THE BRITISH STABLE

THE BRITISH STABLE

GILES WORSLEY

photography by

WILLIAM CURTIS ROLF

Published for

THE PAUL MELLON CENTRE FOR
STUDIES IN BRITISH ART

by

YALE UNIVERSITY PRESS
NEW HAVEN AND LONDON

Second printing 2005

Designed by Gillian Malpass

Printed in Italy

Library of Congress Cataloging-in-Publication Data
Worsley, Giles.
The British stable / Giles Worsley.
p. cm.
Photography by William Curtis Rolf.
Includes bibliographical references and index.
ISBN 0-300-10708-0 (cl : alk. paper)
1. Stables–Great Britain. 2. Architecture–Great Britain. I. Rolf, William Curtis. II. Title.
NA8340.W67 2004
728'.92--dc22

2004014768

A catalogue record for this book is available from The British Library

Page i Roger or William Hurlbutt, entrance to the stables of *c.*1670 at Tredegar House, Gwent, before 1680 (William C. Rolf, 1999).

Frontispiece Alfred Waterhouse, the stables of *c.*1870–82 at Eaton Hall, Cheshire (William C. Rolf, 1999).

In memory of my mother

ACKNOWLEDGEMENTS

The British Stable is a project that has stretched on for much longer than it should have done. My thanks are due to three people for ensuring that it was finally brought to completion: John Nicoll of Yale University Press, for recognising that this could make a good Yale publication; Brian Allen of the Paul Mellon Centre for Studies in British Art, without whom far fewer good architectural books would be published; and my photographer, Bill Rolf. To Bill my special thanks must go, not only for his wonderful photographs, which have been critical to this book, but for his astonishing tolerance as other projects kept pushing its completion just out of reach.

The long gestation of *The British Stable* means that the following acknowledgements can be only partial. If anyone has been omitted I ask for their forgiveness; sadly some of those mentioned have since died. I am very grateful to all the owners of stables who allowed me to poke round their buildings and who suppressed their bemusement at my excitement over apparently arcane matters of drainage and ventilation, and above all to those who agreed to allow Bill Rolf to photograph their buildings. My profound thanks also go to my erstwhile colleague at *Country Life* Rupert Uloth, who cast the eye of a former cavalryman over the text to make sure that there were no obvious equine howlers.

In particular I thank David Adshead, Malcolm Airs, Finch Allibone, Jill Allibone, Clive Aslet, Frances Bailey, Andrew Barber, the Duke of Beaufort, Anne Bennet, John Berkeley, James Bettley, Ray Biggs, Neil Bingham, Marcus Binney, D. J. Bonney, Heter Borron, David Bowark, Peter Boyden, Ian Bristow, E. M. Brooks, Randle Brooks, Edmund Brudenell, E. Bulmer, S. Bywater, C. F. Cattermole, Mr and Mrs Guy Cavenagh-Mainwaring, John Chesshyre, John Chichester-Constable, the Marquess of Cholmondeley, Russell Cleaver, Sir John Clerk Bt, Sir Howard Colvin, W. J. Connor, Alan Cook, the late John Cornforth, Dr John Crook, Professor J. Mordaunt Crook, Elizabeth Cuthbert, Marilyn Dalton, Graham Damant, C. R. Davey, Peter de Savery, the Duke of Devonshire, Captain H. Drummond of Megginch, Peter Drury, Anna Eavis, Barry Futter, Julian Gibbs, Mark Girouard, John Goodall, Margaret Gray, Jane Greenacombe, Jenny Greene, Lord Gretton, Pat Grover, Christopher Hagon, Michael Hall, R. Hall, Robin Harcourt-Williams, Eileen Harris, John Harris, Richard Haslam, Lord Hastings, Robin Herbert, the Lady Hesketh, Mr and Mrs Tony Hide, S. D. Hobbs, Helen Hudson, Gordon Higgott, Rosemary Hill, Neil Hynd, Gervase Jackson-Stops, the Earl of Jersey, Simon Jervis, Mr and Mrs David Johnson, Jean Kennedy, P. I. King, G. A. Knight, Tim Knox, Peter Leach, Valerie Lee, James Lees-Milne, Tom Lloyd, Peter McKay, Edward McParland, John McVerry, Peter Meadows, Jim Meads, Lord Methuen, James More-Molyneux, Jeremy Musson, John Newman, The Earl of Normanton, the Duke of Northumberland, Emma O'Reilly, Robert Oresko, Anthea Palmer, Lord Palmer, Stephen Parissien, J. M. Paterson, Jeremy Pearson, the Earl of Pembroke, A. Piper, Brian Playne, Giles Quarme, Anthony Quiney, Peter Reid, Barbara

Rich, Margaret Richardson, the Duke of Richmond, the Hon. Jane Roberts, Mary L. Robertson, John Martin Robinson, J. Rollason, Linda Shaw, Colin Shrimpton, Clare Slemeck, Earl Spencer, Martin Stancliffe, Margaret Stone, Elizabeth Steele, David Starkey, Sir Richard Storey, Sir Roy Strong, Lord Sudeley, R. H. G. Suggett, Tim Tatton-Brown, the Marchioness of Tavistock, Professor F. M. L. Thompson, Dr Simon Thurley, Thomas Tuohy, William Wake, Margaret Walker, Dr Alastair Ward, K. T. Ward, Valerie Watson, J. H. Richmond-Watson, the Duke of Westminster, Anthony Wharton, S. Whitbread, Helen Whittow, C. Wilkins-Jones, Barry Williams, Hilary Williams, J. Wilson, C. M. Woolgar, Dr Lucy Worsley, Sir Marcus Worsley and the Earl of Yarborough.

My thanks are owed also to all those at Yale University Press who have made this book possible, in particular my editor, Gillian Malpass, who designed the book and is always a delight to work with; Paul Sloman, who copy-edited the text; Barry Kew, who proofread it; Sarah Faulks, who kindly stepped in to sort out the picture collection; and Hazel Hutchison and Katie Harris in the publicity department.

Finally, I thank Charlotte Hoyle, my daughters, Alice, Emma and Lucy, and above all my wife, Joanna, for applying just the right degree of pressure to make me finish this book.

CONTENTS

1 Brocklesby Park: the pack at check.

Go anywhere in England, where there are natural, wholesome, contented, and really nice English people; and what do you always find? That the stables are the real centre of the household.
George Bernard Shaw, *Heartbreak House* (Act III)

INTRODUCTION

NO ANIMAL HAS BEEN SO FAVOURED BY ARCHITECTS AS THE HORSE. Ever since Francesco di Giorgio built the famed ducal stables at Urbino in the 1480s, and probably long before, horses have been associated with fine buildings. Jean Aubert's stables at Chantilly for the Prince de Condé of 1719–40 were so grand that foreign potentates could dine there without realising their practical function. The Baroque stables of Christianborg Palace in Copenhagen, though built for a relatively minor monarch, are on such a scale that they appear to curve around to infinity. In England, Roger Morris's stables at Althorp, Northamptonshire, are markedly more imposing than the house they sit beside, and it is a moot point whether James Paine's lapidary stables at Chatsworth, Derbyshire, more than stand their ground beside the rather confused ducal residence they serve. To study these and similar buildings, such as the Wellington Lines at Aldershot, built in the 1850s to hold 1,584 cavalry horses, or the massive four-storey stables housing six hundred horses, built in the 1870s at Paddington Station in London, is to realise that horses deserve a special place in the history of architecture.

And yet, the stable seems to have passed the architectural historian by. The planning of medieval cloisters, and the positioning of halls and apartments in medieval castles have been studied in detail, but historians have rarely asked where the horses, so critical to medieval life, were housed. Georgian London would not have been possible without the invention of the mews, but mews are not even discussed in Sir John Summerson's highly influential *Georgian London*.

Where stables have been discussed it has been in passing. The assumption has been that stables were fundamentally utilitarian buildings, one step up from the farm. The fact that stables were places of display for princes, noblemen and even nineteenth-century brewers, and were often lavishly fitted up, goes unmentioned. So does the physical relationship between the stable and the country house, a key determinant in understanding attitudes towards the house and its landscape. Quite where the estimated one million horses needed in Victorian London were housed has not been a particular concern to urban historians.

The fact that stables have received little attention from architectural historians is hardly surprising, for a similar lack of interest in matters equestrian is common among academics. The work of George Stubbs aside, equestrian art has always been treated as second-rate. It is only recently that the centrality of the hunt to Court life, with all its implications, has come to be acknowledged by historians. The same is true of economic and social history. In 1978 Joan Thirsk noted that 'when we

2 Thomas Worsley, the riding house of 1767–9 at Hovingham Hall, Yorkshire (William C. Rolf, 1999).

consider the age-long dependence of man on the horse coupled with the dramatic expansion of economic activity in the early modern period, it is remarkable how little interest historians have shown in the way that horses were made available to meet more insistent and fastidious demands'.[1] A quarter of a century on, her words still ring true.

There have been detailed investigations of almost every building type from churches to country houses, farms to market halls, barracks and dockyards to ice-houses and windmills, and yet the stable has been largely neglected. This book is an attempt to redress that neglect. It focuses on British and primarily English stables. Though Continental stables were often even grander and more impressive than those in Britain, there was a traditional belief that the British had a particular affinity with the horse and that this was expressed in the quality of their stables. As François de la Rochefoucauld noted, 'The English have an extraordinary affection for the horse – a passion which is common to the whole country.'[2] It is an attitude summed up by Sir Osbert Sitwell, writing of his Edwardian childhood: 'Horses – horses and carriages more than people – gave London, as I remember it then, its particular distinction, for it was the capital of an island where the horse was still god, where stables were often finer than the house, racecourses, better than parks.'[3]

Horses do not need stables to live. Wild horses show only a passing interest in shelter. But for men, the stable has been as fundamental a tool of equine domestication as the bit, the stirrup, the saddle or the harness. Practicality dictates that working horses should be kept in a stable, at least for much of the time, so that they can be fed efficiently and be to hand when needed. Such shelter prevents horses wasting energy by growing a thick coat in winter and permits the neat, short coat that smart society has always preferred. Simple prudence has always valued the security of the stable, for there is nothing novel about the fabulous sums achieved by modern racehorses.

Yet, though practicality explains the need for stables, it cannot of itself justify the energies and resources that have been poured into their building, often far beyond functional requirements. To understand this, it is necessary to remember that until the rise of the motor car rendered it largely obsolete, the horse lay at the centre of western culture and the stable was the physical expression of its importance. A handsomely appointed stable was often as much a setting for horses as a shelter.

Much of the attention of the book, particularly in the earlier chapters, is directed at country-house stables. Until the late eighteenth century these were (with the exception of a handful of royal and military examples) the largest and most architecturally impressive stables, as well as the most technologically advanced. In the late eighteenth and the nineteenth centuries, the focus widens to include other types of stable, especially military and commercial stables, which faced new problems as the number of horses needed to service an industrialised society rose exponentially. Special attention is paid also to the largest single building in the stable complex, the riding house, and with it the uncharted history of *haute école* in Britain, which largely determined the fortunes of the riding house before the nineteenth century.

Though surviving medieval stables are rare, they are often well documented, so the scope of this book runs from the twelfth century through to 1914. That year was a watershed for equestrian history, the moment when the increasing acceptance of the internal combustion engine fatally undermined the principal function of the horse as a means of transport and haulage. Horses continued to serve an important economic role in the 1920s and 1930s, but their decline was swift and the need for new stables had gone. Since 1945 horses have continued to be popular for hunting, racing and general recreation, but with a handful of exceptions there have been few stables of architectural interest.

I

PROLOGUE: THE MEDIEVAL STABLE

The medieval stable remains something of a mystery. The horse was of supreme importance during the Middle Ages, for war, hunting and transport, and it is known from archives that the grandest medieval stables rivalled their successors in scale and elaboration. And yet it is very hard to identify any physical remains with certainty. Few medieval stables survive in any form; none remains intact, and very few have been positively identified by archaeologists. Indeed, where they have been, the identification needs to be treated with caution as it sometimes seems to be based on no more than assigning a name to an otherwise unexplained building. Excavations in the late 1960s at Elstow nunnery, Bedfordshire, for example, uncovered a building that was identified as a stable even though it was only ten feet deep, too tight to be used by horses.[1] At Bolton Castle, Yorkshire, two ground-floor rooms have been identified as stables despite the fact that this would have meant that large, unwieldy horses had to share a narrow door with all the traffic of the castle to the Guardhouse and Great Hall beyond (including food from the kitchens), and despite the two sharp right-angle turns the horses would have needed to make before they could enter their stables.[2] Even with intact sites such as Bodiam Castle, Sussex, or Nunney Castle, Gloucestershire, stables cannot be found. Medieval stables, it seems, are an archaeological black hole.

Quite why this should be so is a matter of debate, but the answer is probably very simple. Nobody has found stables because nobody has really tried to look for them, and nobody has tried to look for them because they have not been considered important by archaeologists. Despite their self-evident centrality to medieval aristocratic life, stables have largely been ignored because they have never held the status accorded to churches, fortifications or medieval halls. As Helen Clarke noted with respect to monastic life, the interests of late nineteenth-century archaeologists 'lay mainly in the recovery of the ground-plans of monastic churches and monks' dwelling-quarters rather than in the exploration of monastic sites in their entirety . . . their concentration on the central, "most important", areas of monastic complexes resulted in an almost total disregard of the outer portions of the precincts . . . we still know very little of the structures of medieval monasteries which were not directly associated with church or cloister'.[3]

Clarke's words were echoed by Glyn Coppack: 'The [monastic] outer court is very much an unknown quantity, spurned by the historian and the architectural student alike, for in theory it is little more than a farm yard, its buildings crude and functional, and with little apparent relevance to religious life.'[4] Neil Rushton could have been talking about stables when he noted that 'Understanding of almonry sites has been hindered by the lack of archaeological excavations. This has been due largely to the archaeological neglect of the outer precincts of monastic houses

3 Boucicaut Master and workshop, *Sale of Horses to Arabs* (detail of fig. 9; larger than actual size).

where almonries were invariably located.'[5] These comments on monasteries hold true for castles and manor houses. Even Anthony Emery's recent comprehensive study of the medieval great house does not engage with the siting of stables.[6]

In the absence of a solid body of documented remains, it is necessary to rely on written sources, accounts, inventories and building contracts, which though scattered over several centuries allow a general, if not very precise, picture of the medieval stable to emerge. One useful, though problematic, source is medieval manuscripts, which provide some illustrations of stable fittings and which, if approached with care, can be informative. But inadequate though the documentary record is, it shows that medieval stables could be as large, well equipped and solidly built as those erected in subsequent centuries and it also allows the opportunity to establish some idea of where stables were sited in medieval complexes. What can be learnt will establish a solid foundation for the study of later stables, and perhaps a foundation that archaeologists might use to conduct a more thorough search for medieval stables.

Horses were critical to a royal and aristocratic way of life that was based on constant movement, a passion for hunting and the military power of mounted knights. They were also expensive to buy and maintain, and therefore needed to be carefully looked after. In 1315, when the daily wage of a skilled man was 4d per day, it cost 6–7d a day to keep one of the king's horses at Kennington in Surrey. Their impact on limited agricultural economies was also significant as horses ate an estimated two tons of oats per year.[7] For this reason horses were often split up into small numbers around the country, especially in winter. But when they were needed, it could be in large numbers. To take one typical example, when the Earl of Warwick stopped at Berkeley Castle, Gloucestershire, in early March 1421, hay and provender, and therefore accommodation, had to be provided for seventy-two horses and thirty-eight hackneys. His wife, Elizabeth Berkeley, Countess of Warwick, travelling to London later that month, was accompanied by fifty-seven horses and eleven hackneys. Of these, seventeen horses and four hackneys were being fed at Berkeley Castle. Her chief groom brought a further twelve from elsewhere, while eight were borrowed from the Bishop of Gloucester, five from the Prior of Llanthony and six from the Master of the Hospital of St John the Baptist in Warwick.[8]

Horses were also vital for war. When the kings of England periodically invaded France, vast numbers of horses had to be mobilised, as in 1359 when more than 3,245 horses were shipped from England to Calais.[9] Not all the horses were needed for knights or men-at-arms. Mounted archers and the baggage train also had to be provided for. Nevertheless, so many horses, especially the prized great warhorse or destrier, could not be produced swiftly without their careful maintenance in large numbers during peacetime, and that required stabling. Constant border skirmishing with Scotland also meant that substantial numbers of horses had to be housed in the northern castles. In 1325 there were thirty mounted troops or hobelars and twenty men-at-arms stationed in Bamburgh Castle, Northumberland, whose horses had to be stabled. At Norham Castle a survey under Henry VIII records permanent stabling for fifty horses with temporary stabling in an ox-bier for fifty more and a further twenty in a room under the chapel.[10]

Then there was hunting, the passion of all medieval monarchs and one that needed large numbers of expensive horses at hand. Some idea of the scale of stabling required to support this activity comes from the building accounts of the King's Works which record that Edward I erected stables for two hundred horses at a cost of £104 8s 5d at the royal hunting lodge at Clipstone, Nottinghamshire, in 1282. Few post-medieval stables, even royal stables, could have rivaled them in size.[11]

Equestrian studs were established to maintain a sufficient supply of horses, the best known of which were the monastic and royal studs. In King John's reign early in the thirteenth century at least thirteen sergeants or keepers moved around the country from park to park and stud to stud with horses and grooms. Thomas de Landa, the chief keeper, for instance, was at Worcester in August 1212 with twenty-four horses and twenty-four grooms.[12] The stables at Pickering Castle, Yorkshire, chiefly used as a royal stud, were 240 ft long and when first mentioned in 1322 housed fifty horses.[13]

Equally important were the monastic studs. Beaulieu Abbey in Hampshire had 112 horses, 63 mares, 120 affers, 46 colts and 48 foals distributed between 17 granges in 1269–70.[14] Some of these would have been for domestic use but others would have been bred commercially as the quality of the monastic studs were renowned. Sir A. Darcy writing to Thomas Cromwell at the dissolution of the monasteries in 1537 described Jervaulx Abbey in Yorkshire as 'one of the fairest churches that I have seen, fair meadows, and the river running by it, and a great demesne'. It was an ideal spot, he suggested, for a royal stud: 'for surely the breed of Jervaulx for horses was the tried breed in the north'.[15] Monasteries also had their own horses to stable. Samson, Abbot of St Edmund's Abbey from 1182 to 1211, had twenty-six horses and would also have needed to provide stabling for travellers' horses.[16]

Because so little evidence survives, it is hard to be certain whether stables were built for architectural effect, though the scale of the stables at Clipstone, for instance, must have been impressive. The clearest evidence available that at least some patrons were aware of the architectural impact of their stables comes at Warwick Castle, Warwickshire. The late fifteenth-century chronicler William Worcestre recorded that Richard Beauchamp, Earl of Warwick, 'had a stable of great size built at Warwick of plaster of Paris, at a cost of 500 marks' (that is £333 6s 8d). The fact that Worcestre thought the stables, which were built in 1423, worth mentioning suggests how impressive they must have been, as does their extraordinary cost. The reference to plaster of Paris is enigmatic. It may simply have been used to emphasise the whiteness of the panels of what was almost certainly a timber-framed building – perhaps over a stone base. But it is quite likely that it refers to decorative plasterwork, like pargeting, on the exterior of the building. Remarkably, an architectural drawing of the stables survives, albeit part of a general survey of the castle made by Robert Smythson, probably in about 1605 (fig. 4). This shows a long, symmetrical building with a projecting central porch in a prominent position facing the main castle gates. There can be no doubt that Beauchamp, renowned across Europe for his prowess at jousting and therefore presumably a proud horse owner, was making a deliberate statement of display in these stables, an approach that was to become common in great stables in the sixteenth and seventeenth centuries.[17]

It is possible that Warwick was exceptional. Generally, no attempt seems to have been made to isolate stables from other buildings, nor even necessarily to group all the stables on one site together. At Windsor Castle, stables are variously referred to

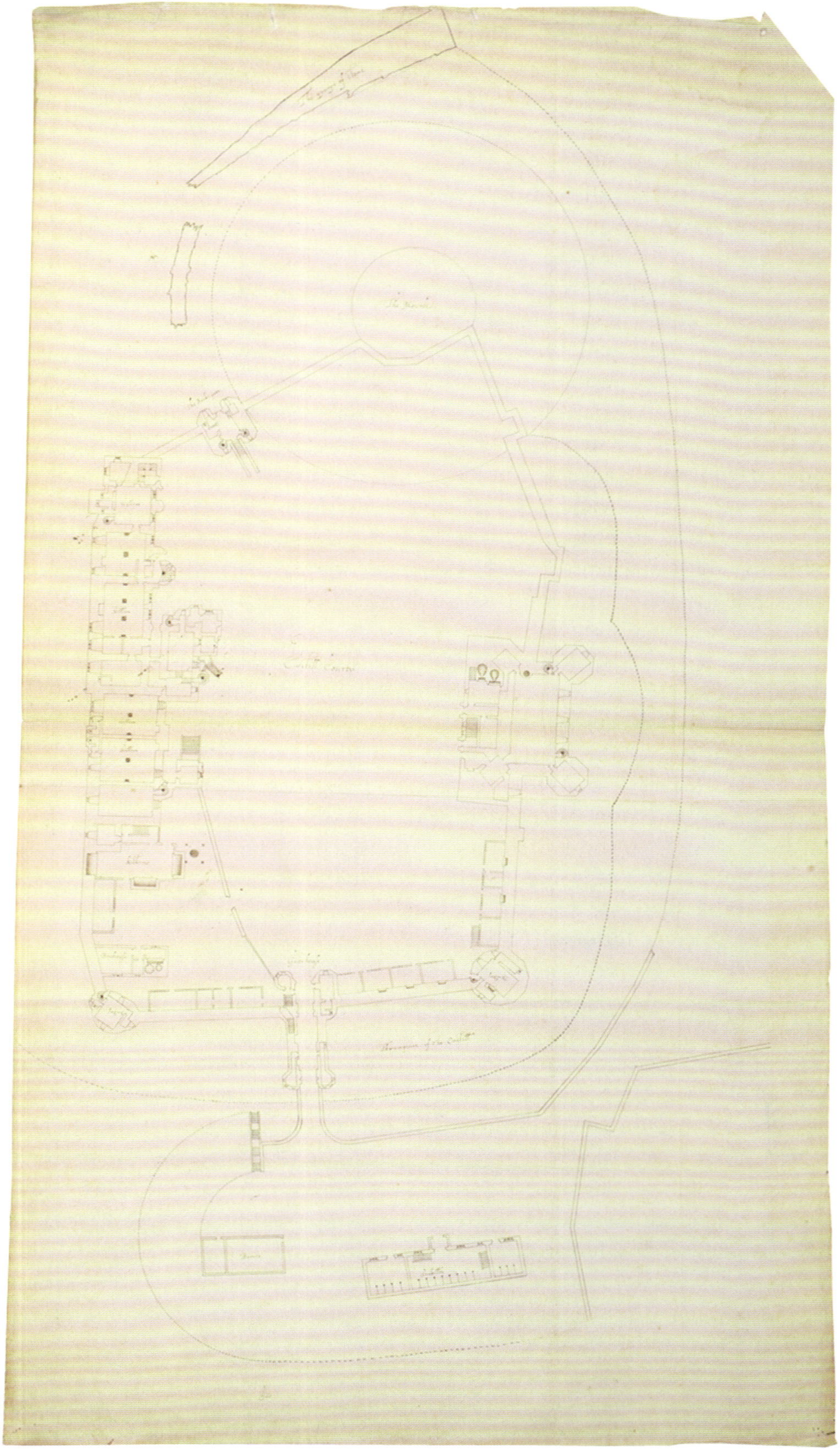

4 Plan of Warwick Castle, Warwickshire, by Robert Smythson, *c.*1605. The symmetrically designed stables face the main entrance to the castle at the bottom of the drawing. The depiction of the stalls is probably representative rather than accurate (RIBA Library Drawings Collection, Smythson Collection I/21).

in the inner bailey, the middle bailey and the lower bailey, and outside the castle, near the garden. At Alnwick Castle, Northumberland, a survey of 1567 mentions six different stables: three normal stables and three in the ground floors of towers. Instead stables were often organised according to the different households a building held. In royal palaces the king's and queen's stables were usually kept separate. At Bury St Edmunds as well as the general stable there were stables for the abbot, the prior, the sacrist and the cellarer, reflecting the range of establishments of that great abbey. Even smaller monasteries often made a distinction, as at Thicket Priory in Yorkshire, which had a guest and a workhorse stable.[18]

Where a house had two courts, the stables were usually in the outer court, along with the lesser offices, as at Dartington Hall, Devon, Croydon Palace, Surrey, and Eton College, where the founder's design stated that the outer court 'shall be edified with diverse housing necessary for the bakehouse, brewhouse, garners, stables, hayhouse, with chambers for the stewards, auditors'.[19] The Cistercian statutes specified that the stables should be within the precinct walls, and in monasteries the general stable at least would have been in the outer court, as at Mount Grace, Yorkshire.[20] An obvious site was by the gatehouse, or as the lease of Minster Lovell to Alexander Umpton in 1536 put it 'the stable next to the gate'.[21] This was certainly the case at Durham Cathedral where before the Dissolution children were educated by the almonry in a loft over stables on the north side of the abbey gates. After the Dissolution, the stable was converted into a kitchen for Stephen Marley's lodging and still survives.[22] Another probable medieval monastic survivor is the Priory Stabling at Winchester Cathedral (figs 5 and 6). This 100ft-long, timber-framed range, which was probably built about 1490–1500, lies just inside the Close Gate and is now part of the Pilgrims' School. It was certainly the 'common stables' of the Dean and Chapter of Winchester from the early sixteenth century (the earliest reference to 'common stables' is in 1541) until the early twentieth century and is described as such in the parliamentary survey of 1649. Though documentary proof is lacking because of lacunae in the cathedral's pre-Reformation archives, the building's size (it is 25 ft deep), date and position make it likely that it was used as stables before the dissolution of St Swithun's Priory in 1538.[23]

Castles with an outer court or bailey usually had their stables there, as can be seen in a pictorial survey of Castle Hedingham, Essex, of about 1610 (fig. 7), and in a survey of Maxstoke Castle, Warwickshire, of 1521, which details that 'without the said castell is a goodly and a large bace corte with roomes for stables and also there are 2 fair barnes for hay, all walled with stoone and coverde with slate'.[24]

Even where no outer bailey is known to have existed it might nevertheless make sense to look for the stables in a similar position. At Bolton Castle, Yorkshire, it is assumed by P. A. Faulkener that everything lies within the body of the castle: 'Its [Bolton Castle's] advantages as a subject lie, not only in its completeness, but in the fact that it must, being self-contained, of necessity include within one building all the permanent accommodation demanded, not only the halls and chambers but stables, mills, stores and kitchens.'[25] But there is no documentary evidence to support this assumption and the practicalities seem to argue against it. As has been noted above, access to the site assigned to the stables would have been cramped and inconvenient. The number of horses the space could have housed would have been very limited for a castle where hunting would have been a principal occu-

5 The Priory Stables, Winchester Cathedral, Hampshire, probably pre-Reformation (John Crook).

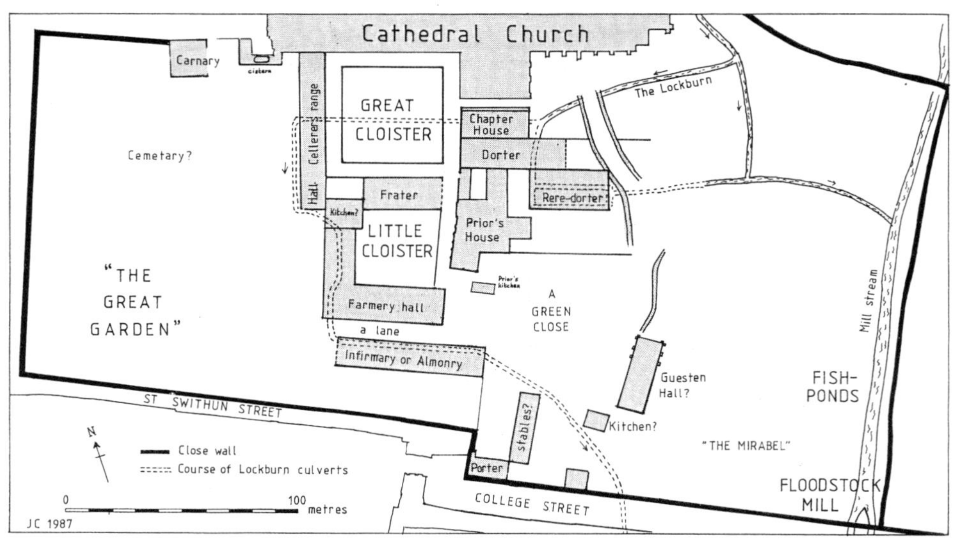

6 Plan of the precinct of Winchester Cathedral. The Priory Stables lie on the south side of the precinct, by College Street (John Crook).

pation. And at the very least there must have been a haybarn outside the confines of the castle. It seems likely that the stables would have been there too.

At Southwell Palace, Nottinghamshire, a regularly planned manor house following the model of the concentric castle, the stables were probably outside the courtyard. But even some important manors, such as the Bishop of Durham's manor

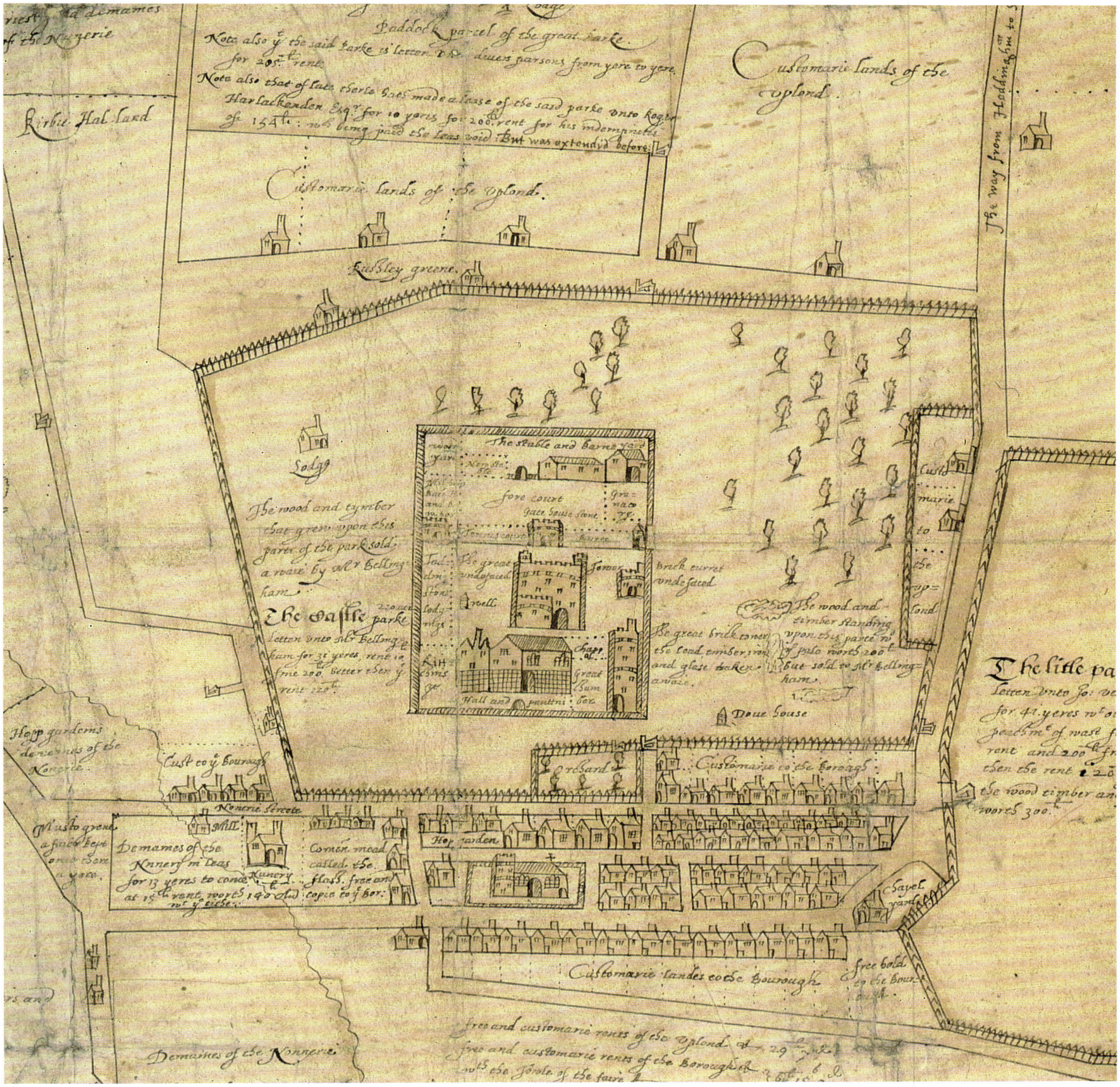

at Howden, Yorkshire, were built around a single courtyard with little separation between the stables and the domestic quarters.[26]

The quality and durability of the materials from which stables were made seem to have been determined less by date than by the importance of the relevant stable. It is hard to see any absolute improvement in the quality of stable design or progress

7 Detail from a survey of Castle Hedingham, Essex, *c.*1610. The stables lie in the outer court (Essex Record Office D/DmhPI, survey of *c.*1610).

between the thirteenth and sixteenth centuries. Instead the stables of more important establishments were more likely to be solidly built regardless of date.

The most substantial stables had ground floors of brick or stone with a timber-framed first floor, examples of which can be found throughout the Middle Ages. The new stables at the Royal Mews at Charing Cross in London were built with a stone base and half-timbered superstructure in 1287–88, as were those at Westminster in 1342 and at Sheen, Surrey, in 1369. At Langley Abbey, Norfolk, a stable range that probably pre-dates the Reformation survives with a brick base and a wooden first floor. The stable at Tattershall Castle, Lincolnshire, was exceptional in being two storeys of brick in 1439–40. This was still a rare material to be used on any building. Its use on a stable was partly a mark of the high status of the horse, but perhaps more an example of conspicuous extravagance by its builder, Ralph Cromwell, the Lord Treasurer who used brick extensively at Tattershall.[27]

Well-built stables had tiled or slated roofs, a sensible precaution given the threat of fire in a building already full of straw. At St Edmund's Abbey, Abbot Samson 'ordered that the stables and offices roundabout the court which were formerly covered with reeds should be reroofed and covered with tiles' at the beginning of the thirteenth century. By the early fourteenth century important stables were often tiled. The great stable at Windsor Castle was tiled in 1311–12; the new stable at the Palace of Westminster, in 1342; and the Black Prince's stables were retiled in 1355. But even important stables might be thatched long after tiled or slated roofs were common, as was the case with King John's hunting lodge at Writtle, Essex. In 1419 this had a thatched roof, even though the domestic buildings were tiled. At Warkworth Castle the stable was rethatched in 1488, though it had a slate roof by 1537. As late as the dissolution of the monasteries of the nine Yorkshire priories whose surveys mentioned the condition of their stables, only Wykeham Abbey had a slate roof, as opposed to thatch.[28]

A solid roof and walls were important for a well-built stable, but equally important was the quality of the floor on which the horse stood, as horses' feet always need careful attention. Evidence for such flooring becomes available only in the fifteenth century. At Farleigh Castle, Somerset, excavations revealed that what appear to be the stables, erected after 1425, probably had a floor of natural bedrock made even where necessary with stone slabs. Excavation at Kirkstall Abbey, Yorkshire, showed that the guest-house stable was paved with stone flags when the building was converted from domestic use in the late Middle Ages. The building accounts for Tattershall Castle reveal that the stables there were paved in brick in 1439. Elsewhere there is extensive evidence for stalls being floored with wooden planks, a system that would remain popular into the seventeenth century. A survey of 1440 reveals that the stables of the Old Deanery at Salisbury had a plank floor, and accounts for Great Milton, Oxfordshire, show that the stables were built with a planked floor in 1457, as were those in a house in London, which were described as 'new planked' in 1485. It is possible these references are evidence of an innovation in stable design in the fifteenth century. Certainly the evidence points to the widespread use of planked floors in all types of stable during the century, although many stables still had earth floors, as must have been the case with Water Hall at Wixoe, Suffolk, where the otherwise very detailed building contract of 1473 makes no mention of flooring material.[29]

By at least the fifteenth century some stables were being carefully planned for drainage. Four stables from that century have left evidence of gutters running down the centre of the stable, apparently connected to streams or drains beyond. In the Kirkstall Abbey guest-house stable was a gutter nine inches wide and six deep, which ran into a drain serving the whole monastic complex. At Farleigh Castle, a natural gully was straightened and deepened with an outflow cut through the base of the curtain wall into the castle ditch.[30]

Some stables even had running water laid on. At Carew Castle in Pembrokeshire, the stable was given its own water point on the pipes that brought water for the entire castle from a source a mile away. Leland, writing in the 1540s, mentions a 'fair stable with a conduit comming to the haul side' at Ravensworth Castle in Yorkshire, and pipes are referred to at the stables at the king's palace at Reading in Berkshire.[31]

There are many references throughout the Middle Ages to wooden 'hekkes' or hayracks, and to mangers for horses' feed and water. The stables at the royal hunting lodge at Clarendon in Wiltshire, for instance, were reported as wanting racks and mangers in 1272.[32] As none survives, it is not possible to be entirely certain of their form, but a reasonable idea can be gathered from illustrated manuscripts. These need to be treated with care as the interesting examples are Continental and often schematic, especially in Nativity scenes, where depictions of the stable may owe more to accepted conventions and the imitation of earlier prototypes than to contemporary examples. But it is unlikely that English stable fittings at this date differed significantly from those on the Continent and as our only visual insight into the medieval stable interior they cannot be ignored. The Adoration of the Magi from *Le Livre d'heures du duc Louis de Savoie*, painted at the beginning of the fifteenth century clearly shows a slanting wooden hayrack level with the horses' heads and a continuous wooden manger below. Similar racks and mangers can also be found in the *Adoration of the New-born Child* by the Boucicaut Master and workshop and in what must be one of the earliest specific representations of a stable, the illustration of Arab horses being sold to merchants from India by the same workshop (figs 8 and 9).[33] Identical fittings continued to be used until the horse fell out of everyday use in the twentieth century and can still be found in many old farm stables. Although it is not possible to be certain when such racks and mangers were first used we know that they were in use at least from the beginning of the fifteenth century and probably much earlier.

8 Boucicaut Master and workshop, *Adoration of the New-born Child*, *c.*1415. The rack and manger suggest the appearance of a medieval stable (Walters Art Gallery, Baltimore, MS 770 fol. 139v).

The form of partitioning, if any, that divided the horses from each other seems also to have been determined by the status of the stable. Though mangers and stalls were repaired at Windsor Castle in 1319–20, most documents that refer to mangers and racks do not refer to stalls, as is the case with the detailed Water Hall contract of 1473.[34] Similarly, though the Nativity of the Follower of the Master of the Saluces Hours of about 1470 shows a wooden partition between people and animals that looks like a stall, the Boucicaut workshop stable shows no form of stall.[35] The evidence suggests that partitions were unusual, as can still be seen in old farm stables. An alternative partition that might have been considered too insubstantial to mention in documents is wattle. Wattle fencing is commonly depicted even in schematic representations of the Nativity, showing that it was commonly associated with stables. Wattle fencing also seems to have been used for temporary sta-

9 Boucicaut Master and workshop, *Sale of Horses to Arabs*, *c.*1413. A rare depiction of a medieval stable (Bibliothèque Nationale, Paris, fr 2801 fol. 92).

bling of horses in the open, as can be seen in the Hunting Book of Gaston Phoebus in which horses are corralled within a wattle fence while the Prince dines before the hunt.[36]

The excavated foundations of stables allow us to establish their size and therefore their spaciousness, although the absence of traces of partitions makes it impossible to establish how much space was allowed between each horse. Most stables had only a single row of horses, generally allowing them between 16 and 20 ft in depth, 18 ft being the most common. Any building narrower than 16 ft is unlikely to have been a stable given the difficulty of access.

Surveys sometimes speak of garners or haylofts over stables, but hay was more unusually kept in neighbouring barns. Grooms or other servants could be accommodated over stables and their rooms can occasionally be identified by fireplaces, as at Dartington Hall, Devon. At Broughton Castle, Oxfordshire, the north range by the gatehouse, built by Sir William Fiennes after 1457, certainly had lodgings on the upper floor and probably stables below. More unusual were the two chambers that Henry III ordered to be included at each end of the new stable at Clarendon, one for the king's and one for the queen's harness, each with a privy chamber. Harness rooms remained rare until the nineteenth century.[37]

Though surviving physical evidence may be limited, it is clear that the best medieval stables were probably as large and well designed as anything built under the Tudors and would compare favourably with surviving farm stables today.

In 1592 the Duke of Wurttemburg toured Theobalds, Hertfordshire. One of the greatest of the Elizabethan prodigy houses, it was a veritable palace built expressly to honour the queen by her chief minister, William Cecil, Lord Burghley. The tour concluded with a visit to the stables, 'in which were kept many fine horses'. Four years later, on a similar trip to England, the Prince of Anhalt visited Cobham Hall, Kent, and was again shown the stables where 'there were standing many fine horses'. The Countess of Dorset, then at Knole, Kent, recorded a similar affair in her diary in October 1617: 'Upon the 4th came Sir Percival Hart and Sir Edward to dine, and after dinner my Lord shewed them his stables and all his great horses.' Stables, at least gentlemen's, nobles' and above all royal stables, were seldom simply the utilitarian quarters of horses in the early modern period.[1]

2

THE TUDOR AND JACOBEAN STABLE

The princely or noble display of the horse was a critical function of stable design, at least as important as its practical management, and to some even more so. Handsome horses were often cited as the embodiment of noble virtues and their ostentatious display in stables was a deliberate act of self-propaganda on the part of their owners. The great stables at Versailles were in their own way as important an embodiment of the grandeur of Louis XIV as the Galerie des Glaces. A nobleman might wish to display his wealth and sophistication to his peers by showing them a handsome gallery full of paintings and sculpture, or by displaying the choice contents of his cabinet, a precious bronze, perhaps, or a rare drawing. In both cases he would have ensured that the contents were shown off to their best advantage. He might equally wish to show off his horses, similarly items of conspicuous expense where connoisseurship was an essential ingredient in appreciating the subtle qualities of what was being displayed. If he did so he would want to be sure that the horses would look at their best in a suitable architectural setting.

Relatively few stables survive from the sixteenth and early seventeenth centuries, and after several centuries of neglect most that do have the air of mildly decrepit farm buildings. It is important to remember that when originally built, the impression would have been very different. But at least some stable interiors survive, together with much more documentary evidence, both written and visual. There was also a vocal school of commentators writing about stables. Together these allow a much more accurate picture to be established of the Tudor and Jacobean stable. What emerges is a marked change in stable design that is profound enough to be described as a stable revolution.

Books devoted solely to stable construction are a nineteenth- and twentieth-century phenomenon, but there is plenty of advice on stables to be found in books on architecture, horsemanship, husbandry and animal health, as well as in manuals of a more general nature. Thomas Blundeville's *The Fower Chiefyst Offices Belonging to Horsemanshippe* of 1565 forms the first considered advice on stable design by an Englishman, including one chapter entitled: 'In What Ayre is Beste for a Horse

10 Interior of the stables of 1654 at Peover Hall, Cheshire (William C. Rolf, 1998).

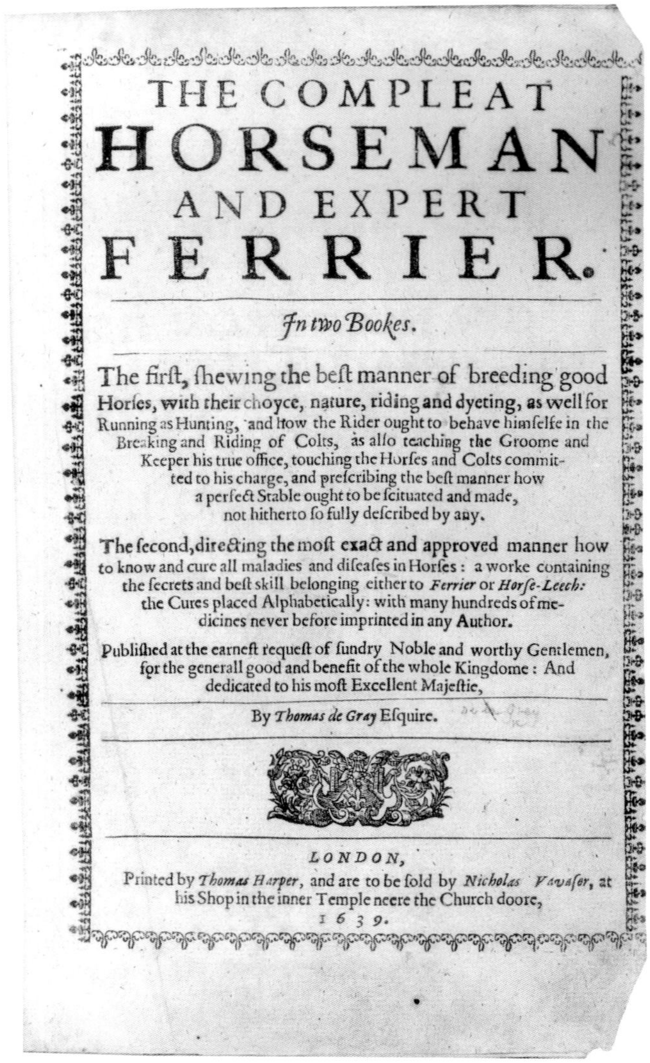

THE COMPLEAT
HORSEMAN
AND EXPERT
FERRIER.

In two Bookes.

The firſt, ſhewing the beſt manner of breeding good Horſes, with their choyce, nature, riding and dyeting, as well for Running as Hunting, and how the Rider ought to behave himſelfe in the Breaking and Riding of Colts, as alſo teaching the Groome and Keeper his true office, touching the Horſes and Colts committed to his charge, and preſcribing the beſt manner how a perfect Stable ought to be ſcituated and made, not hitherto ſo fully deſcribed by any.

The ſecond, directing the moſt exact and approved manner how to know and cure all maladies and diſeaſes in Horſes: a worke containing the ſecrets and beſt skill belonging either to *Ferrier* or *Horſe-Leech:* the Cures placed Alphabetically: with many hundreds of medicines never before imprinted in any Author.

Publiſhed at the earneſt requeſt of ſundry Noble and worthy Gentlemen, for the generall good and benefit of the whole Kingdome: And dedicated to his moſt Excellent Majeſtie,

By *Thomas de Gray* Eſquire.

LONDON,
Printed by *Thomas Harper*, and are to be ſold by *Nicholas Vavaſor*, at his Shop in the inner Temple neere the Church doore,
1 6 3 9.

11 Title page of Thomas de Grey, *The Compleat Horseman and Expert Ferrier . . . prescribing the best manner how a perfect Stable ought to be scituated and made, not hitherto so fully described by any* (London, 1637) (© The British Library).

to Lye and How his Stable Ought to be Sytuated, Made, Kept and Used'. Blundeville's work was carefully considered. A mix of quotation from classical authors, such as Xenophon, Columella, Palladius and Vegetius, together with informed comment about contemporary stables, shows it to be a work of scholarship as well as the manual of a practical man. His common sense can be seen in his recommendation that stables should be placed so that the owner 'may have a delight to come alwayes thyther to see his horse. For according to the olde proverbe that horse is most commonlie fat which is fedde with his own masters eye.'[2]

Four equally useful books followed over the next seventy years: Christopher Clifford's *The School of Horsemanship* in 1585, Gervase Markham's *Cavelarice or The English Horseman* in 1607, Nicholas Morgan's *The Perfection of Horsemanship* in 1609 and Thomas de Grey's *The Compleat Horseman and Expert Ferrier* in 1639. De Grey's book was the first to include the word 'stable' in its long title: . . . *prescribing the best manner how a perfect Stable ought to be scituated and made, not hitherto so fully described by any* (fig. 11).

Between them, these five authors cover every facet of stable design, and de Grey's advice was superseded in print only towards the end of the eighteenth century. For the next 150 years there is little about stables in any published work that they had not already covered. In Morgan's book, the chapter entitled 'Of a Perfect Stable' proved particularly influential and was copied by many subsequent authors.

During the years in which they were writing there was a marked change in attitudes towards stable design. Stables came to be distinguished from the other service buildings with which they had been grouped in the outer court. They were set apart, built of high-quality materials and given a status second only to the house itself. At the same time, their interiors were treated with considerable care.

Because hardly any Tudor and Stuart royal stables survive, it is easy to overlook their importance at royal palaces, but both Henry VIII and Queen Elizabeth spent extensively on royal stables. Indeed, the extent of royal stable building in the early part of Elizabeth's reign led Howard Colvin to note that 'a striking feature of Stockett's surveyorship [Lewis Stockett was Surveyor of the Queen's Works from 1563 to 1579] is the amount of money spent on barns and stables'. To cope with the volume of work, a separate surveyorship of the stables, under the control of the Master of the Horse, was subsequently created. Today it is largely through the building accounts, and occasional plans, that their grandeur can be recaptured. The stables built at New Hall, Essex, in 1517–21, were 145ft long. Those built at the royal palace at Reading in 1570 were 162ft long and cost £1,000. The greatest of all the royal stables, those at the Royal Mews at Charing Cross, where the 'Great Stable, garnetts, barns and hay lofts' were built in 1550–56, cost £6,516 12s 6d, the price of a very substantial country house.[3]

One major royal stable does survive, though much altered: Henry VIII's Royal Mews at Hampton Court (figs 13 and 14). This was built in 1537–38 at a cost of £130 by the king's Master Bricklayer, Christopher Dickenson. As Simon Thurley explains, the stable establishment at Hampton Court was divided into two wings, one for the king and one for everyone else, as was the case with the other departments of the Tudor Court. At Greenwich, for instance, there was 'a partition between ye comyn stabyll and ye master of ye horse Stabyll'. Limiting the number of horses that courtiers brought to Court was as important as controlling the number of people, and in the Eltham Ordinances Thomas Wolsey set down the number of horses each courtier was entitled to stable at the king's cost.[4]

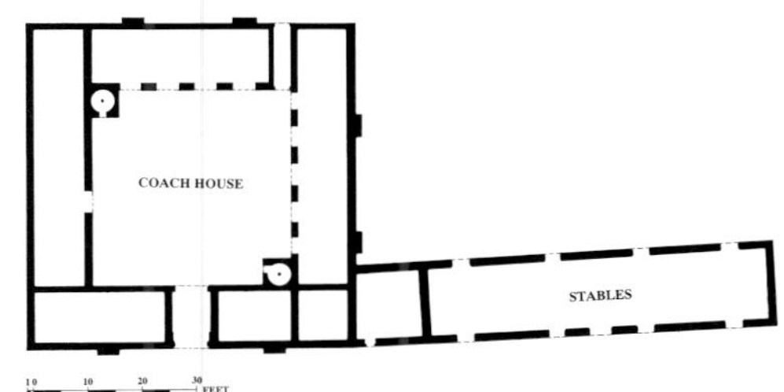

12 Plan of the stables of 1537–38 at Hampton Court, Middlesex (Historic Royal Palaces).

As would be expected at the greatest of this monarch's palaces, the Hampton Court Mews, lying at some distance from the palace itself, is a distinguished building built of brick and red tile, and approached through an impressive stone arch. Unusually, these stables were not designed as a series of separate elements but as a carefully composed quadrangle around a symmetrical two-storey courtyard, with spiral staircases in two of the angles. Similar quadrangles could be found at other major royal palaces. Seventeenth- and eighteenth-century surveys show that the stables at Greenwich Palace, which lay to the west of the main complex, were also built as an elongated quadrangle, with entrances at the north and south. To the south of the stable lay the great barn.[5] No detailed view is known of the stables and as the earliest surveys date from the seventeenth century it is impossible to be certain about their building history, but they can probably be identified with the stables built for the king's coursers by Henry Smith in 1509. If so, they were one

13 The courtyard of Henry VIII's stables of 1537–38 at Hampton Court, Middlesex (William C. Rolf, 2000).

14 (*facing page*) The entrance to the courtyard of Henry VIII's stables at Hampton Court, Middlesex (William C. Rolf, 2000).

of the very first acts of Henry VIII's reign.[6] The stables at St James's Palace are clearly shown as a quadrangle in Faithorne and Newcourt's map, surveyed in 1643–47. There are no building accounts for St James's Palace during Henry VIII's reign, but as no major stable building is recorded after his reign, these stables are likely to be his work.[7] Nor should it be forgotten that the Royal Mews at Charing Cross took the form of a great quadrangle.

No non-royal quadrangular stables are known from the sixteenth century and they remained exceptional in the seventeenth century, but it is perhaps significant

that Henry's great rival, François I, built a stable quadrangle at about the same time at Fontainbleau. This is as exceptional in France as Henry's stables in England. Known as the Grand Chenil, it has since been demolished, but paintings show it to have been an elaborate Classical structure similar in design to the Cour du Cheval Blanc, and therefore probably the work of Sebastiano Serlio who arrived at Fontainbleau in late 1540 or early 1541.[8]

Such quadrangular stables were clear statements of royal prestige. A better idea of what grand stables were more commonly like can be gained from an examina-

15 The Earl of Leicester's stables of *c.*1563 at Kenilworth Castle, Warwickshire (National Monuments Record).

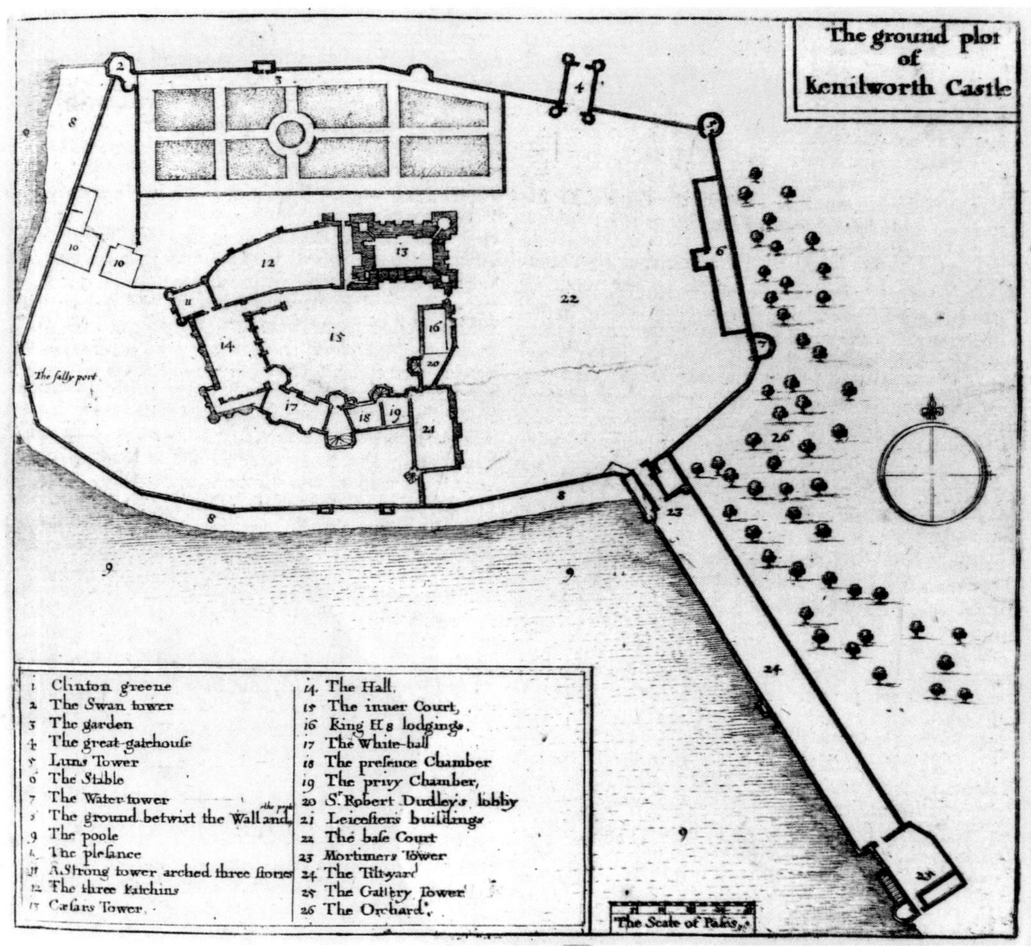

16 Plan of Kenilworth Castle, Warwickshire, from William Dugdale, *Antiquities of Warwickshire* I (London, 1730), p. 243 (© The British Library). The stable lies on the right-hand side of the base court.

17 The stables of *c.*1600 at Bashall Hall, Yorkshire (National Monuments Record © Crown copyright).

tion of the Earl of Leicester's stables at Kenilworth Castle, Warwickshire (figs 15 and 16). The obvious grandeur of Leicester's stables when set against their apparently utilitarian interior and (to twentieth-century eyes) humble status concerned Peter Ellis, who edited the report of the archaeological study into the building: 'the exterior architecture gives little hint that these were the castle stables, promising far more'. He suggests that the building can be seen as disguising its function, 'allowing the stables to be hidden away, playfully altered'.[9] The reality is that Leicester, the Master of the Horse, would have intended the stables for display. Queen Elizabeth's favourite, Lord Leicester, recovered Kenilworth Castle in 1563 and 'having obtained this noble seat, spared for no cost in enlarging, adorning and beautifying thereof'.[10] Leicester entertained the queen at Kenilworth three times. The third visit, in 1575, which lasted for nineteen days, was of legendary splendour. Dudley's improvements included impressive stables 164ft long and 20ft wide, with a two-storey projecting porch and the upper floor framed with cross braces, designed to represent his heraldic device of the 'ragged staff', a motif repeated across the castle. The apparently functional nature of the interior is simply the result of the loss of the timber fittings that would no doubt have set the horses off to their best advantage.

Interestingly, Leicester's elder brother Ambrose Dudley, Earl of Warwick, owned Warwick Castle, less than five miles from Kenilworth. Indeed, Queen Elizabeth visited Kenilworth from Warwick in 1572. The similarities between Leicester's stables at Kenilworth and the great Beauchamp stables at Warwick are marked. Both were large, half-timbered, symmetrical buildings with a central porch and were prominently placed near the entrance to their castle. It has been suggested that there was an element of deliberate archaism in Leicester's use of half-timbering at Kenilworth.[11] Could this have been a deliberate reference to the famed stables at his brother's castle?

Similar, though humbler, stables of about 1600, with a ground floor of stone and an open timber gallery above, survive at Bashall Hall in Yorkshire (fig. 17). Another

18 The stables of *c.*1600 at Baylham Hall, Suffolk (National Monuments Record/English Heritage, 1964).

19 (*facing page*) Anon., *Prospect of Charlecote Park, Warwickshire, from the West* (detail), *c.*1695–1700. The stables of *c.*1558 lie to the right of the house (National TrustPhotographic Library/Derrick E. Witty).

example, this time with a brick base and a half-timbered upper storey, can be found at Baylham Hall in Suffolk (fig. 18). But half-timbered buildings were already beginning to seem unfashionable, even for stables, as can be seen in a late seventeenth-century painting of Charlecote Park, Warwickshire, which had been built in 1558 (fig. 19). That part of the stables visible from the forecourt was built, like the house, of brick, but that hidden from view in the stable yard was half-timbered, presumably as an economy.[12] Though Markham, writing in 1607, considered brick and stone uncommon and recommended them only 'if your ability stretches to it',[13] de Grey did not prevaricate in 1639. Plaster and timber were not mentioned. In his opinion, stables should be either of stone or brick.[14] Brick stables were increasingly standard with new houses built in brick, as can be seen at the now-

20 Anon., *View of the East Front of Burton Constable, Yorkshire*, *c.*1680. The stables of *c.*1600 lie to the right of the house (Burton Constable Foundation).

21 The stables at Osterley Park, Middlesex. Though sixteenth-century in origin, the stables have been much altered, probably in the early eighteenth century (William C. Rolf, 1999).

demolished stables at Burton Constable, Yorkshire, and the surviving, though much altered, stables at Osterley Park, Middlesex (figs 20 and 21).[15]

The rising standard of materials used for stables can also be seen in the growing use of slate and tile instead of thatch, and of glazing for the windows. The stables at Old Thorndon Hall were tiled in 1589, as were those at Nonsuch in 1599. At Grafton Manor in Worcestershire the stables were slated in 1569 and those at Hardwick Hall in Derbyshire, in 1593.[16] Blundeville, writing in 1565, stated that windows should either be glazed or latticed, primarily to prevent birds getting in, whose feathers and dung were considered very dangerous for horses. Similarly, in 1607 Markham stated simply that stable windows should be glazed, as did de Grey in 1637. A survey of Corsham Court in Wiltshire of 1602 referred to the newly built stables as glazed and the building contract for the new stable at Blickling of 1622 also called for the windows to be glazed.[17]

Elsewhere, some Tudor houses, such as Hengrave Hall, Suffolk, completed in 1538, maintained the medieval tradition of placing the stables in an outer court (fig.

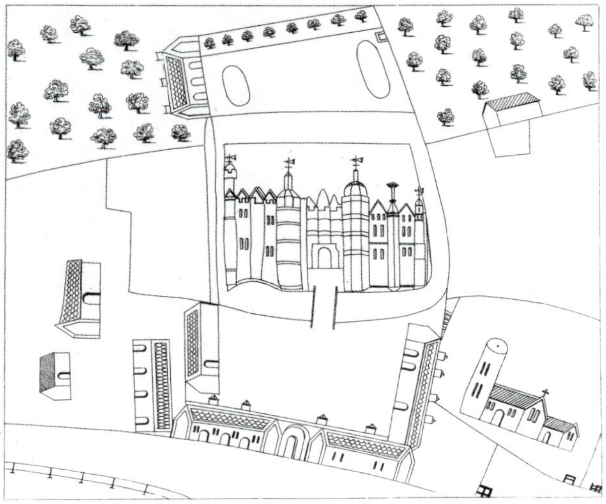

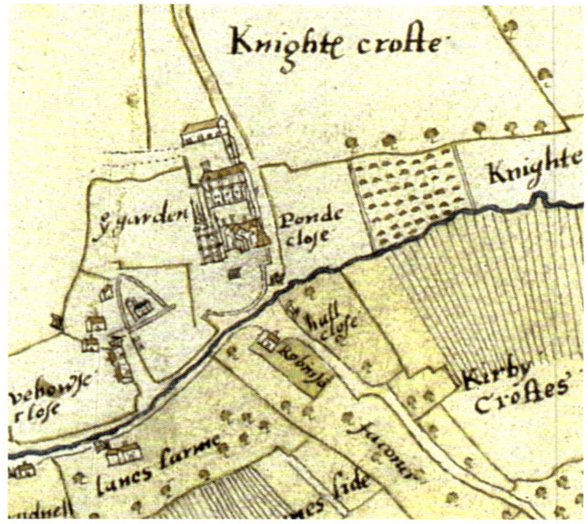

22 (*far left*) Hengrave Hall, Suffolk, showing the now-demolished stables of *c.*1538, from John Gage, *Hengrave Hall* (London, 1822), pl. 1 (detail) (© The British Library).

23 (*left*) Detail from an estate map of Kirby Hall, Northamptonshire, 1586/87. The stables of *c.*1570 lie beyond the house (Northamptonshire Record Office FH 272/8).

22).[18] It was a tradition that continued quite late into the sixteenth century, even at some of the most sophisticated houses of the day, such as Kirby Hall, Northamptonshire, begun in 1570, which an estate map of 1586/7 shows originally had a large outer court (fig. 23).[19] Few such courts survive and their loss has profoundly influenced our appreciation of the setting of these houses. In general though, the sixteenth century saw a shift of the service buildings away from the entrance to the house, for reasons of sanitation and aesthetics. Instead, the builders of Elizabethan houses came to prefer a small walled forecourt containing only a

24 Leonard Knyff and Johannes Kip, Newnham Paddox, Warwickshire, from *Britannia Illustrata* (London, 1707), p. 25 (detail) (© The British Library).

25 Mellichamp, *Chawton House, Hampshire*, eighteenth century. The stables of *c.*1593 lie to the left of the house (Edward Knight).

gatehouse and perhaps corner pavilions. In many cases, however, though the other ancillary buildings were pulled out of sight, the stables were left in a prominent position beside the approach to the house.

A good example of this is the stable at Chawton House, Hampshire, built by John Knight in 1593 (fig. 25). An eighteenth-century painting by Mellichamp shows the stable as the only building flanking an outer court. The stables are thus positioned as a dominant statement on the formal approach to the house and their design is appropriately distinguished. Indeed, the symmetrical building with its projecting gabled wings is so substantial that later writers could not believe that it had been designed as a stable and took to describing it as 'The Old Manor House', a hypothesis disproved not only by its design but by dendrochronological dating of the roof to 1592–93.[20]

Not all sixteenth-century stables were as well designed or carefully placed as those at Chawton House. Many stables, even those for the grandest houses such as Wollaton Hall or Newnham Paddox, Warwickshire, were completely vernacular, with no attempt at any form of architectural distinction (fig. 24). Robert Smythson's ideal plan of Wollaton Hall, Nottinghamshire, shows a courtyard with office buildings, one of them a stable, facing each of the four façades. But, as Siberechts's painting shows, in reality the stables were housed in a huddle of vernacular buildings to the north of the house (figs 26 and 27).

But there was a growing architectural awareness. The better Elizabethan and Jacobean stables, like those at Chawton House, were symmetrical with hipped roofs, gables and small windows. Larger stables, such as those at St Giles's House, Dorset, often had a sequence of gables decorated with finials. Such stables can be compared with multi-gabled houses such as Condover Hall, Shropshire, or Bryanston, Dorset. A particularly fine example was the great stable built probably for the extravagant third Earl of Southampton at Place House, Titchfield, Hampshire, in the first years of the seventeenth century and recorded in two eighteenth-century survey drawings (figs 28 and 30). These show a 106ft-long stable that would have housed twenty horses, an unusually large number, in a symmetrical building with a dominant projecting porch. This also included a kitchen on the ground floor, rooms for bridles and saddles in the mezzanine and a further heated room for the grooms above. There was a separate hayloft, a granary over the stables and another room, perhaps also used for grooms, over the granary. One of the plans shows that the stalls were arranged as a series of bays for architectural effect. These would have been particularly impressive if, as has been suggested, the elegant Ionic capitals in the stables at Cams Hall, Hampshire, originally came from Place House (fig. 29).[21]

There is generally little sign of the advanced ideals of Elizabethan or Jacobean architecture in stable design. Longleat House, Wiltshire, is noted for its incipient Classicism, an impression reinforced by photographs that show it in isolation, but originally visitors would have been very aware of the stables just to one side. Knyff and Siberechts show this to have been a commonplace sixteenth-century gabled building with standard Tudor windows that shared none of the architectural sophistication of the house (fig. 35).

This was also the case at the grandest surviving stables of this date, those at Audley End, Essex, the palatial great house that the Earl of Suffolk is said to have begun in 1603, the year he was created an earl (figs 31 and 32). It was the largest new mansion of the Jacobean era, and the stables were on a comparable scale. Like those at Chawton, the Audley End stables were set apart from the mansion house, flanking the main approach, and were treated as an architectural entity in their own right. In scale they rival many contemporary manor houses, being handsomely built

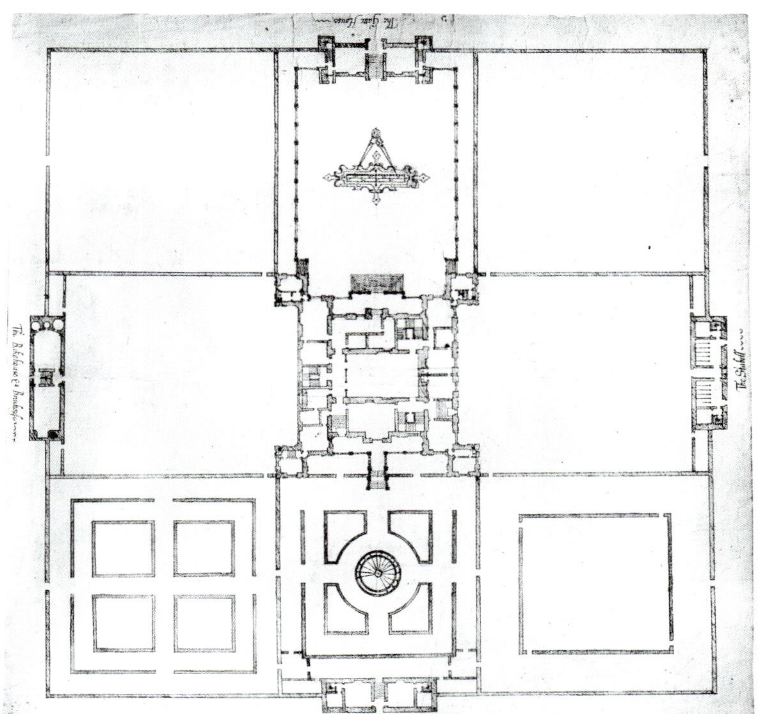

26 (*below left*) Robert Smythson, plan for Wollaton Hall, Nottinghamshire, *c.*1580. The symmetrical service ranges, including the stables on the right, were not built (RIBA Library Drawings Collection, Smythson Collection I/25 (I)).

27 (*below right*) Jan Siberechts, *Bird's Eye View of Wollaton Hall, Nottinghamshire* (detail), 1697. The stables lie beyond the house (Yale Center for British Art, Paul Mellon Collection; photo: Bridgeman Art Library).

28 Early eighteenth-century perspective drawing of the exterior of the early seventeenth-century stables at Place House, Titchfield, Hampshire (Hampshire Record Office, 5M53/1563).

29 Detail of one of the Ionic capitals in the stables as Cams Hall, Hampshire, believed to come from Place House (National Monuments Record © Crown copyright).

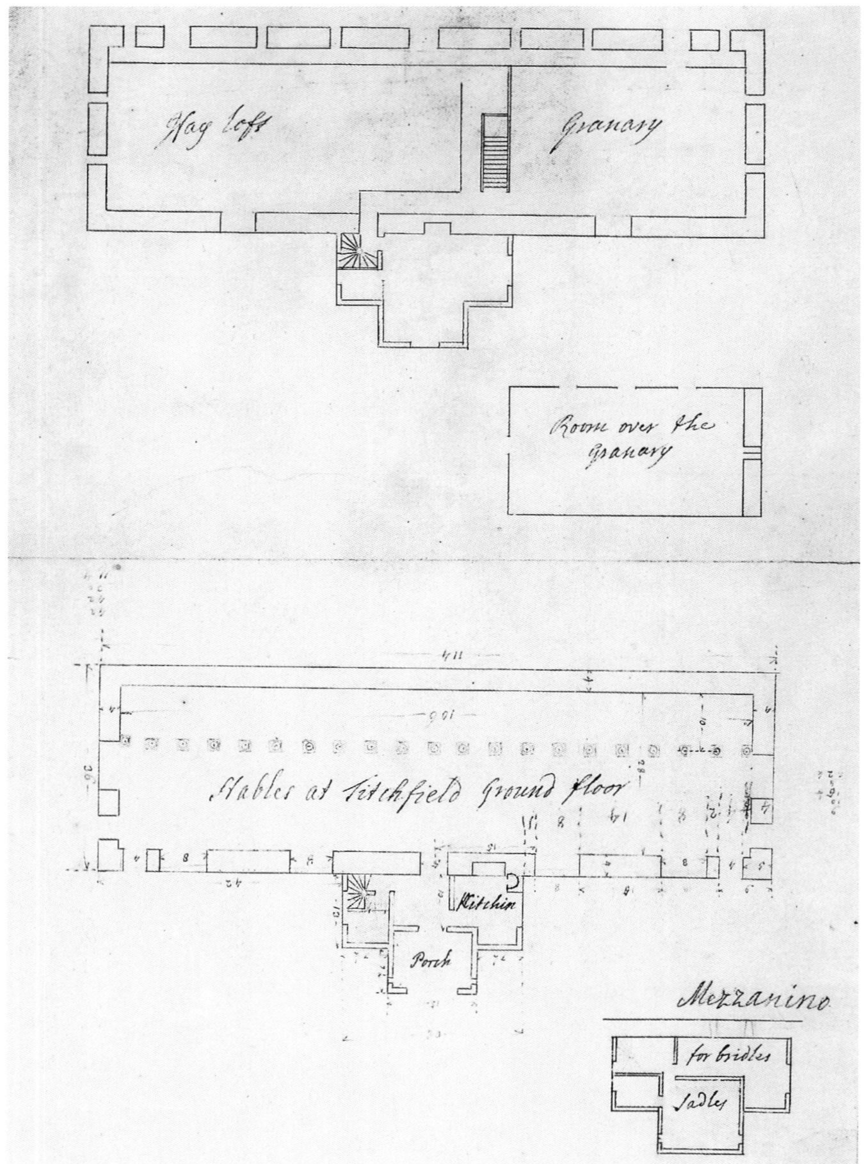

30 Early eighteenth-century survey drawing of the ground-floor, first-floor and mezzanines of the early seventeenth-century stables at Place House, Titchfield, Hampshire (Hampshire Record Office, 5M53/1564).

31 and 32 The stables of *c.*1603 at Audley End, Essex (William C. Rolf, 2000).

33 The early seventeenth-century stables at Wothorpe Lodge, Cambridgeshire (Mark Girouard).

of brick and expensively detailed, with protruding bay windows, gables, stone dressings and an elaborate pilastered and pedimented main doorway. It is easy to imagine the Earl of Suffolk entering it to show off his fine horse to his friends.

But there is a strong contrast between the design of the stables and that of the house, with its large rectangular windows and emphatic cubic design. The steep roof of the stables, the repetitive play of the three large gables, and the small

34 The entrance front of Blickling Hall, Norfolk. The stables of 1622 are on the right (National Monuments Record).

35 Jan Siberechts, *Longleat House from the South*, *c.*1675. The stables, with the added coach houses, lie to the left of the house (reproduced by permission of the Marquess of Bath, Longleat House, Warminster, Wiltshire, Great Britain).

windows with round-headed lights and hood-moulds, all appear strangely old-fashioned. This conservative approach may partly be due to the secondary status of stables, but the demands of practicality were probably more pertinent. The architectural language of advanced Elizabethan and Jacobean houses was not well suited to stable design. Large windows would have let in too much light. Flat roofs allowed owners to walk on the leads of their houses but this was not needed for a stable. Instead, sloping roofs were preferred, which were structurally less complex than flat roofs, and provided room and light for grooms' quarters or a hayloft.

Stables that are stylistically innovative remain rare at this date, but one unexpectedly fine survival is the stable at Wothorpe Lodge, Cambridgeshire, built early in the seventeenth century for Thomas Cecil, first Earl of Exeter, who died

36 and 37 (*top left and right*) Exterior and interior of the stables, built before 1597, at Whitmore Hall, Staffordshire (William C. Rolf, 1999).

38 and 39 (*above left and right*) Front door and window of the stables at Whitmore Hall, Staffordshire (William C. Rolf, 1999).

40 Interior of the stables at Whitmore Hall, Staffordshire (William C. Rolf, 1999).

41 Interior of the stables of 1654 at Peover Hall, Cheshire (William C. Rolf, 1998).

in 1623 (fig. 33).[22] The sophistication of Wothorpe Lodge, built on the plan of a Greek cross with four octagonal towers, is well known. Less familiar is the large stable range, with its elegant Doric-pilastered doorcase and surprisingly sophisticated windows, modelled on a simplified version of those found in Serlio. This source, and the careful Classical mouldings of the door and windows, makes Wothorpe, which was probably used as a hunting lodge, the first surviving Classical stable in Britain.

In the early seventeenth century, the idea that architecturally impressive stables might enhance the impression of the house was taken one step further by the creation of symmetrical wings flanking the forecourt. At Blickling Hall, Norfolk, the approach to the house was originally flanked by a single range of offices, giving the asymmetrical appearance common in the sixteenth century (fig. 34). In 1622–25 Sir Henry Hobart contracted to match this with a new stable range. This was 'to be built . . . to keepe a forme to the ould building in the Courte opposite against it'. In other words, the original range was to be copied to create a grand and symmetrical approach to the house. The pages of Kip and Knyff's *Britannia Illustrata* include several examples of symmetrical flanking ranges of stables and offices contemporary to those at Blickling.

42 The ceiling of the stables at Peover Hall, Cheshire (William C. Rolf, 1998).

The contract for the stables at Blickling Hall, Norfolk, of 1622–25, called on the carpenter to devise the interior, 'according to the plote with pillers of dorica and arches over the Capitals with arkatrive, freese and Cornish . . . not much differing from your honnours stable at Highegate'.[23] The 'plote' must have been similar to that attached to the contract between the Earl of Salisbury and Thomas Avis to convert the hall of the Bishop's Palace at Hatfield into stables in 1628.[24]

These stables do not survive, but two comparable stable interiors survive, both built by the Mainwaring family. The date of the stables at Whitmore Hall, Staffordshire, is uncertain, but they are believed to pre-date 1597 when they appear on an estate map. Those at Peover Hall, Cheshire, were built, according to the inscription over the door, as a gift from Mrs Ellen Mainwaring to her son Thomas in 1654. Though that date seems late for their essentially Jacobean form of the detail, their similarity to the chapel at Cholmondeley Castle, Cheshire, dated 1655, make it plausible.[25]

The Whitmore stable is a small but well-built two-storey stone building with a massive stone lintel over the door (figs 36–40). Though not large, it is typical of the individual stables commonly built at country houses at this time. What is remarkable is the surviving arcading along the rear of the stalls and the arcaded rack at

43 Interior of the stables of *c.*1617 at Dunster Castle, Somerset (William C. Rolf, 1999).

the horses's head. The design is probably similar to that intended for the stables at Blicking and is of considerable distinction for a country squire.

The original ceiling of the stable at Whitmore has been lost, leaving bare boards, but at Peover it survives, revealing elegant plasterwork that combines strapwork and floral decoration (fig. 42). As at Whitmore, a timber arcade runs behind the stalls and above the rack, though the Peover arcade is the more elegant of the two. Even today, with its slightly scruffy sense of faded grandeur, the sophistication of the stable at Peover Hall is evident. The treatment is what one would expect of a domestic interior, not a utilitarian space. There is no reason to believe that the Peover stables were exceptional and they clearly demonstrate that display was an important factor even in gentry stables.

Stables such as those at Whitmore and Peover, together with documentary evidence and printed sources, allow us to gain a fairly full understanding of the interior of the late sixteenth- and early seventeenth-century stable. It is here that the clearest evidence of the marked improvement in stable design can be seen: in the form of the divisions between horses; in the design of the rack from which the horse ate his hay and the manger and trough from which he ate his feed and drank his water; in the materials used for flooring the stall; and in the trouble taken to arrange the drainage.

One issue that was much debated at the time was whether or not to divide the horses with fixed partitions. Such partitions were expensive, both in terms of cost of the materials used and in the greater space a stable needed, and remained unusual in the sixteenth century. 'They that lysted to bestowe more cost', declared Blundeville in 1565, could board up the partitions so that they were fixed and so prevent horses biting, striking or seeing each other, but he argued that 'suche cost is not greatly commodious, onles they might cary their stables with them'.[26]

A more common solution at this date was the use of hanging bales: wooden bars hanging on a chain or leather thong between the manger and the heel post at the foot of the stall. These had been used in Italy since at least 1515, when Leonardo da Vinci's notes for the Medici stables referred to poles being placed between each horse, and were recommended by Blundeville in 1565.[27] They can be found in the 1600 inventory of Ingatestone Hall in Essex and in the specifications for the stables at Blickling Hall of 1622–25, which include 'heel peeses [posts] for the horse and bars between them'.[28] Hanging bales continued to be used in the best stables throughout the seventeenth century. Roger Pratt included them in the stables at Clarendon House, London, in 1666, the most prestigious building project of the day. Similarly, a design by Wren for a stable for Charles II was clearly intended to have hanging bales as no partitions are shown in the section.[29]

The Whitmore stables would have had hanging bales, though they do not survive. Perhaps the earliest that do are those in the handsome and important surviving stables, probably built for George Luttrell by William Arnold at Dunster Castle, Somerset in about 1617 (figs 43 and 44). Though the walls have been stripped down to bare brick and stone, giving the stables a rather scruffy, abandoned air, the grandeur of the fittings and the scale of the interior is impressive even today. At Dunster, the arcade familiar from Whitmore and Peover is replaced by a series of heelposts or elaborately carved columns shaped rather like chess-pieces, at the foot of each stall. Heelposts brought an element of sculptural display into the stable, the

44 Manger in the stables at Dunster Castle, Somerset (William C. Rolf, 1999).

repeated rhythm of the heelpost down the long run of the stalls providing the most striking image in many stables. They were obviously cheaper than arcading, which grew increasingly uncommon from the middle of the century. A large timber construction combining rack and manger runs the whole length of the stable, divided into individual stalls by elegantly scrolled, short partitions, each terminated by another column. In a number of cases, bales still hang between the two columns.[30] The short fixed partition at the head of the stall was recommended by Roger North, who preferred bales to partitions: 'for the horses doe not love to be recluse, but hanker after the enjoyment of their company, which is by the eye, and so have pleasing converse'. But, he continued, 'a post and partition at the manger is good, because it prevents snapping, and unequall feeding, as will be, if they can come at each other's meat'.[31] The disadvantage of bales was that they did not prevent horses from disturbing, and in particular biting, each other. The problem was particularly acute with stallions, as the Earl of Pembroke commented in 1781: 'they are devilish troublesome in strange stables, where there are not stalls, but only barrs'.[32]

By the beginning of the eighteenth century, bales had fallen out of favour in the better class of British stable, though they continued to be used in military stables, where economy and space were more critical, into the nineteenth century.[33] On the Continent, however, bales continued in standard use in grand stables long after they had been abandoned in England, although Mansart's late seventeenth-century stables at Versailles had elegant partitions.[34] The French writer de Garsault commented in 1741 that, while the French used bales, the English separated their horses more securely using partitions, and Briseux still took bales for granted in his *L'Art de bâtir des maisons de campagne* of 1761.[35] In Vienna, the elegant stables of Prince Eugene of Savoy's Belvedere Palace, built in 1721–23, had bales, and fine examples survive at the great stables at the Villa Pisani at Stra near Venice, built around 1735.[36]

As for width, Clifford recommended in 1585 that a stall should be 5ft wide, and this seems generally to have remained standard in the seventeenth century. In the 1660s, the Earl of Clarendon's grand stables at Clarendon House, London, designed by Sir Roger Pratt were 5ft wide, as were those at Kingston Lacy, Dorset. Roger North writing at the end of the century also preferred stalls that were 5ft wide.[37]

The second element in the stable revolution was the transformation of the hayrack that lay at the horse's head. In 1585, Clifford announced that he had devised a new rack for Sir Thomas Scott's stables at Scott's Hall, Kent, and had perfected it at Sir Bernard Drake's stables at Ashe House, Devon. He explained that it was superior not only to the sloping rack, formed of a series of wooden bars set at an angle to the wall and held in place by a further bar along the top (familiar from medieval stables as well as from old farm stables), but to every other rack, including the Italian, German, Spanish and French racks. Clifford's rack was a much more substantial timber construction, with the bars placed vertically, at a distance of 1½ft from the wall. The manger, which ran below, was equally substantial, 1½ft wide and deep and running the whole length of the stable, 3–4½ft off the ground. The advantages of Clifford's rack were that it made it easier for the horse to eat and kept the dust out of its eyes.[38] Its disadvantage was its extravagant use of space and materials, which made it considerably more expensive. Despite this, Clifford's rack became standard in the better class of seventeenth-century stables, with many surviving examples, including those at Whitmore Hall, Peover Hall and Dunster Castle (figs 40–44). An ingenious variation was recorded at Richard Fermor's stables at Tusmore by Plot's *Natural History of Oxfordshire* of 1677. Here, a tube connecting the loft to the manger supplied the correct amount of horses' feed.[39] Fermor's idea was picked up and repeated by several authors in the eighteenth century, but there is no evidence that it was ever acted upon.

The third element in the design of the stall was drainage and flooring. Here again, the late sixteenth and early seventeenth centuries saw substantial change. The need for drainage in stalls for stables to remain healthy is self-evident. As William Horman put it in 1519: 'I have made in every place gutters and synkis to avoyde filthe.'[40] The basic principles changed little. Urine ran either directly, or along a smaller gutter under the stall, into a larger gutter at the horses' rear, which could be open or covered. This main drain had to be kept clean by regular scouring, although in 1635 the grooms at the princely stables of the ninth Earl of Northumberland at Petworth in Sussex could turn on a tap to wash it clean.[41]

Blundevil made use of the Classical Roman authors Palladius and Vegetius to support his argument that wooden planks – the common flooring for stalls at the time in Britain – were the best material. Thomas Tusser's *500 Points of Good Husbandry* of 1557 reveals their ubiquity:

> Sawn slab let lie/ for stable and stye

and

> A stable well planked with key and with lock,
> Walls strongly lined to bear off a knock;
> A rack and a manger good litter and hay,
> Sweet chaff and some provender every day.[42]

Documentary sources record planks exclusively in the sixteenth and into the seventeenth century. Inventories refer only to planks, and they are commonly mentioned in building accounts, where few references can be found to other materials. In 1528, John Newman was paid 4s 4d for planking the stable at Hengrave Hall; in 1602–07, Thomas Taylor of Chobham received 11s 6d for 344ft of planks for the stables at Ashley House, Surrey; the new stables at Penshurst Place, Kent, were planked in 1611–12; the specification for the stables at Blickling of 1622–25 included planking, and the stables at Naworth Castle in Cumberland were planked in 1628.[43] Numerous probate inventories of those who were not particularly wealthy demonstrate that planked stables were ubiquitous at every level of society, as can be seen in the inventories of people like Elinor Pare, late wife of Thomas Pare, saddler of Oxford, in 1582, or Luke Cooke of Bedford, malt miller in 1619.[44] No planked stalls are known to survive, possibly because of the natural deterioration of wood under foot, but more probably because they fell out of favour during the seventeenth century.

The alternative to wooden planks was paving, with stone, small cobbles or brick. Blundeville, who considered paving an unsuitable Italian habit, knew of only one set of paved stables in England when he wrote in 1565. These had been built for Sir Thomas Cheyne at his great house at Shurland in Kent. But Cheyne, who owned many fine horses, was exceptional as he had travelled extensively on the Continent, visiting Rome in 1513–14 and France in 1522 and 1526.[45] Clifford, writing twenty years after Blundeville, was more open-minded, explaining that he had shared the common English prejudice against paved stables until he was forced to use them abroad and discovered that his horses did not suffer from the experience. He recited the standard English objections to paving: that if the horses rest on the ground it would soon turn to mire, but that if the stall were paved it would be very cold and soon wear down the horses shoes; that it would cause them and their hoofs to break more often; and that it was twice as expensive in terms of litter. These he dismissed, explaining instead how to lay a floor of gravel or small stones so that it would remain hard. He argued that although paving might occasionally break a shoe and use more litter, it was much better than planking because it was not so slippery and cost only half as much to maintain.[46]

By the seventeenth century, the tide was turning. Although Markham favoured planking in 1607, both Morgan in 1609 and de Grey in 1639 attacked it vociferously, condemning planks as unhealthy, slippery, liable to shrinkage which would leave dangerous cracks, and to dry out during the summer and break when the horse returned from summer pasture. The trenches below the planks meant that the 'horse lyeth over a moyst and dampish place and vault; and beside that, the evill savour of horse-pisse will be evermore in their nose'. In their eyes, paving was more durable, much cheaper, hardened the horses' feet and did not cause them to slip, and any danger of cold was prevented by having plenty of litter.[47]

Evidence of the move away from planking comes in the royal *Proclamation for the Preservation of Grounds for the Making of Salt-Peeter* of 1624. This declared that 'no stables [are] to be paved, pitched or gravelled where the horse foot used to stand but planked only and neither gravel, stones nor pitch be used deceitfully (as heretofore hath been used) laid under the planks, nor before the planks next the manger, purposeth to destroy the growth of petre and to keep the petre men from working, but only good earth and such as is fit and apt to breed and increase the mines of petre'.[48] Saltpetre was a vital national asset because of its use in making gunpowder and, as stables were an important source, saltpetre makers were allowed by law to dig up their floors to remove it. That this was no idle threat is shown by the entry in the Countess of Leicester's Household Book for 13 March 1635: 'given the saltpetre men not to break up the stable'.[49] The proclamation is revealing because it shows that by 1624 the move away from planked floors was sufficiently marked for even the government to have noticed it.

The proclamation reveals the wide variety of material – gravel, stones and pitch – then in use. Clifford describes a compact surface of gravel, sand and stone. De Grey recommended the stall be pitched in flint. Flint cobbles were popular and can be found at the earliest surviving stable interiors. Whitmore Hall, Peover Hall and Dunster Castle all had cobble floors, although the passageway of the first two, like that at Llanvihangel Court, was flagged in stone. By contrast, at Dunster the whole stable was cobbled (figs 37, 41, 43 and 97). The disadvantage of cobbling was that it was difficult to clean properly as dirt caught in the cracks. It could also present an uneven foothold to the horse.

Bricks were a better, more even solution. In 1635 Sir William Brereton visited Dublin Castle and described the new stables of the viceroy, the Earl of Strafford:

'no planks [are] made use of, but Holland bricks placed upon edges, whereon the horses lie and you walk; these as easy to walk upon as to lie upon, and these are made of Holland earth, which is harder and more durable than our clay: with these the streets are paved in Holland'.[50] Such bricks were known as clinkers and Brereton clearly considered them a novelty. They seem to have become more common around the time of the Restoration. Sir Roger Pratt recorded that the stables at Horseheath, Cambridgeshire, were pitched with the best Holland brick at 3s per yard. They remained expensive. Pratt noted that pitching with flint at Kingston Lacy and Farnham Castle cost 2d per yard if materials were provided and 7–8d if not. What are probably the original clinkers can still be seen in the stables at Powis Castle.[51]

Horses were not the only valuable items kept in stables. Saddles, harness and other trappings also needed to be looked after, particularly those used for display which could be extremely expensive. These could be kept in chests or on pegs behind the horses, but were most conveniently and safely kept in separate saddle rooms. The finest medieval stables, such as the royal stables at Clarendon, had separate saddle rooms, but it was not until the sixteenth and seventeenth centuries that these became common, further evidence of rising standards in stable design. Blundeville, writing in 1565, suggested that though bridles could be hung on the heelposts behind the horse, close-presses should be set up to store saddles, harnesses, stirrups and bits either in the horse-keeper's chamber or in a secure place divided from that chamber. Morgan agreed that grooms' lodgings were the place to keep saddles and bridles, but de Grey thought that they should be kept in a separate saddle room, which could also be used for storing grain. The importance of security was stressed by Markham who was quite happy for ordinary saddles to be kept on a strong piece of timber placed between the windows in the stable, but who argued that the better harnesses and saddles should be kept secure in a separate room.[52] Old Thorndon Hall had a saddle room in 1587–88; Nonsuch Palace, in 1599; Hengrave Hall in Suffolk one in 1603; and Ludlow Castle one in 1631. John Smythson's drawing for the stables at Clifton Hall, Nottinghamshire, of 1632 has a room marked 'sadeles' adjoining the grooms' lodging on an attached flap, which suggests that these were on the first floor (fig. 45), as does the survey of the early seventeenth-century stables of Place House, Titchfield, in Hampshire (fig. 30). The survey drawing of the stables at Place House, Titchfield, shows separate rooms for bridles and saddles on the first floor above the porch.[53] The Petworth inventory of 1632 records that George Browne's chamber contained one 'horse' on which to set saddles.

A constant refrain of those writing about stables from Alberti onwards was the importance of a groom sleeping nearby, both for security and in order to hear if the horses were disturbed in the night. As de Grey put it, a 'convenient lodging chamber for your Groomes whose nearnesse together with their care and vigilancy might prevent many dangers and inconveniences which may accrued to your Horses by night'.[54] Senior grooms might have had the privilege of a separate chamber, but others would be expected to sleep with the horses. Thus the 1598 inventory of Gilling Castle in Yorkshire lists a bed 'in the chamber over the stable' and records a cupboard, bedstead, mattress, boulster, woollen blanket and two coverlets 'in the stable'.[55]

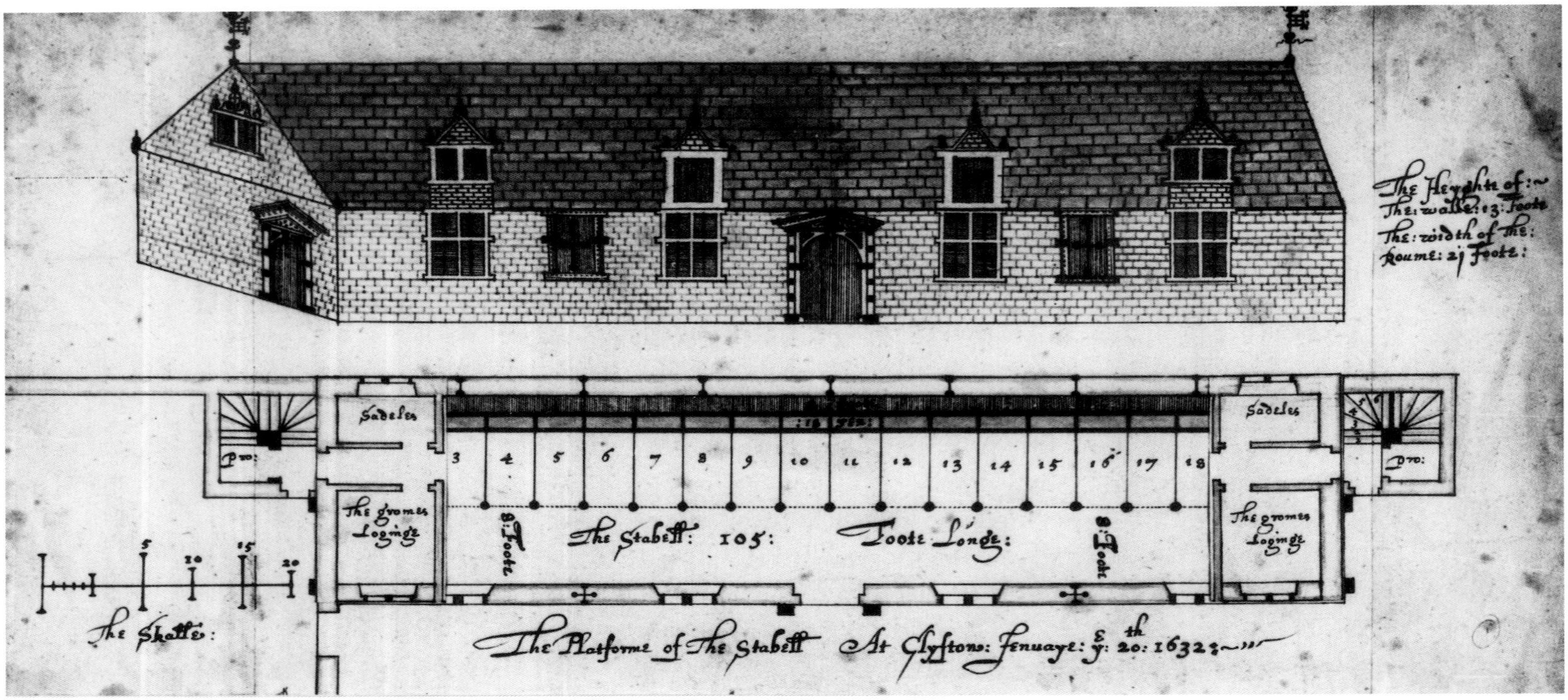

45 John Smythson, plan and elevation of the stables at Clifton Hall, Nottinghamshire, 1632. The grooms' lodging rooms and saddle rooms, drawn on flaps, were on the first floor (RIBA Library Drawings Collection, Smythson Collection III/2).

Alternatively, hay could be kept above the stable. The best Elizabethan and Jacobean stables often included a barn for hay, either on one side of a quadrangle, as at Theobalds in Hertfordshire, or separate, as at the Royal Mews at Hampton Court, Middlesex. Here, the upper floor of the stable quadrangle was occupied by the Grooms of the Stables, the assistants of the Master of the Horse. But for lesser stables where there was no need for an entire barn it was increasingly common to add another storey above the stables for hay, as Clifford recommended.[56] In the seventeenth and eighteenth centuries this became standard.

The introduction of bales, and subsequently fixed partitions; the introduction of new forms of racks; the move from wooden planks to paved, preferably clinker floors; and the growing use of such ancillary features as saddle rooms, together with the higher status accorded to stables in their architectural design and placement, all point to a marked change in stable design in the late sixteenth and early seventeenth centuries. It is a change that can be seen in the writings of stable theorists and is confirmed both in documentary evidence and surviving stables. There is some suggestion in the works of Blundevil and Clifford that the spread of foreign ideas may have been partly responsible: paved floors, which were common on the Continent when English stables were planked, eventually became standard in England. But Continental practices were not always followed, even when known. The Italian ground-level manger recommended by Blundevil was never accepted in England. Other improvements such as Clifford's rack were invented in England and in some areas, such as the preference for partitions over bales and the dislike of sloping stalls, England led the Continent. Instead, the reason for this revolution in stable technology probably lies in the introduction of a new, expensive breed of horse which merited greater comfort than existing breeds.

The Great Horse, the large, heavy-boned horse traditionally used in the Middle Ages for war, had always been very expensive and comparatively rare. Its use had been restricted to battle and tournaments. In England its supply was particularly problematic after the royal studs, the main source for native Great Horses, were run down after 1360. In 1495 Henry VII complained of the shortage of good horses and mares and banned their export. By then Great Horses were largely obsolete in their traditional medieval military form, but they took on a new, courtly role as the fashion grew for riding them on occasions of state and for including displays of horsemanship or *haute école* at court festivals.[57] *Haute école* was taken up with enthusiasm by Henry VIII in the second decade of the sixteenth century and though it declined in popularity later in his reign, it was revived under Elizabeth and flourished under James I and Charles I.

Haute école required large, well-built but nimble horses and as the native English Great Horses were not only scarce but slow, unhandy and lacking in endurance, new sources had to be found abroad, the best being from Andalusia in southern Spain. Henry VIII made great efforts to improve the quality of the native Great Horse and the royal accounts and correspondence include many references to gifts or purchases of Great Horses from the Continent. The Emperor Charles V, for instance, gave him twenty-five Spanish horses. Such imports significantly improved English horse stock but the numbers involved were relatively small as *haute école* was largely restricted to court circles. The stable revolution cannot be put down to them alone. A more profound impact was played by a major social revolution: the introduction of the coach.

'Of old time coatches were not knowne in this Island . . . but now of late yeares the use of coatches is taken up, and made so common, as there is neyther distinction of time, or difference of persons observed.' Stow's censorious words of 1598 on the introduction of the coach to Britain were echoed twenty years later by Fynes Moryson: 'Sixtie or seventy yeeres agoe, Coaches were very rare in *England*, but at this day pride is so farre increased, as there be few Gentlemen of any account (I mean elder Brothers) who have not their Coaches, so as the streetes of *London* are almost stopped up with them.'[58]

The coach, a rarity in the middle of the sixteenth century, had become quite usual by its end. In 1556, two years before Elizabeth became queen, Sir Edmund Hoby offered to let Lady Cecil ride in his coach, but it was Queen Elizabeth's example in acquiring a coach that made them fashionable. The earliest coaches were largely restricted to royal circles, but 'after a while divers great ladies, with as great jealousie of the Queene's displeasure, made them coaches, and rid in them up and down the countries, to the great admiration of all beholders; but then by little and little they grew usual among the nobilitie and others of sort'.[59] Within eight years, so many people had coaches that there was an outcry against the evil effect that they were having on the art of riding (and on the use of boats on the Thames). The rapid rise in the popularity of the coach is borne out by a sudden wealth of contemporary references. In 1573, Sir Thomas Kytson of Hengrave bought a 'coche, with all the furniture thereto belonging except horses' for £34 14s. The 1575 inventory of Lambeth Palace taken after the death of Matthew Parker, Archbishop of Canterbury, included two coaches. And in his will, dated 1569, although not proved until 1580, Sir Giles Alington of Horseheath in

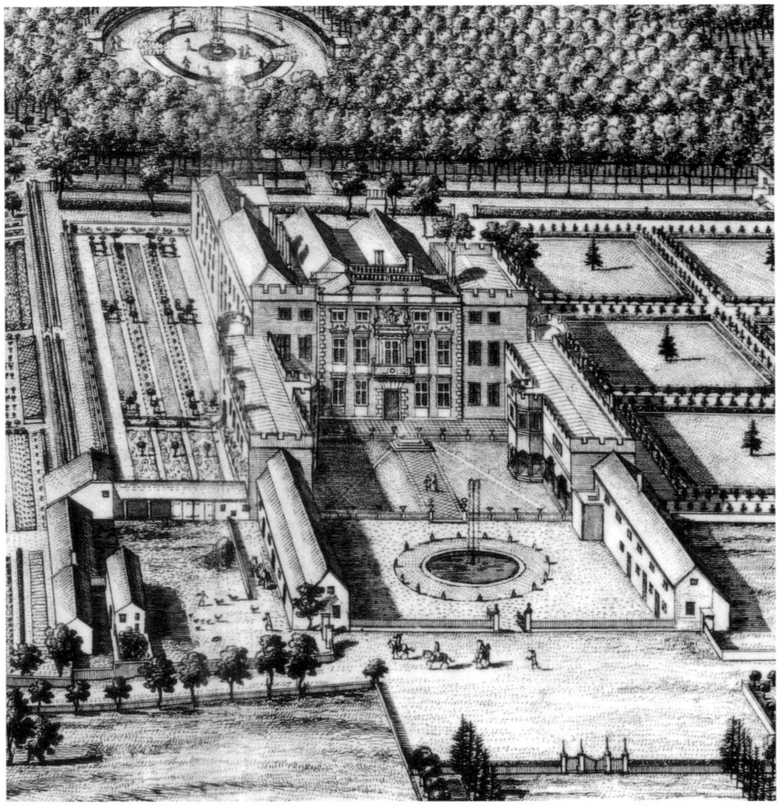

46 (*above left*) Leonard Knyff and Johannes Kip, Hutton Hall, Cumberland, from *Britannia Illustrata* (London, 1707), p. 59 (detail) (© The British Library).

47 (*above right*) 'Master of the Uninhabited Garden', *The North Front of Newburgh Priory, Yorkshire*, *c.*1695–1700. The coach houses lie to the right of the house (*Country Life*).

Cambridgeshire left his wife the coach she was accustomed to use. Thus by 1569 even the wife of a knight had grown accustomed to riding in a coach.[60]

Although the coach was first popular in London, it soon spread to the country. The 1585 inventory of Chenies Manor in Buckinghamshire mentions two coaches; the new stables at Old Thorndon Hall in Essex were built in 1589 with two coach houses; and two coach houses were included in the inventory made in 1593 after the death of Sir Henry Widdrington, deputy-governor of Berwick-upon-Tweed. Frequent references can be found to coaches and coach houses in surveys, inventories and building accounts in subsequent decades. At Ashley House in Surrey, the building accounts of 1602–07 mention the erection of a coach house; and at Petworth House in Sussex, where there had been no coach in the 1574 inventory, that taken in 1632 of the ninth Earl of Northumberland's goods included a coach house.[61] Such references became steadily more common as the century progressed.

The introduction of the coach brought a new element to the stable complex. Coaches needed protection, and the accounts of the Office of Works show that a coach house was built at the Royal Mews, Charing Cross, in 1568, soon after Queen Elizabeth acquired her first coach, apparently in 1564. Substantial works were carried out at the stables at Greenwich in 1568–69 and a new complex of barns and coach houses was built at Hampton Court, which was completed in 1570.[62]

The simplest coach house was a lean-to with no walls but a roof held up by wooden supports. A good example of this can be found in Hollar's view of Arundel House in London of about 1646 (fig. 78) and others can be seen in Kip's view of

Hutton Hall, Cumberland (fig. 46).[63] Kip's view also shows two more coach houses, still simple lean-tos but protected by walls and doors, a design found in the surviving coach houses of 1567–70 at Hampton Court Mews. Such coach houses might be added to the end of a barn or stable without thought for their architectural effect, as at Newnham Paddox, Warwickshire (fig. 24), or tacked onto the return side of the range of stables, as at Coley, Berkshire, and Longleat. Alternatively, coach houses could be treated with a certain architectural distinction by giving each of them a gable, as at Newburgh Priory, Yorkshire, where the coach houses were built in stone with the finials on the gables matching those on the house (fig. 47).[64]

But the arrival of the coach did more than just stimulate a demand for coach houses. Tudor proclamations attempting to improve the quality of horses make it clear that most English horses would now be considered ponies, and the palfrey, the standard travelling horse in sixteenth-century England, would not have been strong enough to pull the new coaches. Larger horses were needed. In time an established native breed of heavy horse would be developed, but this was only really achieved between 1640 and 1700. In the meantime, owners of coaches had to rely heavily on expensive imported horses. Coach horses were as much about display as simple practicality and the demand for handsome horses, in particular the matched teams demanded by fashion, made coach horses very expensive. In the survey of Wanstead made in 1588 at the death of Queen Elizabeth's favourite and Master of Horse the Earl of Leicester, the twelve coach horses were worth £87 13s 4d. By contrast, twelve Irish hobbies were worth £41 13s 4d, twelve 'Baye ley' £26 13s 4d and four cart horses and a colt £5 5s. In 1616, when the average price of a gelding in the royal stables was £12–20, £30 a piece was paid for three large grey geldings to pull the king's coach.[65] In the country, at least four and possibly six horses were needed to draw a coach because of the appalling state of the roads. In 1616 the Earl and Countess of Dorset travelled north in two coaches, each with four horses. However, when the earl died in 1624, he left his wife his 'carriage made by Mefflyn, lined with green cloth and laced with green and black silk lace', and his 'six bay coach geldings', suggesting that six coach horses were considered a team. The Gosfield inventory of 1638 confirms this with reference to a fine new coach and a suit of harness for six horses.[66]

It seems unlikely to be a coincidence that the Tudor and Stuart revolution in stable technology happened at the same time as the spread of the new, expensive breed of great horse for *haute école* and pulling coaches. As coach owning spread throughout the nobility and gentry, it was above all the introduction of the coach horse that stimulated new thinking in stable design.

Inventories show how significant an element of a stable coach horses could comprise. At Cockesden in Yorkshire in 1626 there were four coach horses, three saddle geldings, two nags and one mare with colt. At Stapleford Park, Leicestershire, in 1640 there were seven coach horses, one white stone horse, one old bay stone horse (both probably riding horses), one mare, four geldings, two little nags and some foals. At Ham House in Surrey there were fourteen coach horses and ten saddle horses in about 1679. The Stapleford inventory also gives an indication of the relative value of coach horses. The seven of them were worth £58 10s while the other nine horses were worth £50.[67]

48 The mid-seventeenth-century stables at Boundary Farm, Framsden, Suffolk (National Monuments Record).

The need to house these extra horses, and to accommodate the improved stable technology, probably explains the series of new, relatively small but architecturally distinctive stables built at country houses in the seventeenth century, such as those at Whitmore Hall, Peover Hall and Llanvihangel Court, Gwent (figs 36 and 96). These were clearly prestigious buildings, handsomely designed and well constructed to house about eight or ten horses. A good example can be seen in the fine painting of Llanerch in Denbighshire, made in 1662, where a five-bay, two-storey stable with quoins and pediments over the windows was added to the late sixteenth-century house, probably in the early seventeenth century (figs 49 and 50).[68]

One of the most remarkable surviving examples of these new compact stable blocks dates from the middle of the seventeenth century at Boundary Farm, Framsden in Suffolk (fig. 42). This three-bay, two-storey stable was solidly built of brick with elaborate gables at each end and pediments over the windows. The interior was clearly meant to be equally impressive. Sadly today only one large acorn pendant survives, but the fittings were described in detail by a local antiquarian, the Reverend Edmund Farrer, in 1920:

> Running the whole length of the east wall, is a hay rack for the horses, which, I think, may be considered unique in East Anglia. It so exactly resembles the early altar rails that one is inclined to jump to the conclusion that we have here an instance of ecclesiastical spoliation . . . but a further investigation reveals [that they] were made for this old stable. There is, of course, the usual series of rails

through which the horses obtain the hay, but as ornamentation there are five oak pillars with capitals, one exactly in the centre, and those at the sides being equidistant from the walls; below the rails is a running Jacobean pattern, rather suggestive of the earlier years of the 17th century.

The architectural ambition is clear, but what makes these stables remarkable is that they were not built by a wealthy gentry family. James Wythe, who built them, along with a fine summer house, was at the bottom end of the gentry class, only a short step above the level of a yeoman – the title his grandfather had been content to claim not many years before.[69] Fine stables were becoming an essential feature of even the smallest country houses.

49 and 50 (*above and facing page bottom*) Anon., *View of Llanerch, Denbighshire, with St Asaph's Cathedral in the Background*, 1662. The mid-seventeenth-century stables lie in the entrance court, to the left of the house, with a mounting block in front (© Yale Center for British Art, Paul Mellon Collection; photo: Bridgeman Art Library).

THE RIDING HOUSE IS ONE OF THE FORGOTTEN BUILDING TYPES of architectural history.[1] Substantial covered spaces used for the exercise and training of horses, they were perhaps the largest independent structures associated with the country house, and a significant building type even in urban settings. They were also, alongside the real tennis court, one of the very first buildings designed for a particular sport. But, except for a handful of distinctive examples, in particular that at Bolsover Castle, Derbyshire, built by the Duke of Newcastle, riding houses have largely been ignored by architectural historians.

Riding houses were first built for the practice of *haute école*, the schooling of horses in a series of formalised movements that was developed in Italy, particularly in Naples, in the early sixteenth century. *Haute école* is about training riders to

3

THE RIDING HOUSE AND THE FORGOTTEN ART OF *HAUTE ECOLE*

manage their horses with confidence, whether on the parade ground or the battlefield. The intention was to make horses 'supple, flexible and obedient, and to make them lower their haunches, without which a horse . . . will be neither comfortable in his movements nor pleasurable to ride'.[2] But, as anyone who has ever seen the Spanish Riding School knows, it developed into an art of great beauty. The essence of the art lay in the schooling of a horse by a rider or by a man on foot in a series of set movements within a relatively confined area. Exercises were divided between work on the ground, which taught a horse the agility it needed in ordinary riding and provided the foundation for the more advanced parts of *haute école*, and the high airs, where the horse leapt in the air.

From the sixteenth to the eighteenth century, a knowledge and skill in *haute école* was seen as an important characteristic of those of royal, noble or gentle birth. As Castiglione explained in *The Courtier*: 'therefore will I have our courtier a perfect horseman . . . and because it is the particular praise of us Italians to ride well, to manege with reason, especially rough horses, to run the ring and tilt, he shall be in this as among the best Italians'.[3] Gervase Markham developed the point in 1615: 'There can be no greater or better recreation either for health, profit, or renowning of their owne vertues, then the riding of great *Horses*, which in the verie action itselfe speaketh Gentleman to all that are performers or doers of the same.'[4]

Haute école marked a fundamental change in horsemanship from the Middle Ages. It was particularly inspired by the study of Xenophon's *The Art of Horsemanship*, first printed in 1516, which showed that the scientific study of horsemanship had been practised in Classical Greece. But even by then, as Castiglione's *The Courtier*

51 The entrance to the 1630s riding house at Bolsover Castle, Derbyshire (William C. Rolf, 1999).

published in the same year makes clear, *haute école* had displaced tournaments as the favoured recreation of the Italian courts.[5]

Though various cities were important in the rise of *haute école*, primacy has traditionally been given to Naples. 'Ay, that's a colt indeed, for he does nothing but talk of his horse', declared Portia contemptuously of the Neapolitan prince in *The Merchant of Venice*,[6] reflecting the common belief of the time that Naples was the seat of horsemanship. The city's pre-eminence followed its occupation by the Spanish, which led to the establishment of a viceregal court in 1504, and the foundation there of the first academy of *haute école* by Federico Grisone in 1532.[7] The Spanish also brought the gineta, the 'breeding Jennet, lusty, young and proud', as Shakespeare described it in *Venus and Adonis*.[8] The gineta was the classic *haute école* horse, introduced by the viceroy Gonzalo Fernandez de Cordoba as part of his reforms in horsebreeding.[9] It was celebrated in one of the first books on the subject, *De la Cavelleria di la Gineta* of 1551, by the Spaniard Ferdinando Chacon.

Grisone was also the author of *Gli Ordini di Cavalcare*, published in Naples in 1550. Its influence is clear from the speed with which it was translated, first into French in 1559, then English in 1560, Spanish in 1568 and German in 1570. By 1620 there had been sixteen Italian editions and eleven French editions. But Grisone's influence was also personal. For decades, enthusiasts flocked to his academy from all over Europe, while his pupils took the art to foreign courts. Giovanni Battista Pignatelli trained the three key French horsemen of the late sixteenth and early seventeenth centuries, the Chevalier de St Antoine, Salomon de la Broue and Antoine de Pluvinel. Paolo d'Aquino and Bernardo de Vargas Machuca went to Spain, and Claudio Corte and Robert Alexander went to England.

In 1514, the Marquis of Mantua made a gift of horses to Henry VIII which were put through their paces 'in the Spanish fashion' by his envoy Giovanni Ratto.[10] Henry VIII was quick to take up *haute école* as one of the skills that distinguished the cultivated monarch, as the Venetian ambassador Sebastiano Giustiniano revealed in his description of a festive day spent by the Court at Greenwich in May 1515. The king appeared mounted on one of the Marquis of Mantua's horses: 'directly we came in sight, he commenced making his horse curvet, and performed such feats, that I fancied myself looking at Mars'.[11]

Haute école and jousting were often linked and in July 1517 another Italian, Nicolo Sagudino, reported a joust to honour the Flemish ambassadors: 'Between the course, the King and the pages, and other cavaliers, performed marvellous feats, mounted on magnificent horses, which they made jump and execute other acts of horsemanship, under the windows where the most serene Queens of England and Dowager of France were, with all the rest of the beauteous and lovely and sumptuously apparelled damsels . . . The King performed supernatural feats, changing his horses, and making them fly rather than leap, to the delight and ecstasy of everybody.'[12]

Henry tried without success to persuade Ratto to stay.[13] He was more fortunate when Alfonso of Ferrara sent a gift of horses in 1519 accompanied by Antonius Scaticia, a groom described as 'the most skilful rider in Italy', 'to teach their manage'.[14] According to the Duke of Newcastle, Henry also sent for two pupils of Grisone.[15] One of them was Robert Alexander, later celebrated in *The Perfection of Horsemanship:*

Who brought the proudest Coursers to his becke,
And with his hand, spurre, voice and wand, did tame
The stately Steeds that never brookt the checke[16]

The king's success in instilling *haute école* as a necessary part of courtly culture is summed up by Sir Thomas Elyot in 1531 in *The Boke Named the Governour*, addressed to the king: 'The most honourable exercise in mine opynion, and that becometh the estate of every noble personne, is to ryde surely and cleane on a great horse and a roughe, whiche . . . importeth a majesty and drede to inferiour persones, beholding hym above the common course of other men, daunting fierce and cruel beast.'[17]

There is no evidence that Henry built riding houses to practice *haute école*, which was originally practised in the open air, in open maneges. According to Richard Berenger, Pignatelli never used a riding house, and no example is known to have been built in England in the sixteenth century.[18] As Henry VIII often displayed his skill at *haute école* in the intervals between jousts, he may have used his tiltyard rather than a specific manege area, the first record for which dates from the reign of Queen Elizabeth I.

Haute école flourished during Elizabeth's reign, despite the fact that as a woman she was not able to participate. 'Overlong it were to reaccount the names of all the Noblemen and Gentlemen, that in this age have aspired to singular skill in horsemanship', was how Thomas Bedingfield summed up the reign in 1584, adding: 'which I impute cheeflie to the imitation of great personages'.[19] The queen's favourite and the Master of the Horse Robert Dudley, Earl of Leicester, was a great enthusiast, having presumably been schooled by his father, the Duke of Northumberland. Elizabethan *haute école* probably owed its success to his support. Certainly, Thomas Blundeville, author of the first English treatise on *haute école*, *A New Booke Containing the Arte of Ryding, and Breaking Grete Horses*, probably published in 1560 (fig. 52), dedicated his work to Leicester, whom he described as having 'no small skill in the art wherein you are a master as well by knowledge as also by office'.[20] In his dedication, Blundeville appealed to Leicester to encourage *haute école* among the riders of the queen's stable: 'especially now Her Majesty provides for breeding and keeping the Great Horse'. A survey of the royal studs at Tetbury and Malmesbury in 1576 provides evidence for this. The list of stallions included three from Naples, Il Superbo, Non Piu and Grisone, clearly manege horses.[21]

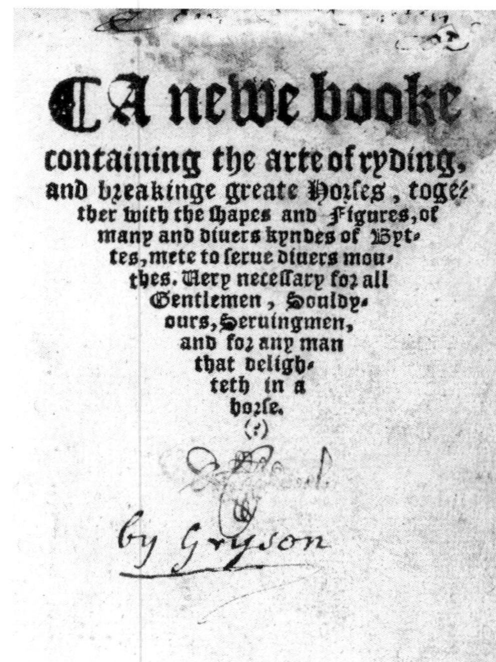
A newe booke
containing the arte of ryding,
and breakinge greate Horses, toge-
ther with the shapes and Figures, of
many and diuers kyndes of Byt-
tes, mete to serue diuers mou-
thes. Very necessary for all
Gentlemen, Souldy-
ours, Seruingmen,
and for any man
that deligh-
teth in a
horse.
by Gryson

52 Title page of Thomas Blundeville's *A New Booke Containing the Arte of Ryding, and Breaking Grete Horses*, *c.*1560 (© The British Library).

'Breeding and keeping the Great Horse' included establishing customised open maneges, though not covered riding houses. In 1576, when Prospero had just arrived at the mews, a 'riding place' 300ft long and divided off by posts and rails was created. This was probably intended for *haute école* as was the 'Rode or manneginge place' for the queen's great horses formed in 1582–84.[22]

Leicester's nephew Sir Philip Sidney was another distinguished Elizabethan enthusiast for *haute école*. As part of his Continental travels from 1572 to 1575 he visited the emperor's court at Vienna with Edward Wotton, where 'we gave our selves to learn horsemanship of Jon Pietro Pugliano'. So taken was Sidney by Pugliano's praise of horsemanship that he confessed: 'if I had not bene a peece of a *Logician* before I came to him, I thinke he would have perswaded me to have wished myself a horse'.[23] Sidney's love of horsemanship can be seen in his poetry, in his sonnets, in *The Countess of Pembroke's Arcadia* and in *Astrophel and Stella*, where he compares love and the manege:

> I on my horse, and *Love* on me doth trie
> Our horsemanships, while by strange work I prove
> A horseman to my horse, a horse to *Love*:
> And now man's wrongs in me, poor beast, descrie.
> The raines wherewith my Rider doth me tie,
> Are humbled thoughts, which bit of Reverence move,
> Curb'd in with feare, but with guilt bosse above
> Of Hope, which makes it seem faire to the eye.
> The Wand is Will, thou, Fancie, Saddle art,
> Girt fast by Memorie, and while I spurre
> My horse, he spurres with sharp desire my hart;
> He sits me fast, how ever I do sturre:
> And now hath made me to his hand so right,
> That in the Manage myselfe takes delight.[24]

Sidney not only praised horsemanship in his poetry, but encouraged it in others. To his brother Robert, who was about to travel abroad, he recommended a list of books on horsemanship to read, including those by Grisone and Corte: 'so shal you profit more in a month than others in a year'.[25] For his nephew William Lord Herbert, later third Earl of Pembroke, Sidney brought over an Italian riding master, Signior Romano.[26] Another Italian whom Sidney brought back was 'the celebrated Prospero' from whom Gervase Markham, author of a series of books on horsemanship, learnt much at Master Thomas Story's riding school at Greenwich.[27] Prospero was one of several Italians employed there.

It is a sign of the popularity of *haute école* in Elizabethan England that London seems to have been well provided with schools such as Story's and teachers such as Prospero. 'I may not forget the praise due to those that be teachers of this art in the city of London', wrote Thomas Bedingfield, 'for by their industry and virtue, the number of serviceable horses and horsemen is daily increased.'[28] There was even a proposal in about 1570 by Sir Humphrey Gilbert to set up a more formal academy where the queen's wards and other young men would be taught horsemanship.[29]

An informal version of such an academy was held at Holme Lacy in Herefordshire by Sir James Scudamore (1568–1619), who was renowned for his skill at tilting in front of the queen. 'When he came to reside with *Sr John Scudamore* his Father,' recalled William Higford, 'Himself and some other brave Cavalliers, and some of their Menials and of his suit, to manage every morning Six or more brave well-ridden Horses, every Horse brought by his Groom to such decency, order and honour, that *Holme Lacy* at that time, seemed not onely an Academy, but even the very Court of a Prince.' Higford was so taken by Scudamore's prowess that when he came to list the virtues of a gentleman in his *Institutions or Advice to his Grandson*, he declared: 'the first that I shall propose unto you is the noble exercise of *Riding the great Horse*. A Knight on Horseback is one of the goodlyest sights in the World.'[30] Higford compared Holme Lacy to a royal academy and its model was indeed followed in the next reign by Prince Henry at Richmond and the Earl of Northampton at the Court of the Lord President of the Council of Wales in Ludlow.

Haute école was certainly well enough established in the Elizabethan consciousness for Shakespeare to use it as a frequent source of metaphor. In *Henry VIII*, Bishop Gardiner reveals an unexpected familiarity with *haute école*:

> For those that tame wild horses
> Pace 'em not in their hands to make 'em gentle
> But stop their mouths with stubborn bits and spur'em
> Till they obey the manage.[31]

Similarly, the horse skilled in *haute école* is one of the key images in *Venus and Adonis*:

> Sometimes he trots, as if told the steps,
> With gentle majesty and modest pride;
> Anon he rears upright, curvets and leaps,
> As who should say 'Lo, thus my strength is tried,
> And this I do to captivate the eye
> Of the fair breeder that is standing by.

To cater for all this interest, a series of books on horsemanship were published. The first was Thomas Blundeville's *A New Booke Containing the Arte of Ryding, and Breaking Grete Horses*, probably published in 1560. Encouraged by its reception, Blundeville published an expanded version in 1565, *The Fower Chiefyst Offices Belonging to Horsemanshippe*, which by 1609 had run into six editions. Two further works were published in 1584, Thomas Bedingfield's translation of Claudio Corte's *The Art of Riding* (fig. 53) and John Astley's *The Art of Riding*. Christopher Clifford's *The School of Horsemanship* was published in 1585 and in 1593 Gervase Markham published his book on the subject, *A Discourse on Horsemanshippe*. It becomes clear from these books that there was a closely knit circle of enthusiasts at work, exchanging and training each other's horses, while dedicating books to each other. Blundeville notes how one of Leicester's horses had been trained by Astley's man Abraham.[32] It was Astley who persuaded Blundeville to translate Grisone. Both Bedingfield's and Astley's books were dedicated to Henry Mackwilliam, one of the Gentleman Pensioners, who had been given a special responsibility by Henry VII for improving horsebreeding and seem to have been a centre of interest in *haute école*. Sir James Scudamore, who ran the informal academy at Holme Lacy, was Standard Bearer of the Gentlemen Pensioners.

The Italian dominance of *haute école* came to an end with the rise of Pignatelli's three French pupils, de Pluvinel, de la Broue and St Antoine. As the Duke of Newcastle later put it, they 'filld *France* with *Frenchmen Horse-men*; which were fill'd were *Italians*'.[33] The most important of the three was de Pluvinel, who founded the first French riding academy in Paris in 1594 with the support of Henri IV.[34] With its establishment, French *haute école* could be said to have come of age. Before 1594, young Frenchmen had to travel to Italy to study horsemanship, but following de Pluvinel's success other academies sprang up in Paris and in the rest of France, attracting both Frenchmen and foreigners. French dominance in *haute école* was reinforced by the publication of a series of books, beginning with de la Broue's *Preceptes principaux que les bons cavelerises doivent exactement observer en leur escoles* of 1593 and de Pluvinel's *Maneige royale* of 1623 and *L'Instruction du roy en l'exercise de monter a cheval* of 1625.

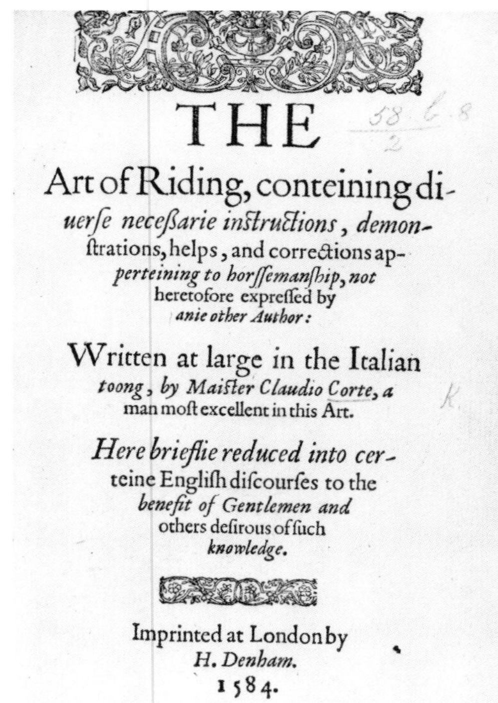

THE

Art of Riding, conteining di-
uerse necessarie instructions, demon-
strations, helps, and corrections ap-
perteining to horssemanship, not
heretofore expressed by
anie other Author:

Written at large in the Italian
toong, by Maister Claudio Corte, a
man most excellent in this Art.

Here brieflie reduced into cer-
teine English discourses to the
benefit of Gentlemen and
others desirous of such
knowledge.

Imprinted at London by
H. Denham.
1584.

53 Title page of Claudio Corte's *The Art of Riding*, translated by Thomas Bedingfield, 1584 (© The British Library).

Although the Duke of Newcastle later criticised de la Broue for being tedious and derivative, de Pluvinel's more progressive approach was a considerable improvement on that of the Italians. De Pluvinel considered the horse a highly developed animal that must be taught to think and believed that this was to be achieved by a system of rewards and a sparing use of the whip and spurs. In this way the rider would arrive at a closer unison with the horse and achieve 'grace and elegance, in addition to science and ambition'. The book was well illustrated with plates of de Pluvinel teaching the young Louis XIII and took the novel form of question and answer between the young king and his master. The prestigious nature of the art could hardly have been more clearly stated. As Walter Liedtke puts it, 'for the first time a particular monarch, indeed an ambitious young King much on the minds of his royal neighbours, was acclaimed as a master of horsemanship'.[35]

When compared with Italy, the convenience of France, especially Paris, and the glamour of the French court made it almost inevitable that the French academies should prove attractive to young Englishmen anxious to acquire a gloss of sophistication. Among the first was Henry, Lord Clifford, son-in-law of the Earl of Salisbury, who joined de Pluvinel's academy in 1609. Others soon followed him. 'Though I never knew yet of any Englishe noblemen or gentlemen undertake this cours of the Academy before,' wrote his tutor, William Becher, to Lord Salisbury, 'yet my Lord's example hath already brought one more which is Mr Puckering.'[36]

The importance of *haute école* is perhaps most clearly shown in Henri IV's coronation gift to James I in 1603. This comprised six beautiful and richly caparisoned horses and the Chevalier de St Antoine, 'an esquire chosen by my hand to help my dear nephew your son ride'.[37] The duc de Sully who presented the gift described the horses as 'parfaitement beaux et bien dressés, et richement caparaconnés' and de St Antoine as 'le plus excellent homme de cheval, quoi connut'.[38] The royal stables continued to be enriched by such gifts during the reigns of James I and Charles I, to the great advantage of English horsemanship and horsebreeding, as when the Archduke Albert sent a dozen mares in foal, four horses and eleven stallions from Italy in 1605. The Prince of Wales's consolation prize after his unsuccessful visit to Madrid in search of the Infanta's hand included twenty-four horses from the royal stud at Cordoba. His travelling companion the Duke of Buckingham was given a further dozen horses.[39]

St Antoine ensured that the previous Italian dominance of the royal mews was replaced by French control. In 1610, three French esquires were listed in the establishment of Prince Henry's stables: St Antoine, M. le Coste and M. Beauclare.[40] The Duke of Newcastle mentions all three in his history, adding a fourth, M. Fountenay, whom he believed was St Antoine's nephew or natural son.[41] In 1611, when the Earl of Salisbury wanted a riding master for his eldest son, Lord Cranborne, he followed the royal example and asked Sir Thomas Edmondes to find him a French écuyer. Edmondes was clearly successful as payments to M. le Fevre, the rider, appear in Lord Cranborne's accounts the following year.[42]

French dominance was not total. Richard Zinzano, presumably from the well-established Zinzano family of horsemen and inevitably known as Alexander, kept his position at the mews and obtained the reversion of an annuity for riding the King's Great Horses in 1624.[43] John Evelyn, whose horse was trained by Zinzano, mentions him as the king's *cavallerizzo* in 1643.[44] Nor was Italy abandoned as a place of study, often being visited by those who had first been taught in France, like Sir Thomas Puckering and Alan Percy who rented a house in order to take riding lessons from the renowned Rustico Picciardini in Florence.[45] Such direct contact meant that Italians were still brought back to England. Lord Howard de Walden, eldest son of the Earl of Suffolk, who spent some time in Naples, had a Neapolitan, Signior Hannibal in his stable.[46]

Though James I had no interest in *haute école*, preferring to spend his time hunting, his elder son Prince Henry's love of *haute école* was attested by the author 'W. H.': 'He took great delight in riding of great horses, and laboured to have the best and rarest that were to be found, and had such care taken of them that he went often to the stables to see them and accounted them to be part of his best jewels.'[47] According to Sir Charles Cornwallis, 'For practice, hee used in a manner daily to ride and manage great horses, with which he had his stables most excellently furnished, oftimes to runne at the ring, and sometimes at Tilt, both with he so well and dextrously performed, and with so great comelinesse, as in those his first years, he became second to no Prince in Christendome, and to many that practised with him much superiour.'[48]

Henry's enthusiasm for the manege may well have been inspired by his father's lack of interest, just as his dislike of hunting was probably influenced by James's passion for it. As early as October 1606, when Henry was only twelve, the French ambassador reported that 'he is a particular lover of horses and what belongs to them, but he is not fond of hunting'.[49] In May 1611, the Venetian ambassador reported that 'The Prince will retire to his house at Richmond, the King allowing him as he knows that the Prince does not care for such assiduity in the chase.'[50] Henry's occupation at Richmond is confirmed in a later report: 'The remainder of the morning was spent by the Prince tilting at the ring and in his *manege* in which he made a fine show.'[51]

To accompany the prince, a select group of young nobles was gathered as an informal academy. Lord Cranborne was one of those closest to the prince. His six months in Paris in 1609–10 included tuition from the best French riding masters, and on his return to England his training was continued at Richmond. In September 1610 the Hatfield accounts refer to the cost of keeping Lord Cranborne's Great Horses at Richmond for a month and in March 1612, 'Monsieur St Antwaine' was paid £30.[52] St Antoine accompanied another of the prince's intimate circle, the Earl of Essex, on a tour of France, Germany and the Low Countries in 1608.[53]

The best known of Prince Henry's riding companions was William Cavendish, later created Earl and ultimately Duke of Newcastle, the author of *Le Methode et invention nouvelle de dresser les chevaux* of 1658 and *A New Method and Extraordinary Invention to Dress Horses* of 1667 (fig. 54). Cavendish's wife subsequently recalled that he was sent by his father as a boy of ten 'to the *Mewse*, to *Mons. Antoine*, who was then accounted the best Master in that Art'.[54] Clarendon later described Newcastle as 'a very fine gentleman, active and full of courage, and most accomplished in those qualities of horsemanship, dancing and fencing, which accompany a good breeding; in which his delight was'.[55]

The shameless self-promotion of the duke and his duchess means that much more is known about him than about any other horseman of his generation. But Newcastle, whom de la Gueriniere allowed had been the most skilful horseman of

54 The Marquess (later Duke) of Newcastle instructing in the art of horsemanship, from William Cavendish, *Methode et invention nouvelle* (Antwerp, 1658) no. 18 (© The British Library).

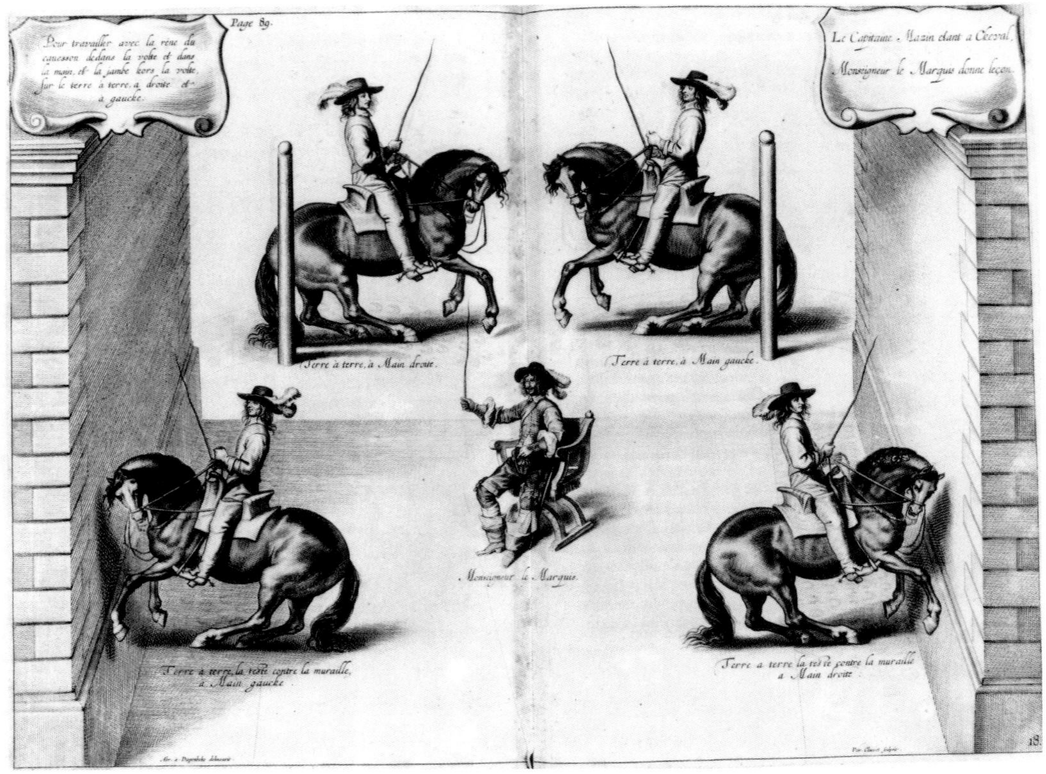

his age[56] – a telling comment from a Frenchman given that the French considered themselves the masters of the art – deserves his fame. His two books form invaluable sources of information about the state of *haute école* both in his own time and in the previous century.

Ben Jonson was equally flattering:

> When first, My Lord, I saw you backe your horse,
> Provoke his mettall and command his force
> To all the uses of the field, and race,
> Me thought I read the ancient Art of *Thrace*,
> And saw a Centaure, past those tales of *Greece*:
> So seem'd your horse and you both of a peece.[57]

Another member of the informal academy was Lord Percy, eldest son of the Earl of Northumberland. The fact that his father spent much of James I's reign in the Tower of London for his supposed involvement in the Gunpowder Plot does not seem to have spoilt Lord Percy's chances at Court. The Petworth archives include payments for stabling one of the earl's Great Horses at the Royal Mews in 1612–13.[58]

Not everyone was entirely positive about the influence of *haute école*. Writing about the value of exercise in his *Anatomy of Melancholy* of 1621, Robert Burton speaks of 'Riding of great horses, running at rings, tilts and turnaments, horse-races, wilde-goose chases, which are the disports of greater men, and good in themselves, though many Gentlemen by that means gallop quite out of their fortunes.'[59] At the universities it was considered a dangerous distraction. When Henry Slingsby, whose brother William had gone to de Pluvinel's academy in Paris, was at Queen's College, Cambridge, in 1619, he passed his spare time fencing and riding the great horse.[60] The authorities were sufficiently concerned by such distractions that in 1637 the Archbishop of Canterbury sought to ban Mr Crofts from teaching *haute école* at Oxford: 'for certainly it cannot be fit for the university; though the exercise itself be exceedingly commendable'.[61] The Inns of Court appear not to have shared his qualms for in 1634 a Mr Crofts, presumably the man later banned from Oxford, wrote to then Earl of Newcastle, 'We are here in great expectation of a horse masque and a foot masque of the Templars, and opinions are divided which will prove the best. I am for the horse masque because I think the horses will dance under them much better than they can when they are on their own legs.'[62]

It is to the informal circle of horsemen around the Prince of Wales that we owe the first English riding houses. An open manege on its own may have made sense in balmy Italy – they can often be found in seventeenth-century Italian paintings – but in the colder, wetter climate of northern Europe, the advantages of a covered area in which to practise *haute école* were clear. Not that the advantage of the riding house was just shelter. As Berenger later explained: 'In the early days of the *modern Manege*, horsemen had not the advantage of covered buildings, expressly erected for the purpose of riding and breaking horses; but from the want of them were exposed to great inconveniences, and obliged to have recourse of various shifts and contrivances; for besides the shelter which a roof affords both man and beast, the *Walls* of the riding-house are immediately necessary to assist the horseman by awing, guiding and confining the horse.'[63]

The first English riding houses were probably those built for Prince Henry. In 1611, the Landgrave of Hesse visited the prince at his palace at Richmond, the event being recorded by Johann Georg Dehn-Rotfelser, the secretary of one of the Hessian ambassadors: 'In the morning of 26 July we saw 22 fine horses from several nations, also the Prince and the Duke of York riding in the Riding School. The stable is in an old monastery called Sheen, the riding instructor was called de S. Anthoin, a Frenchman.'[64] The stables at the royal palace of Richmond were in the adjoining former Carthusian monastery at Sheen, and the parliamentary survey of 1649 confirms that there used to be a riding house there: 'one other barn heretofore used for the riding of the King's Great Horses', as well as a great horse stable.[65] The riding house was almost certainly built as part of the major improvements to the palace carried out by Prince Henry at a cost of £10,000 between 1610 and 1612.[66]

Prince Henry's other riding house, at St James's Palace in London, is well known from John Smythson's survey drawing taken soon after the building was completed (fig. 55), as well as from Kip's engraving of St James's Palace.[67] This was built by William Browne and James Hutchinson in 1607–9 for

55 Robert Smythson, survey plan, elevation and section of the riding house of 1607 at St James's Palace, London, 1609 (RIBA Library Drawings Collection, Smythson Collection I/14).

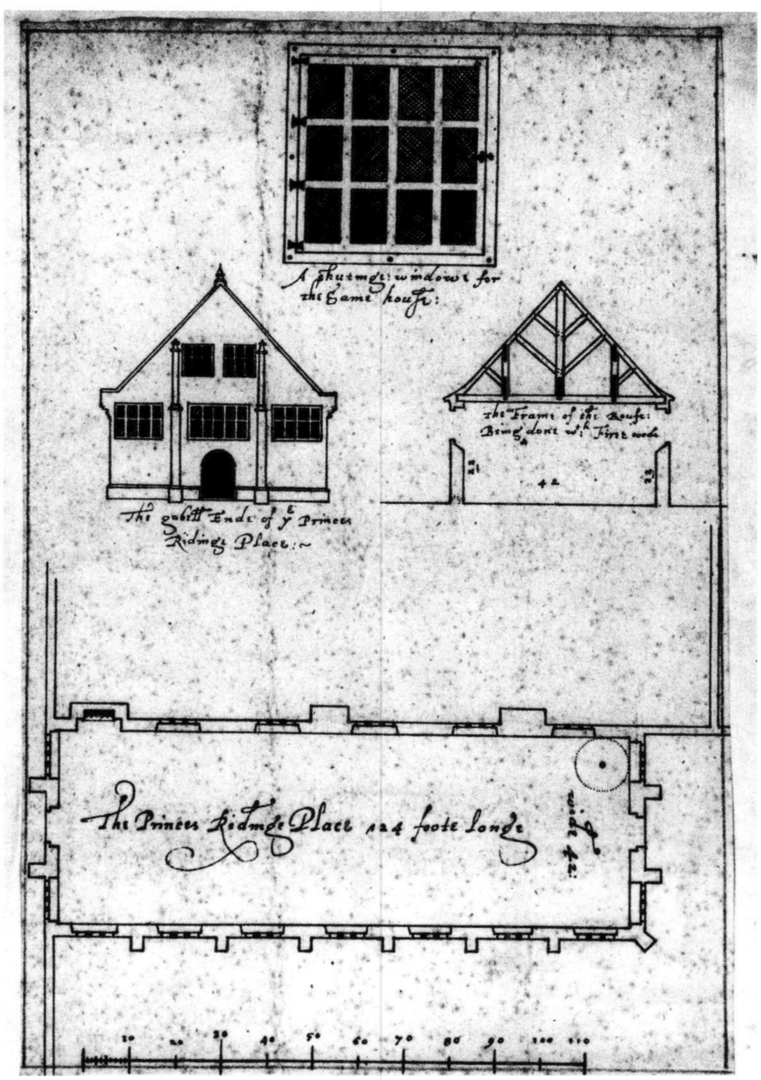

£149 18s. Smythson shows the building to have been a free-standing rectangle 124ft by 42ft, built of brick, with a simple gable and buttresses and minimal architectural detail. There was a round-headed door at each end, and the interior was well lit by seven windows on the long side and another five at the gable end. These were multi-paned and opened on hinges. Their novelty led Smythson to draw one in detail. At one end of the riding house was a viewing gallery heated by a fireplace.

Several of Prince Henry's companions also built riding houses. There was a riding house at Petworth House in Sussex for Lord Percy (fig. 68), and another was planned (although probably not built) for Syon House (figs 63 and 64). Hammond, visiting Petworth in 1635, considered it a great wonder that one of the four sides

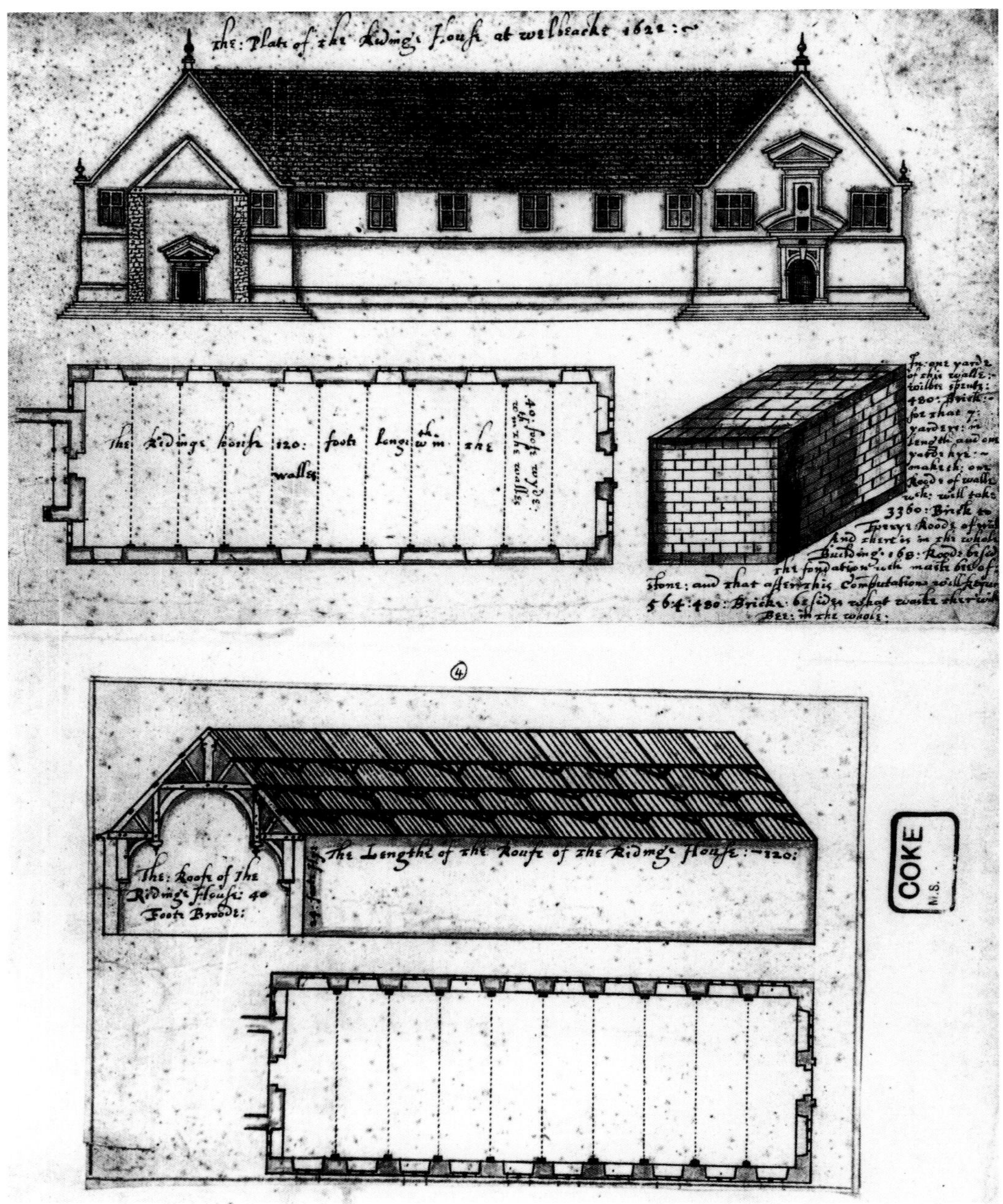

56 John Smythson, plan, elevation and section for the riding house at Welbeck Abbey, Nottinghamshire, 1622 (RIBA Library Drawings Collection, Smythson Collection III/15 [3–4]).

57 The riding house of 1623 at Welbeck Abbey, Nottinghamshire, in William Cavendish, *Methode et invention nouvelle* (Antwerp, 1658), no. 8 (© The British Library).

58 (*below left*) S. H. Grimm, view of the interior of the 1623 riding house at Welbeck Abbey, Nottinghamshire, eighteenth-century (© The British Library, Add. MSS 15545 fol. 68; photo: Conway Library Courtauld Institute of Art, University of London).

59 (*below right*) The 1623 riding house at Welbeck Abbey, Nottinghamshire, *c.*1895. This shows the riding house with later windows on the side elevations but before additions to the western (near) end obscured the original door-case (private collection; photo: Lucy Worsley).

of the magnificent stable quadrangle was 'onely built for a stately Riding House for the greate horse to ride in all weathers and all seasons whatsoever'.[68] William Cavendish built two riding houses, one at Welbeck Abbey, Nottinghamshire, in 1623 (figs 56–59), the other at Bolsover Castle, Derbyshire, probably in the 1630s (figs 51, 60 and 61).[69] Inigo Jones also designed two riding houses. The first was

60 The exterior of the 1630s riding house at Bolsover Castle, Derbyshire (William C. Rolf, 1999).

61 The interior of the 1630s riding house at Bolsover Castle, Derbyshire (William C. Rolf, 1999).

62 Inigo Jones, elevation of a riding house, probably for Hatfield House, Hertfordshire, *c.*1610 (RIBA Library Drawings Collection).

63 (*below*) Inigo Jones, elevation of a riding house, perhaps for Syon Park, Middlesex, *c.*1620s (Worcester College, Oxford).

64 (*bottom*) Inigo Jones, section of a riding house, perhaps for Syon Park, Middlesex, *c.*1620s (Worcester College, Oxford).

doubtless intended for Lord Cranborne at Hatfield in about 1610, on his return from studying *haute école* in France. It would have had an elaborate pedimented façade with a tall, arched entrance into the riding house, which would have been flanked by lower stable ranges (fig. 62). The second, which dates from the early 1620s and may perhaps have been intended for Lord Percy at Syon House, was less elaborate but followed the same model. Neither was built (figs 63 and 64).[70]

Prince Henry died suddenly and to great lament in 1612, but his death did not affect the course of English *haute école*. His brother Charles, the new Prince of Wales, was equally enthusiastic – the anonymous editor of *The Works of King Charles The Martyr* declared him to be the most graceful manager of the Great Horse in the three kingdoms – and took on St Antoine as his écuyer.[71] He seems to have been particularly keen on the way *haute école* could show off his not over-prepossessing stature at its most regal. The most impressive of Van Dyck's state portraits painted in 1633 shows him mounted on a manege horse with St Antoine at his side. Le Sueur's equestrian statue of him, originally commissioned by Lord Weston for his house at Roehampton in 1630 but now at the top of Whitehall, was considered particularly appropriate: 'In his yonger daies, His pleasures were in Riding and sometimes breaking the great Horse; and he did it so gracefully, that He deserved that Statue of Brass which did represent Him on Horse-back.'[72] Two new riding houses were soon built, one at the Royal Mews in Charing Cross in 1614–15, the other at Newmarket in 1615–17.[73]

London was not the site of the only court in Britain in the seventeenth century. At Ludlow in Shropshire, the president of the Council of Wales held his own quasi-royal court as the king's representative. The Lord President in 1618 was the Earl of Northampton who, on inheriting an enormous fortune from his father-in-law in 1610, was said to have spent £72,000 in eight weeks, 'most in great horses, rich saddles and plays'.[74] Northampton saw the model of Prince Henry's academy as an excellent way to bring a little civilisation to his rough Welsh gentry. In a letter to the deputy lieutenants in Flintshire he set out his intention to found such an

65 The riding house at Castle Ashby, Northamptonshire, *c.*1610s. This was subsequently divided up and is now the estate office (William C. Rolf, 2000).

academy at Ludlow with a riding house and a gentleman to teach horsemanship. The cost was to be borne by the deputy lieutenants and by those assessed to supply a horseman to the trained bands.[75] Perhaps significantly, Sir James Scudamore, who had run the informal equestrian academy at Holme Lacy, Herefordshire, served on the Council of Wales and Northampton appointed his son John Scudamore as Captain of Horse in the county.[76] (Northampton's project makes an interesting comparison with Mazarin's project for the College des Quatre Nations in Paris, announced in his will in 1661. This was also intended to aid the assimilation of peripheral regions, in this case the four 'nations' recently incorporated into France, the inhabitants of Flanders, the Catalans of Roussillon, the Germans of Alsace and the Italians of Pignerol. A riding academy was a central part of the original scheme.[77]) The riding house is noted in an inventory of Ludlow Castle of 1631 and again in a parliamentary survey of 1650.[78] Northampton also built what appears to be the earliest surviving riding house at Castle Ashby in Northamptonshire (fig. 65). About 35 ft wide by 70 ft long, under a single, sharply pitched roof like Prince Henry's riding house at St James's and placed in a formal relationship with the house, this was clearly a building of some distinction, although it is now divided up and used as the estate office, with garages below. It is too wide to have been a stable and as the side wall seems to have been open cannot have been a barn. Given its dimensions, its similarity to the St James's riding house and Northampton's known fascination with *haute école*, a riding house is the most likely purpose.

Although the erection of the first purpose-built riding house followed the arrival of Monsieur de St Antoine and so it might be assumed that this was a French innovation, there is no evidence of riding houses in France before they were built in England. There are tennis courts but no riding houses in the plans of French houses published in Jacques Androuet du Cerceau's *Les Plus Excellents Bastiments de France* of 1576–79. Instead the French seem to have made do with open maneges. An engraving of Fontainbleau by du Cerceau not included in *Les Plus Excellents*

Bastiments notes a *cavalleriza* or open manege to the north of the palace on what later became the site of the Conciergerie.[79] This had been built over by 1614 and was replaced by another open manege beside the Chenil or Petit Ecurie, on the far side of the lake. This can be seen in Dan's *Le Tresor des merveilles de la maison royale de Fontainbleau* of 1642 and appears more clearly, and in use, in Jean-Baptiste Martin's view of Fontainbleau in the collection of the earl of Radnor at Longford Castle.[80] There is no evidence of a covered riding house at Fontainbleau, despite its pre-eminence among French palaces.

In Paris, a long, narrow open manege was created beyond the Grande Ecurie, which was built in 1568 by Philibert de l'Orme. The Venetian ambassador described this in 1578 as 'une piste, une carriére a picquer les Chevaulx'.[81] Pascal Lievaux has suggested that a riding house was built at the far end of this open manege under Henri IV, and a building is shown in this position in Nicolay's map of Paris of 1609, Merian's map of Paris of 1615 and the Plan Gomboust of 1652. Lievaux identifies this with the wooden riding house described as ruined by Desgodetz in 1694.[82]

However, each of these maps shows the building as very narrow, less than half the width of the open manege, apparently no wider than a stable. This is presumably the open timber structure shown in open manege by de Pluvinel. To be practical a riding house would have needed to cover most of the width of the open manege, as was the case with the riding house subsequently built by Robert de Cotte on the site in 1720–22. Moreover, Henri Sauval in his detailed description of the Grande Ecurie of 1660 makes no mention of a covered manege.[83] This suggests that the building shown by de Pluvinel, Nicolay, Merian and Gomboust may have been a prototype riding house but could have had some other use.[84]

We do know that the riding academy founded by the great equestrian Antoine de Pluvinel in 1584 used part of the lower range of the Grande Gallerie of the Louvre as a covered manege, but this was not a purpose-built riding house.[85] It would thus appear that though the first English riding houses were erected under the influence of M. de St Antoine, there was no specific French model to follow.

There was, however, a more distant precedent for the riding house as a shelter from the more extreme weather of central Europe in the winter riding school built of timber at the Hofburg in Vienna in 1572. No permanent stone riding house was erected until 1735, although plans were made to replace the temporary wooden shed in 1681.[86]

Haute école was above all a courtly art. All the early English riding houses were built by men closely connected with the court. All, except those for the Earl of Northampton, were built for Prince Henry's contemporaries in the informal academy at Richmond. It flourished when the monarch, a royal prince, or a favourite encouraged it, for this was the high point of English *haute école*. But in England there was a rival equestrian attraction tugging at the interest of monarch and nobleman alike. By 1639 Thomas de Grey, author of *The Compleat Horseman,* was already worried about the health of *haute école*:

> Since the laying aside of the great Saddle and Cannon, and the neglect of the Horse of the Menage, since the applying of our Breed only to Racing or (as I may better say) in furnishing our selves with Horses of speed to runne away

from our Enemy, the most ancient honour of Horseman-ship peculiar to this our Kingdome, and for which all other Nations highly esteemed us, is now almost vanished and lost.[87]

What worried him was not any waning of the popularity of *haute école*, of which many boasted their knowledge, but the inability of his countrymen to breed and train manege horses. But it may be symptomatic that the only English book on *haute école* published in the first half of the century was Gervase Markham's *Cavelarice or The English Horseman* of 1607, itself a reworking of his *Discourse on Horsemanshippe* of 1593.

What killed Stuart *haute école* was the outbreak of the Civil War, when all the Great Horses bred in theory for war found themselves at war in practice, like John Evelyn's manege horse which was sent to the king. The Royalists' defeat meant the destruction of a court that had so encouraged *haute école*. There appears to be no evidence that it was supported under the quasi-court of Oliver Cromwell, although the third Earl of Bridgwater, son-in-law of the Duke of Newcastle, did spend the years of his Interregnum retirement practising *haute école* in the riding house that he built for himself at Ashridge House in Hertfordshire. (He justified the expense of his recreation as an innocent but noble occupation which kept him busy when the difficulty of the times meant he was not allowed to travel abroad.)[88]

Bridgwater's riding house was the last private manege to be built in England for nearly a century. England after the Restoration was not fertile ground for *haute école*. The Duke of Newcastle, who had been the new king's governor and taught him *haute école*, described Charles II as a fine horseman capable of managing horses too fierce for lesser men in the dedication of his *New Method* of 1667. But this standard flattery of kings was to no avail. Monconys reported the king as setting no value whatever on maneged horses and thought his stables 'fort mal garnies'.[89] His interests centred instead on horse racing, a sport that has nothing in common with *haute école* and where the great horse was valueless.

Deprived of its support at court, it was only in the circles of the pre-war enthusiasts that interest in *haute école* survived. Thus, ironically, it was only after English enthusiasm for *haute école* waned that Newcastle's two books, the finest English books on the subject, were published, in 1658 and 1667. Sir William Hope's 1696 translation of de Solleysel's *The Compleat Horseman* shows interest survived, but the title of his second chapter reveals a defensive note: 'That it is a very foolish thing, and a token of great ignorance in the Art, to think the mannage useless.' The evidence of riding houses supports the suggestion of a marked decline in interest. Only a handful of private riding houses are known to have been built in the British Isles in the eighty years between 1660 and 1740. The most prominent of these was built in 1705 at Dublin Castle for the Lord Lieutenant, the Duke of Ormonde, who kept his own quasi-royal court.[90] Another was erected for the dukes of Bolton at Hackwood House, Hampshire, probably between 1699 and 1725. At Windsor, an open manege was created at Ranger's Lodge, probably by the third Duke of Marlborough in about 1735.[91]

Haute école retreated instead to the academies. The most famous of these were abroad in places like Paris, Angers, Strasbourg, Geneva and Turin. A spell in an academy was an important, though now largely overlooked, aspect of the Grand

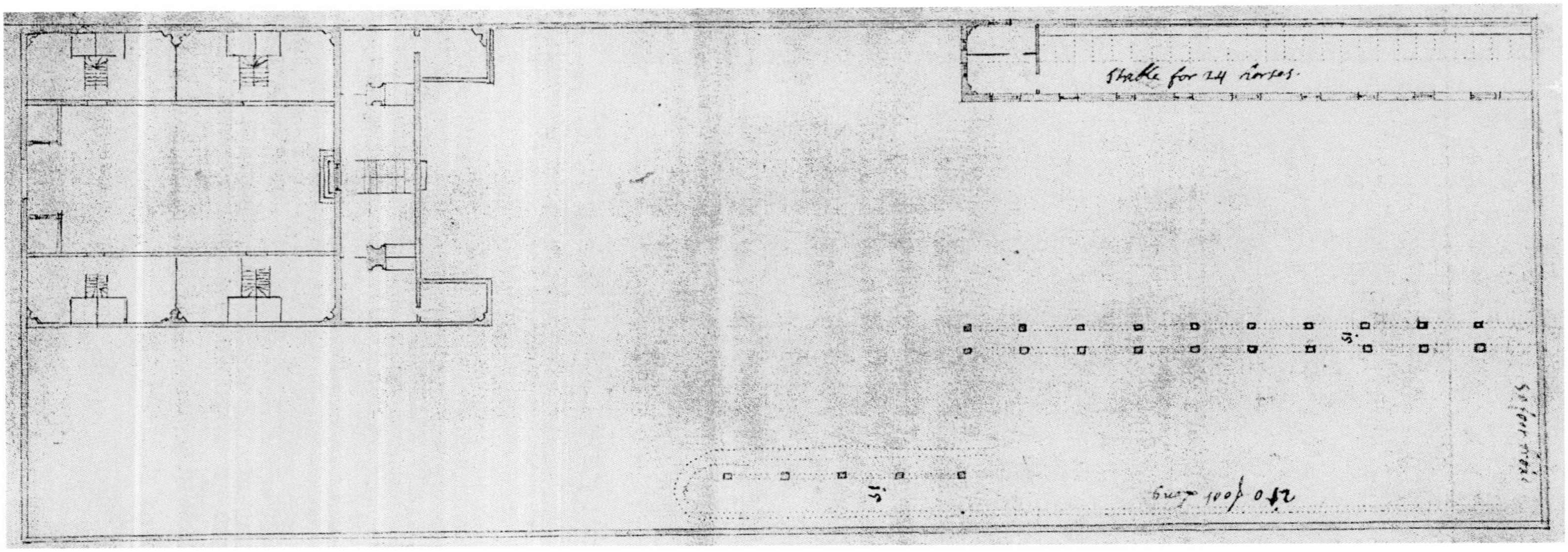

66 Plan of a manege yard, stables and house, perhaps for the Duke of Newcastle's proposed riding academy in Clerkenwell, n.d. (Balliol College Archives MS B21.24; photo: Lucy Worsley).

Tour. But foreign academies were not without their critics: 'There is scarce a Prince in Europe', noted John Evelyn, 'but what have been scholars in the French Academys and which by consequence has leven'd them all with the mode as well as the language of France, and disposed them to an undervaluing of their owne countrys with infinite prejudice to the rest of Europe.'[92]

In this climate it was not surprising that a series of attempts was made to establish academies in England. Sir Balthazar Gerbier had a short-lived academy in Bethnal Green in 1649–50 and it is possible that the Duke of Newcastle considered creating another at his house at Clerkenwell after the Restoration (fig. 66).[93] More successful was the academy established by Henry Foubert after the closure of the Protestant academies and his expulsion from France in 1679.[94] Foubert found it difficult to establish himself and it was not until 1684 that the academy was launched. From 1691 until Foubert's death in 1696 his academy was supported by an annual grant of £500 from William III, a sign again of the importance of royal patronage.

Foubert made do with an open manege, probably in the fields adjoining Brewer and Sherwood Streets where he was visited by John Evelyn in 1683.[95] His rival, Lewis Maidwell, was more ambitious.[96] Establishing himself at King Street in Soho in 1687, Maidwell built a handsome brick house, stables and riding house for £2,000. This can be seen in the centre of the block between King Street, Cross Street, Carnaby Street and Tyler Street, in a survey made by William Dickinson (probably as a potential site for one of the Fifty New Churches) now in Lambeth Palace Library.[97] Dickinson's survey shows the riding house was 100 ft long by 35 ft wide. As well as the usual Classics curriculum, Maidwell taught the sons of the nobility and gentry French, Italian, maths, accounts, navigation, astronomy, geography, history, chronology, dancing, fencing and horsemanship. However, despite its fashionable clientele, the academy cannot have been profitable and in 1699 Maidwell petitioned the House of Commons to turn it into a public institution. Though the school was given a royal charter in 1702 as the Royal School of King William

III, the idea was blocked by the universities. Thomas Hearne noted that 'it was stopped because, without doubt, it would have utterly obstructed all true learning'.[98] Frustrated, Maidwell closed the school in 1704.[99]

Maidwell's site was sold to Henry Foubert, son of the emigré, who had tried unsuccessfully to obtain a grant of land to build a riding house near Old Spring Gardens in 1701. It was perhaps in Foubert's academy that the artist John Vanderbank, 'the better to execute his ideas', was a 'disciple in [the] Riding Schools and purchased a fine Horse as Model for his pencil' while preparing *Twenty Five Actions of the Manege Horse*, published in 1729.[100] Henry Foubert continued to run his academy until his death in 1743. It was then taken over by his nephew Solomon Durrell, under whom it declined into an ordinary riding school. It was finally closed by his son-in-law Thomas Evans in 1778.[101]

4

THE STUART STABLE

Of all those who wrote about stables in the seventeenth century, none showed a greater concern for display than Sir Balthazar Gerbier, Huguenot, courtier, diplomat and one of the leading architects of the early Stuart court. Gerbier played an important cultural role at the courts of James I and Charles I in the 1620s and early 1630s, particularly in the household of the royal favourite, the Duke of Buckingham. In his old age he published two short works on architecture, *A Brief Discourse Concerning the Three Chief Principles of Magnificent Building* in 1662 and the more soberly titled *Counsel and Advise to All Builders* of 1663. The two books shed light on the thinking that must have prevailed at the early Stuart court and in particular on the importance placed on stables as settings for display.

In Gerbier's eyes the perfect stables were those built by Prince Thomas of Savoy, that is Tommaso Francesco, Prince of Carignano (1596–1656), youngest son of Carlo Emanuele, first Duke of Savoy, 'all of marble, the racks, manger and upright posts all of copper, richly wrought'.[1] For Gerbier, who believed that nothing should 'impede the Accesse of a Cavallier to the horses', magnificence was an ideal in building, and in stables quite as much as palaces.[2] It was an ideal shared by others at Court, as can be seen in a series of architecturally ambitious stables whose architectural importance in the early Stuart revolution has never been fully appreciated.

In the early seventeenth century, for the first time, a number of stables built for courtiers attracted national renown, in particular those at Petworth House, Sussex, Burley-on-the-Hill, Rutland and Fonthill House, Wiltshire.[3] The Petworth stables, which were built for the ninth Earl of Northumberland, possibly between 1616 and 1623, were celebrated above all for their size (fig. 68). Thomas Fuller considered them the finest stables of any subject in Europe and Daniel Defoe, lamenting their demolition a century later, referred to 'those noble Buildings: I mean the Meuse or Stables – the finest of their Kind in all the South of England, and equal to some Noblemen's whole houses'.[4] It is a mark of their renown that they were recorded in one of the first paintings to focus on a stable building. In the painting, made in about 1680, the stables stand proudly in the foreground, with the old house set dismissively behind them.

The most detailed description of the stables comes from Lieutenant Hammond, who visited Petworth in 1635: 'first I will begin with that I desir'd first to see, the like whereof I am sure not a subject in all England hath to show: The stabling, lofty, faire built Stables, which are 80 of my Paces in length, wherein I viewed 2 such Rankes of hansome, proper, hunting Geldings, neat brave Saddle nags for the Road, and lusty, stout Coach-Horses, 60 in all in each Ranke, a Troop of Cavalry worth the viewing, the like not in this land'. As well as the stabling, there was a riding house, a large haybarn, convenient lodgings of the Gentleman of the Horse, grooms, coachmen, riders and postilions and 'a fair large armoury bothe for horse and foot'.

67 Jan Siberechts, *Cheveley Park, Cambridgeshire* (detail), 1681 (Belvoir Castle; photo: Bridgeman Art Library).

68 Jacob Knyff (attr.), *The Stables at Petworth House, Sussex* (detail), *c.*1680 (Syon House).

Built around a quadrangle, the Petworth stables were perhaps the largest non-royal stables to be built before the eighteenth century. That the stables took the form of a quadrangle is unlikely to be chance. Though stable quadrangles were to become commonplace in the eighteenth century, they were exceptional before that date. Henry VIII had built the first at Hampton Court in 1537–38. An even grander quadrangle was built at Theobalds in Hertfordshire as a key part of the alterations made in 1607 to 1610 after James I acquired the palatial house of the Cecils.[5]

The importance of the house at Theobalds has long been appreciated. The stables, by contrast, have been overlooked, despite their extraordinary scale and architectural ambition. Fortunately, they were so impressive that Robert Smythson recorded the plan, probably on his journey to or from London in 1609 (fig. 69).[6] No other visual record of the stables exists, but the essential details of Smythson's drawing are confirmed by the parliamentary survey made of the building after the Civil War.[7] This notes that the stables took the form of a great quadrangle 119 ft square, with a central gate, flanked by a pair of stables. Further ranges of stables 119 ft long and 23 ft wide were placed on the flank elevation, and the far end of the quadrangle was closed by a large barn.

Smythson's drawing does not entirely correspond with the parliamentary survey. In particular, some of the dimensions vary and he places further stables at the far

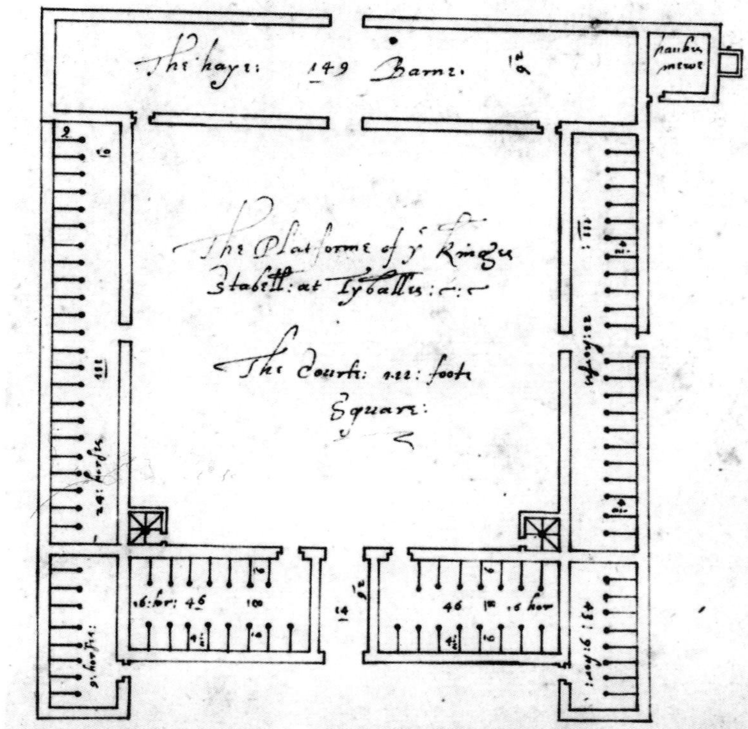

69 Robert Smythson, plan of the stables of 1607–10 at Theobalds Palace, Hertfordshire, 1609 (RIBA Library Drawngs Collection, Smythson Collection 1/4).

ends of the entrance range, where the survey records houses for the saddler and the farrier houses. This implies that he copied a design for the stables, rather than surveyed them himself, and that the building was varied in execution. This suggests that the stables were planned as one of the first improvements to be made to Theobalds after it became a royal palace, but that these improvements had not been completed by the time of Smythson's visit two years later. Theobalds lay in good hunting country, which explains the exceptional size of the stables. According to Smythson, they would have housed eighty-eight horses. Presumably, the king's horses would have been housed in the pair of stables with double rows of stalls on either side of the entrance. Sadly the stables seem to have been demolished along with the rest of Theobalds under the Commonwealth, but they were almost certainly the model that the Earl of Northumberland had in mind when he commissioned the new stables at Petworth.

Though the scale and architectural form of the Petworth stables was remarkable, its architecture, with its array of gables, was traditional. The other two renowned private stables of the early seventeenth century were smaller but stood out for the advanced Classicism of their design.

Thomas Fuller, writing about George Villiers, Duke of Buckingham, in his *History of the Worthies of England* in 1662, highlighted the stables he built at Burley-on-the-Hill in Rutland as one of his particular achievements. He described these as superior to any other stable in the country: 'where the horses (if their pabulum so plenty as their stabulum stately) were the best accommodated in England'.[8] Buckingham, favourite of James I and then of his son, Charles I, was a keen horseman who had studied *haute école* at Duplessis-Mornay's academy at Angers in

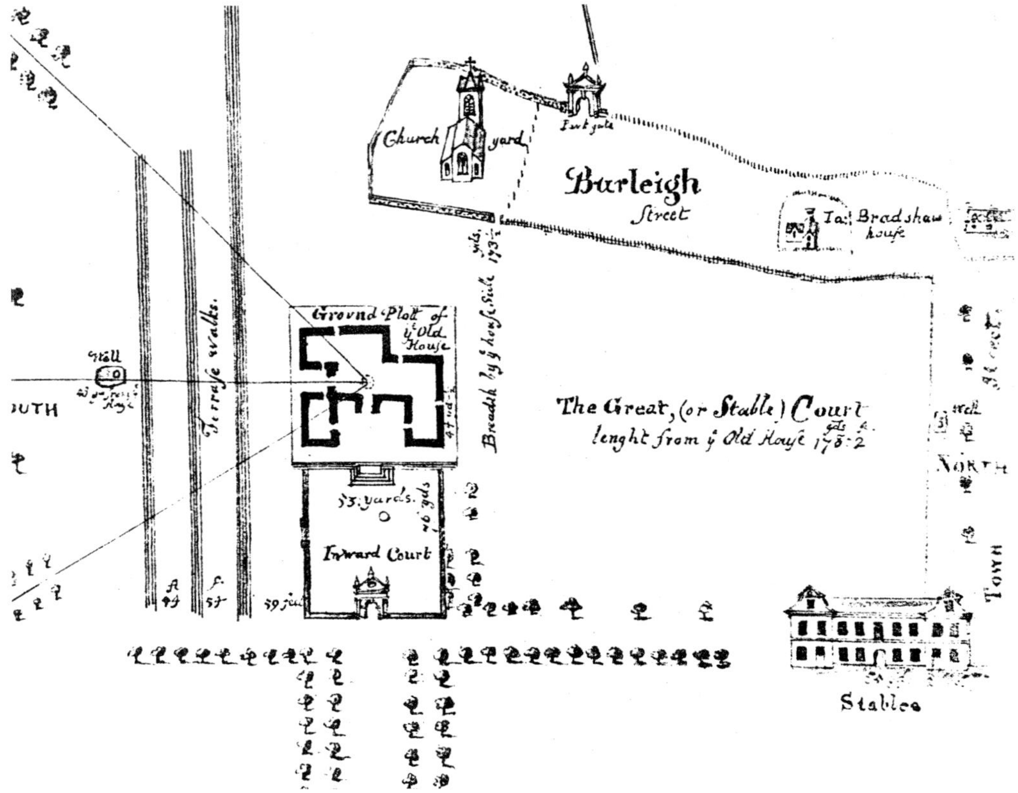

70 Survey of Burley on the Hill, Rutland, *c.*1689/90, showing the Duke of Buckingham's 1620s stables to the right of the house, from Pearl Finch, *History of Burley-on-the-Hill* (London, 1901), pp. 6–7 (detail).

1613 and was appointed Master of Horse in 1616. He was also a famed connoisseur of art and architecture who vied with Charles I to establish Classical tastes in England, collecting paintings and sculpture and commissioning the country's most sophisticated architects to design buildings that set new standards of Classical design. Inigo Jones worked for him at Whitehall in 1619–20 and at New Hall, Essex, in 1622–23, as did Sir Balthazar Gerbier, who was employed at New Hall and York House in the Strand in 1624–25.[9]

Buckingham bought Burley-on-the-Hill in around 1621, entertaining the king there that year, and used it principally during the hunting season. He is said to have much improved the house, but as this was destroyed by fire during the Civil War, the details of these changes are uncertain. What does survive, though altered after a serious fire in 1705, is the impressive stable range, which, according to a survey by Commonwealth commissioners, held forty horses in two stables of twenty each. No documentary evidence survives to date these precisely, but they must have been built before Buckingham's murder in 1628.[10] Their original appearance is known from a copy of a survey of the park, perhaps made in 1689/90 (fig. 70). This shows a two-storey building with casement windows, a hipped roof and curved 'Dutch' gables (probably removed after the fire) over slightly projecting wings.[11] For its date, somewhere between 1621 and 1628, the use of such gables, a hipped roof and casement windows, is highly innovative. The gables can be compared to that on Inigo Jones's Brooke House in London of circa 1617 and even more pertinently with Andrea Palladio's Villa Barbaro at Maser, where they are also used to emphasise the ends of single-storey office ranges.[12] The whole composition recalls Sir Roger

Townshend's Raynham Hall, Norfolk, designed in 1622, which had a pair of similar gables.

The lack of documentation means that there is no record of the architect for the stables at Burley-on-the-Hill, but the purity and innovation of the design and Buckingham's position as royal favourite suggest a court-based architect. It is hard to see who that might be at this date except Jones or Gerbier, with Jones, who designed a number stables in his early career, including one for camels at Theobalds in 1622–23, the most plausible candidate given the links to Brooke House.[13]

The stables at Fonthill were built by another of Charles I's courtiers, Lord Cottington, Chancellor of the Exchequer from 1629 to 1642, who bought the estate in 1631. Six years later George Garrard wrote to the Earl of Strafford of his recent visit to Fonthill: 'He [Cottington] hath built a Stable of Stone, the third in England, Petworth and Burleigh on the Hill only exceed it.'[14] The Fonthill stables have since been demolished but are known from an eighteenth-century painting by George Lambert and from engravings (fig. 71).[15] These show a nine-bay building, with a five-bay central range flanked by two-bay projecting wings, a central pedimented doorcase, mullion and transom windows on the ground floor with *oeil-de-boeuf* windows above and a hipped roof. But for the *oeil-de-boeuf* windows, which would have lit and ventilated the hayloft over the stabling, Cottington's design might be mistaken for an innovative country house of the 1630s, such as West Woodhay, Berkshire, built for Sir Benjamin Rudyerd, a friend of the fourth Earl of Pembroke, in 1635. Examples of the newly fashionable astylar Classical manner of the early Stuart country house are rare, which is why West Woodhay is important.[16] What has not been appreciated is how courtier stables, such as those at Burley-on-the-Hill, Fonthill and those discussed below, significantly increase the number of known examples of early Stuart Classicism.

71 George Lambert, *Fonthill House*, 1740 (formerly in the collection of the Property Services Agency). Detail showing the stables, 1630s, at Fonthill House, Wiltshire.

The architect of West Woodhay and of the Fonthill stables is unknown, but given Cottington's office it was probably a court-based architect. By the 1630s this is unlikely to have been Inigo Jones, who was too busy to design Wilton House, Wiltshire, for the Earl of Pembroke, despite his close connections with Charles I. Edward Carter, chief clerk in the Office of Works, has been suggested as a possible architect for West Woodhay and would be a plausible name for Fonthill. But a candidate with a stronger claim is Isaac de Caus.

De Caus was probably the architect of very similar stables built just before those at Fonthill at Bedford House on the Strand in London (fig. 103).[17] These stables, which were built for the fourth Earl of Bedford in about 1628, are known from a number of surveys and from Wenceslaus Hollar's bird's-eye *West Central View of London*. They lay to the east of the garden at Bedford House in a separate stable court and from the surveys it is clear that they formed a single range about 160 ft wide, with projecting wings, like the stables at Fonthill. Hollar's view shows a nine-bay central range (as opposed to the five bays at Fonthill) with a pedimented central doorcase flanked by two-bay wings. Hollar's details of buildings need to be treated with care and he certainly elides the west wing of the Bedford House stables with an adjacent range. The position of the ground-floor windows, which appear to be set above the level of the pedimented central door, is also curious. Though this is reminiscent of the way thermal windows were placed above blank walls to light stables in the eighteenth century, it has no parallel in seventeenth-century stable

72 and 73 (*right and facing page*) Isaac de Caus, the stables, 1630s, at Wilton House, Wiltshire (William C. Rolf, 2000).

design and would have made for an immensely tall room behind. It is more plausible that, as at Fonthill, the windows were level with the door. What is convincing about Hollar's drawing are the *oeil-de-boeuf* windows on the first floor, below a hipped roof with dormers. Such a detail could only have been used for a stable and again parallels the stables at Fonthill.

For London, where space was always at a premium, these were stables of remarkable grandeur, in their scale (they rivalled Bedford House in length), in the size of the courtyard in front of them and in their fashionable architecture.[18] As Dianne Duggan has shown, the fourth Earl of Bedford inherited the Bedford estates from his cousin in 1627 when the family's fortunes were at a very low ebb. Duggan suggests that the earl extended Bedford House and laid out a large ornamental garden, as well as developing the Piazza at Covent Garden, in a deliberate bid to re-establish his family's status and prestige. She does not discuss the stables, but their size and architectural novelty suggest that they should also be seen as part of that deliberate self-aggrandisement.

No documentary evidence survives to name the architect of the stable. Though Edward Carter is known to have had at least an executant role at Bedford House, Isaac de Caus, who was joint surveyor with Carter on the Piazza and almost

certainly designed the Bedford House gardens, is the more plausible designer. Parallels with the stables at Wilton, which de Caus is known to have designed, make him the probable architect of the Bedford House stables.

John Aubrey was particularly impressed by the Wilton stables, describing them as being 'of Roman architecture, built by Mons. de Caus, [they] have a noble avenu to them, a square court in the middle; and on the four sides of this court were pictures of the best horses as big as the life, painted in severall postures, by a Frenchman' (figs 72 and 73).[19] They presumably date from the 1630s as work began under Isaac de Caus's direction on the gardens in 1632 and on the house in 1636.[20] They may date from 1639 when, as Aubrey mentions, the earl much enlarged his stud of racehorses and hunters. The quadrangle incorporated a fourteenth-century barn on one side (which could have been used as a riding house, though there is no docu-mentary evidence for this).

As Hampton Court and Theobalds demonstrate, a stable quadrangle had a particularly royal connotation. This might have made it seem appropriate for Wilton for according to Aubrey, Charles I 'did love Wilton above all places: and came thither every Sommer', making it a quasi-royal palace. The principal elevation of the stables, of brick with stone details, faced Wilton House across the grand formal gardens

74 Isaac de Caus (attr.), the stables, *c.*1630s, at Ramsbury House, Wiltshire (National Monuments Record).

and was treated with considerable architectural ambition. Two-storey wings with casement and *oeil-de-boeuf* windows and balustrading flanked an elaborate rusticated arcade (since filled in) with roundels between each bay. Aubrey's description, and the comment by Evelyn that it was also adorned with heads of Caesars, shows that the courtyard was equally grand.[21] The life-size paintings of horses in the courtyard, which do not survive, seem to have been unique in England and were presumably inspired by similar paintings at the Palazzo Te in Mantua.

The general air of decay today, alongside the slightly crude handling of the Classical details and the way the gable of the barn breaks over the balustrade, makes it easy to overlook the importance of the Wilton stables. But for their date these were among England's most architecturally ambitious buildings of any type.

De Caus continued to be paid a pension by the Earl of Pembroke until his death in about 1656,[22] which makes him a plausible candidate for two other mid-seventeenth-century stables with *oeil-de-boeuf* windows associated with the Herbert family. Ramsbury Manor, Wiltshire, which lies on the edge of Savernake Forest, would have been used by the fourth Earl of Pembroke as a hunting lodge (fig. 74). The stables there are of beautifully laid brick, with eight tall, oval windows in stone, separated by brick quoins, lighting the ground floor. After the fourth earl's death in 1650, the house was occupied by his widow, then let. It was sold to Sir William Jones in 1676, who rebuilt the house to the designs of Robert Hooke. As the stables stylistically pre-date Hooke's work and there is no reference to them in the building accounts, and as stables of this grandeur are unlikely to have been built for a widow or a tenant, they are likely to be the work of the fourth earl before the Civil War. If so, Isaac de Caus is again the probable architect. The stables at Tythrop House, Buckinghamshire, which were illustrated by Henry Winstanley in 1680 (fig. 75), were no less grand, seven bays wide and four bays deep, with a ped-

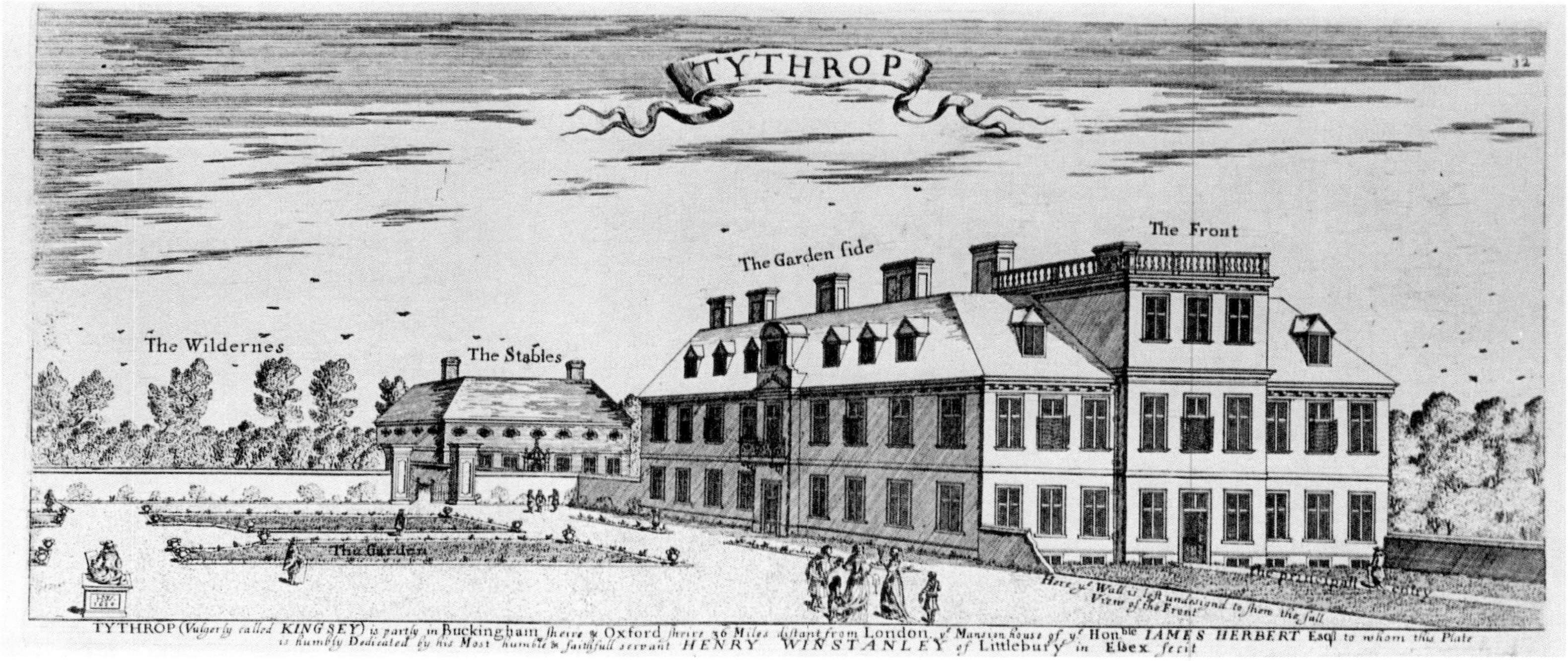

75 (*above*) Isaac de Caus (attr.), the stables, *c.*1650s, at Tythrop House, Buckinghamshire, in Henry Winstanley, *Tythrop*, engraving, 1680 (John Harris).

76 (*left*) Sebastiano Serlio, 'Elevation of the Ninth House outside the City', from *On Architecture*, Book VII (Frankfurt, 1575), p. 21.

imented doorcase and *oeil-de-boeuf* windows under a hipped roof lighting the attic, as at Fonthill (fig. 71). They were probably built by the fourth earl's sixth son, James Herbert, who acquired the property through marriage in 1649 and sold it in 1675. It has been assumed that the Interregnum would not have been a propitious time for building and that the work must post-date 1660, but there is no reason why this should be the case. The Herbert family prospered under the Commonwealth, and Wilton House was expensively rebuilt in 1648–circa 1651 after a fire, so there is no reason to believe that work at Tythrop should not have begun soon after Herbert and his wife succeeded in 1649. If so, the stables could still have been designed by de Caus.[23]

This use of *oeil-de-boeuf* windows on stables was a successful example of adapting Classical detail to practical use. Haylofts needed lighting and ventilation, but casement windows were too large. *Oeil-de-boeuf* windows, ideal in scale for an attic

77 Elevation of the interior of the stables of 1638–40 at Holland House, London, from Isaac Ware, *Designs of Inigo Jones and Others* (London, 1735), pls 52–53.

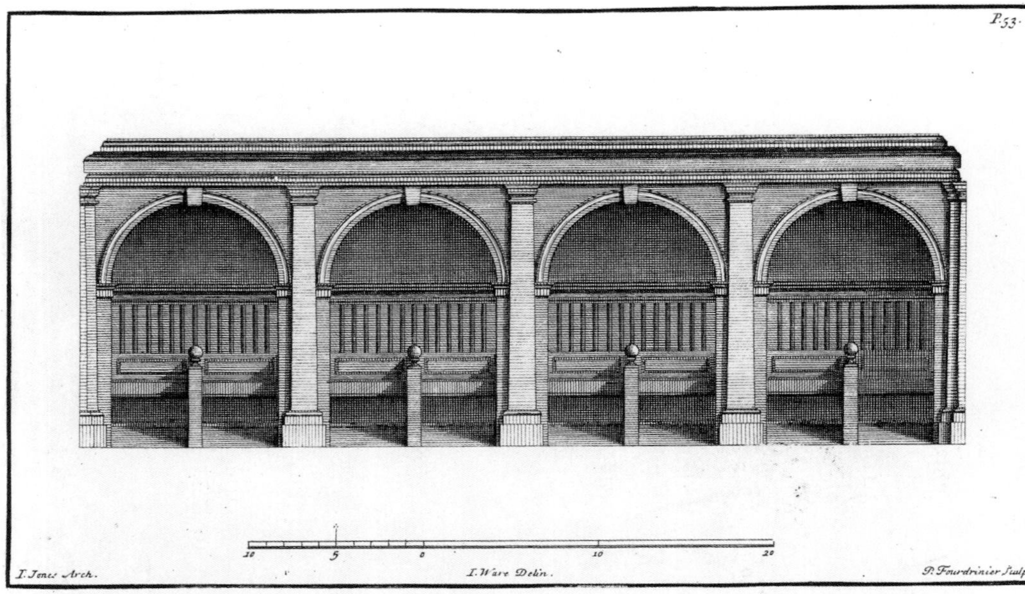

floor, solved the problem. The source for this consistent use of *oeil-de-boeuf* set above casement windows is clearly Sebastiano Serlio's seventh book, published in 1575, where they appear repeatedly, perhaps most pertinently on his Ninth House (fig. 76). A more immediate source may be the stables Inigo Jones designed for the king's Great Horses at Newmarket in Suffolk in 1615–17. It is known that these had two great oval windows in stone but as no illustration of them survives it is impossible to be certain of their form.[24]

Oeil-de-boeuf windows were used on another courtier's stables, those at Holland House in Kensington, which were sufficiently impressive to be attributed to Inigo Jones by Horace Walpole and to be among the early seventeenth-century Classical buildings recorded by Lord Burlington's circle. Two copies of *The Designs of Inigo Jones and Others*, edited by Isaac Ware, annotate the interior of an otherwise anonymous Classical stable ascribed to Inigo Jones with the note that they were the stables for Holland House (fig. 77).[25] These stables were part of major improvements to Holland House carried out by Henry Rich, first Earl of Holland in 1638–40. Building accounts show that Lord Holland spent the considerable sum of £1,800 adding a new wing to Holland House and the enormous sum of £4,000, the equivalent of a complete country house for some less ambitious courtier, on new stables and coach houses.[26] Only a much-altered fragment of the elevation of the stables survives, but the accounts, together with the engraving, whose detail they confirm, allow us to reconstruct the stables and accompanying coach houses in considerable detail. These were 240 ft long and 34 ft wide. Built of brick, they were two storeys high, with six arcaded coach houses balanced by stables behind a matching blank arcade with a central projection between them. The first floor had thirteen oval windows with an attic above lit by dormer windows. The parallels with the arcaded main elevation of the Wilton stables, and in particular with the end pavilions with *oeil-de-boeuf* windows over arcading, are obvious.

Inside, the stable was divided into two compartments, one with eight stalls within four arches (as shown by Ware), the other probably with a further four stalls. The

78 Inigo Jones, the stables of *c.*1618 at Arundel House, London, from Wenceslaus Hollar, *The Courtyard at Arundel House*, engraving, 1646 (British Museum). The stables lie on the left-hand side of the courtyard, with the coach houses beside them.

stalls had a double rack and manger 'made of the best oken timber and the rack with a deep architrave freeze and cornish and having strong shutters to keepein the strawe under the manger and turned bayles betweene the horses with substanciall posts with great turned balls at the heel turn'. The walls and piers were lined with moulded panelling and all the woodwork, including the stable furniture, the great stable doors, the coach house doors and panelling were painted a 'sad [subdued] Timber color in oyle'. There was a 'Great cornish within and without the stables' (reminiscent of Bolsover) of moulded brick, and architraves 'about the Arches and windowes' were similarly treated.

For his workmen, Holland relied heavily on the Office of Works. Thomas Baldwin, comptroller of the Board of Works, was paid for auditing the bills and making payments, and other members of the Office of Works were also involved, including William Dodson, the master bricklayer, William Mason, the deputy mason, Thomas Bagley, chief glazier, and Hugh Justice, seargent plumber. Sadly, the accounts make no payment to a designer, although Holland's close connections with the Office of Works are clear. Malcolm Airs suggests that the attribution to the surveyor-general Inigo Jones made by Walpole and Ware may be possible, perhaps in partnership with Isaac de Caus, though parallels with Wilton may mean that de Caus was working alone.

Not all the courtiers' stables in the new style included *oeil-de-boeuf* windows. Two evocative engravings by Wenceslaus Hollar survive of the outer court of Arundel House on the Strand, the London home of his patron the Earl of Arundel (fig. 78).[27] The yard is shown busy with people, horses and coaches, and a chaotic mix of timber-framed buildings, except along its east side. Here, in marked contrast, is a six-bay range that stands out for its up-to-date architectural language: ordered, astylar, two-storeys with an emphatic cornice, platband and dormer windows. This is the familiar manner of Jonesian stylistic revolution. That it was the stable is shown by the horses gathered outside it and the coach houses beside it. Both engravings make it clear that the range was unfinished, suggesting that it was only the first

79a and b John Smythson, plan and elevation of the stables at Welbeck Abbey, Nottinghamshire, 1625 (RIBA Library Drawings Collection, Smythson Collection III/5–6).

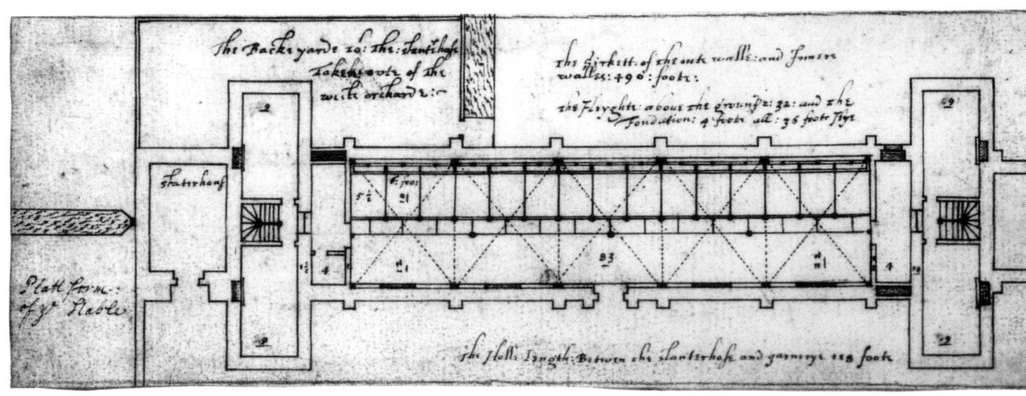

stage in an intended complete rebuilding of the entrance courtyard in the new and fashionable style.

The fact that this work was at Arundel House should come as no surprise. The Earl of Arundel was one of the leading virtuosi and greatest art collectors of his day with an astonishing collection of paintings and sculpture; he was also the owner of two chests of drawings by Vincenzo Scamozzi, the Italian architect who took up Andrea Palladio's mantel.[28] As a young man he had made an extensive tour of Italy in 1613–14, accompanied by Inigo Jones, for whom this journey was fundamental to his emergence as England's leading Classical architect. Arundel's patronage of Jones is well known. Drawings survive of the 'Italian' gate at Arundel House, which was built in 1618, for a house for Lord Maltravers, Arundel's son, at Lothbury, dated 1638, and for a more substantial development associated with it. He also presumably designed the garden gate recorded by John Smythson in 1619 and is generally credited with the impressive gallery wing known from Hollar's bird's-eye view of London.[29]

Hollar's engraving was made in 1646 using an older drawing. Arundel left England never to return in 1642 and Arundel House was used as a barracks by the Parliamentarians. These stables could well have been part of Arundel's major remodelling of Arundel House around 1618, for which Jones was the architect. A direct comparison can be made with the elevation of the building erected for the architects and masons at Newmarket House in 1619–20, recorded by John Webb in the 1640s.[30] If this range is to be dated 1618, as seems plausible, then these stables are one of the first accurately recorded examples of the astylar architectural language, more familiarly known as Georgian, that was one of Jones's great legacies to English architecture.

The grandest stables to survive from before the Civil War were those erected by the great equestrian enthusiast William Cavendish, Earl, subsequently Duke, of Newcastle. Cavendish was educated at the Prince of Wales's informal academy at Richmond and subsequently served as governor of the Prince of Wales. His first new stables were built at Welbeck Abbey, Nottinghamshire, to the designs of John Smythson in 1625.[31] Though now demolished, these were illustrated in Newcastle's *Methode et invention nouvelle de dresser les chevaux* of 1658, and can be examined in more detail in Smythson's drawing, the first surviving detailed drawing of an English stable (figs 79 and 80). This and the inscription concerning the stable in the *Methode et invention nouvelle* reveal how carefully thought out, and indeed how extravagant, the design was. The stable was vaulted in stone, a useful protection against fire. The heelposts were also of stone, as were the mangers, which were described as 'a l'Italienne'. There were heated chambers for grooms or harness, and adjoining areas for preparing the horses' feed, with shoots from the hayloft above, and running water passing under the stable to take away waste. There was even what the *Methode et invention nouvelle* described as a 'petite chiminée' or small ventilation shaft in front of each stall to take away the horse's exhaled breath, which could be opened or closed depending on the temperature outside.[32]

80 The stables of 1625 at Welbeck Abbey, Nottinghamshire, from William Cavendish, *Methode et invention nouvelle* (Antwerp, 1658) (© The British Library).

Though the Welbeck stables included alternating round and segmental pediments over the windows that were presumably a reference to Inigo Jones's recently completed Banqueting House in Whitehall, the design otherwise owed little to the Italianate Classicism then coming into favour at the court. Indeed, the buttresses, arrow-loop windows ventilating the hayloft, and ogee roofs over the pavilions have more in common with Robert Smythson's Little Castle at Bolsover, Derbyshire, built for Newcastle's father Sir Charles Cavendish around 1612–14. John Smythson's other surviving design for a stable, for Sir Gervase Clifton at Clifton Hall, Nottinghamshire, of 1632, is little more advanced, again using arrow-loop ventilators for windows (fig. 45).[33]

The stables and riding house that Newcastle built at Bolsover Castle, Derbyshire, probably in the late 1630s to the designs of Huntingdon Smythson, were more monumental in their ambition (fig. 60). Built entirely of stone, with a series of gabled dormer and casement windows with Classical architraves, the composition is dominated by the massive, rusticated doorcases to the riding house and stable, each with a broken segmental pediment. These details reveal a knowledge of the latest architectural fashions from London and the Continent that is missing at Welbeck. Lucy Worsley and Tom Addyman point out similarities with Jones's work in London in the gables over the windows and in the form of the windows. A particular parallel can be made with Brooke House, though comparison could also be made with the stables at Burley-on-the-Hill (fig. 70). The central doorcase is taken from Alexandre Francini's *Livre d'architecture* of 1631. The interiors of both riding house and stable have been much altered, but enough survives within the stable to show that it was originally decorated with a handsome plaster cornice.[34]

The high status accorded to stables is clear from this list of large, expensive and architecturally advanced examples built by important courtiers in the 1620s and 1630s. As Gerbier suggests, stables were built for display and prestige, quite as much as houses were. Indeed, stables had a particular advantage as a building type in which to express the new architectural fashion. For monarch and courtiers alike, the first half of the seventeenth century was a period of grand but unrealised or incomplete architectural schemes, the reality of which was more limited and often largely restricted to internal improvements. Stables, self-contained architectural units whose scale and simplicity made them well suited to an essay in the new style, which could be built for less cost, and yet were prestigious buildings in their own right, were a chance to realise in miniature these thwarted architectural ambitions.

Given the classical origins of *haute école* in Xenophon's *The Art of Horsemanship*, at a time when such horsemanship was keenly practised at Court, contemporaries may even have seen a particular appropriateness to housing horses in Classical buildings, a reference emphasised by the busts of the Caesars in the courtyard of the stable at Wilton. This body of stables, which shows a consistent approach to design, is an important addition to the few buildings by Inigo Jones and the handful of other houses that provide the evidence for the pre-Restoration Classical revolution in architecture. The importance of the stable in the development of pre-Civil War Classical architecture should not be ignored.

Display was obviously important in the early Stuart stable, but an innate English practicality prevented it from being taken too far. In Gerbier's view, display was best achieved by placing horses in a double range of stalls along both sides of the stable: 'the disposing of a stable into a double range hath been affected by some who would see all their horses at once'. It was equally important that horses should stand on a marked slope. Gerbier favoured a slope of six inches in an eight foot stall, not for comfort or convenience but because 'the higher a Horse stands towards the Manger, a better sight it is, especially when the Lights of the Stable strikes on the Horse their backs'.[35] This had long been the tradition on the Continent. Leonardo da Vinci's design for ideal stables and Raphael's Farnesina stables in Rome, both of the early sixteenth century, had sharply sloped stables.[36]

Both suggestions were heavily criticised. Roger North, writing towards the end of the seventeenth century, presented the matter-of-fact English case against a large stable: 'This looks great, but the master's ostentation is not for the horses' health; the place is cold and noisy, nothing stirs but the room is disturbed, and an horse is a watchful creature and hearkens after all that passeth.'[37] The sloping stall was repeatedly condemned. Eighty years before Gerbier wrote his books, Christopher Clifford had asked rhetorically in *The Schoole of Horsemanship* 'whether it be not good to raise my horses room a foote higher than the grounde behind, to make my horses show the better?' His answer was categorical: 'It were good for a horse courser to make horses shew higher than they be indeed, but for the ease of the horse it is stark nought.'[38] Gervase Markham echoed him in 1607, dismissing such a slope as 'of all errors the grossest, not to be tolerated in any place but Smithfield and among horse copers', while Thomas de Grey, writing in *The Compleat Horseman and Expert Farrier* in 1639, was damning in his dismissal: '[if] a horse standeth higher before than he doth behind his hinder legges will swell, and so he becommeth lame . . . His long standing in this uneasie manner begetteth in the poor beast much paine and griefe.'[39] Though in France de Garsault was still recommending markedly sloping stalls in his *Le Nouveau parfait mareschal* in 1741, a book that ran through sixteen editions by 1805, in England the concept of the sloping stall had long since been abandoned.

The Commonwealth saw the consolidation of the Classical revolution inaugurated under the early Stuarts, which was continued after the Restoration of 1660. The innovative country house of the 1620s and 1630s, with its brick, its stone dressings, its even, symmetrical rows of upright windows, emphatic cornice, quoins, hipped roof, dormers and cupola, became the standard model for the post-Restoration English country house. The more ambitious included a central pediment following the example of Sir Roger Pratt at Kingston Lacy, Dorset, in 1663 and Hugh May at Eltham Lodge, Kent, in 1664. The typical country-house stable of the same date was a smaller version of the same. Such stables were generally designed as a single range (not a quadrangle) with a central door, upright windows and, whether of one storey or two, a hipped roof with dormers. There was usually a cupola on the roof, a diminutive version of that on the house, commonly used for a clock, often a central pediment and perhaps wings. Though some such stables were double pile (or two rooms deep), they were usually only single pile.

The cupola was a particularly distinctive addition to the stable vocabulary and one that was maintained even after it had been abandoned on houses. Though cupolas were a common feature of houses built soon after the Restoration, they had begun to fall from favour by the end of the century. Roger North, writing at the very end of the century, dismissed the 'pompous lanthorne aloft, perpetually

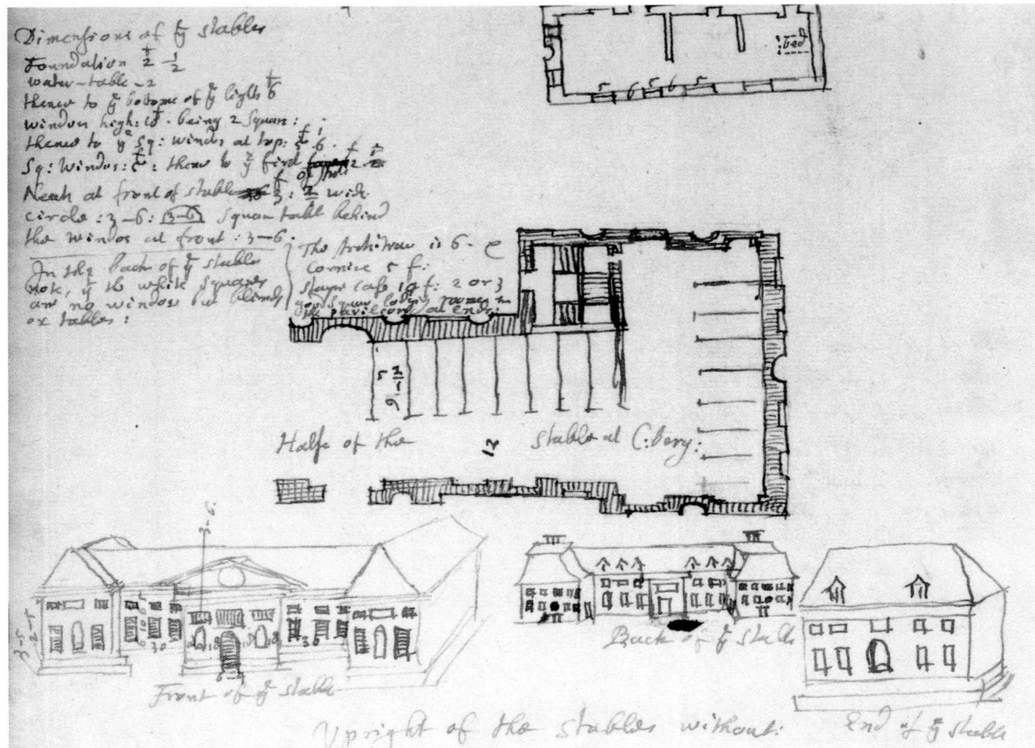

81 John Evelyn, sketch plan and elevations of the stables at Cornbury Park, Oxfordshire, *c.*1664 (© British Library, Add. MSS 78341 fol. 101v).

leaky', adding 'it is irksome even to thinck of climbing so high'.[40] But the cupola survived, in vestigial form, on the stable. Here, it was not used as a viewing platform, and so was smaller than the cupola in a house, but served the purpose of housing a bell and often a clock, as well as acting as a marker in the countryside. 'In the middle [of the stables]', commented Celia Fiennes, when she visited Chippenham Park, Cambridgeshire, 'is a large gate into the ground and built with a high lanthorn where hangs the clock and bell, this stands higher than the house, like a tower . . . this tower I saw 10 miles off'.[41]

One of the earliest, firmly dated examples of a pedimented stable was that at Cornbury Park, Oxfordshire, designed by Hugh May in 1663 and completed by mid-1664 for the Earl of Clarendon, the Lord Chancellor and leading politician of his day. This has been altered, but a sketch made by John Evelyn, who worked with May on the design of the Chapel at Cornbury in October 1664, shows that it was unusually handsome (fig. 81). Fifteen bays wide and two storeys high, with projecting pavilions and pedimented centrepiece, it was built of ashlar stone and embellished with niches in the pavilions and flanking the central door.[42] The Cornbury stables were considered such a good model for a stable that drawings of them were sent to Ireland in 1670 by the Earl of Conway to be used by William Hurlbutt, who was working for him at Portmore in County Antrim.[43]

The Earl of Clarendon went on to build a second set of stables, this time to the design of Sir Roger Pratt in 1664–67, at Clarendon House in London. Prominently sited on London's most fashionable street, and built by one of the country's most powerful men, the importance of Clarendon House as a model for other post-Restoration houses is well accepted. The significance of its stables has not been considered, perhaps because no image of them is known to have survived. Careful study of Ogilby and Morgan's map of London of 1681–82 shows that the stable lay in a flanking wing to the side of the house, with what were probably coach houses behind. In size, its ground plan matched that of Clarendon House without its wings, and like Clarendon House it was a tripartite building with a projecting central section. Analysis of Pratt's architectural notebooks makes it possible to reconstruct the dimensions of the stable and details of its architectural embellishment, which included a wooden modillion cornice costing 5s 8d per foot.[44] Knowing that the façade was treated architecturally, it can be deduced that the stable would have been treated like the house, presumably with a pediment. It was a design that was to became the classic post-Restoration stable.

Such stables predominate in the pages of *Britannia Illustrata*, published in 1707, which records many of the country's principal houses. Often, as had been the case in the 1620s and 1630s, fashionable new stables were added to older houses. They were an easily achievable way for the owner to demonstrate his familiarity with the new architectural fashions without the expense of remodelling or rebuilding his house. Examples of Classical stables added to older houses include Southwick in Hampshire, Margam Abbey, Glamorgan, and Hampton Court, Herefordshire.[45]

The typical post-Restoration stable was thus a miniature version of the contemporary country house, with the grander examples often rivalling smaller country houses. The proposed stables for the Earl of Craven at Hampstead Marshall, Berkshire, for which designs survive, were so grand that they would have had their own pavilions, as if a miniature country house in their own right. That was certainly the impression Celia Fiennes had of the similar, now-demolished stables at Lowther Castle, Westmorland, which she described as 'built like a fine house jutting out at each end and the middle . . . with pillars white and carvings like the entrance of a building' (figs 82 and 83). Joseph Taylor was so impressed by these in 1705 that he exulted that the first stables he saw in England were finer than the royal palace (Holyroodhouse in Edinburgh) that he had just visited in Scotland, while William and John Blathwayt's account describes it as 'a large main building and two pavilions at either end that one would take if one only took into consideration the exterior for a furnished and inhabited house, although it serves for the stables'.[46] The grandest stable of this type was that built by Robert Hooke for the queen, Catherine of Braganza, at Somerset House in London in about 1669–70 (figs 85 and 86). This should be seen as the last and the grandest in the sequence of large stables built for the Strand palaces, following in the footsteps of Essex House, Arundel House and Bedford House. The closest comparison is with Bedford House, which had similarly projecting wings, hipped roof with dormers and the characteristic *oeil-de-boeuf* windows. But, as befitted the stables of the dowager queen, and reflecting the changing fashions of the Restoration court, the Somerset House stables had a central pediment supported by giant Ionic pilasters, with further pilasters on the pavilions, swags between the central pilasters of the pavilions and a rusticated door and handsome balcony in the centre.[47] Few if any late seventeenth-century country-house stables could rival them for grandeur. A handsome, but by no means extravagant, set of stables designed by Sir Christopher Wren for Charles II to house thirty-six horses, for which drawings survive at All Souls, came to nothing (fig. 87).[48] So did a scheme by Wren to rebuild the Royal Mews

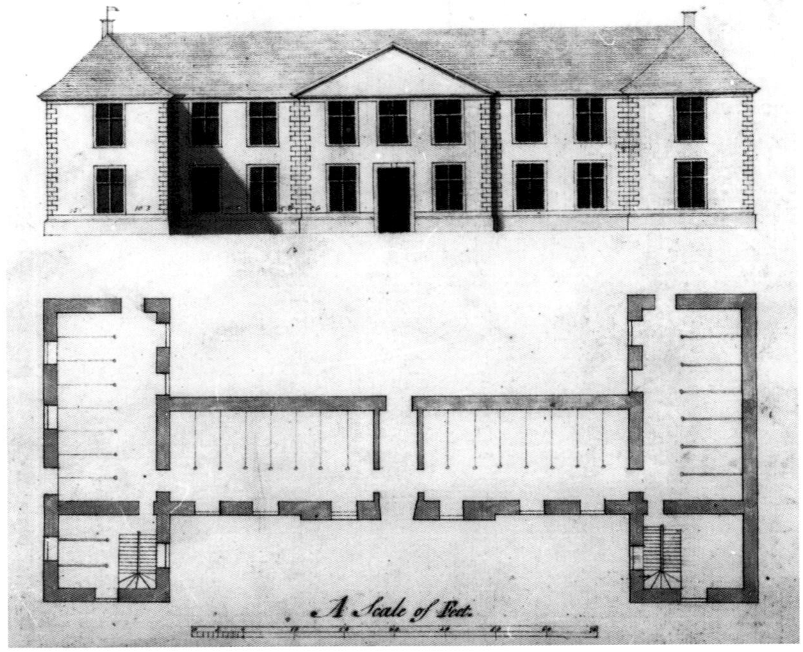

82 Anonymous draughtsman, survey plan and elevation of the stables at Lowther Castle, Westmorland, 1718 (Cumbria Record Office; photo: Conway Library, Courtauld Institute of Art).

83 Leonard Knyff and Johannes Kip, Lowther Castle, Westmorland, from *Britannia Illustrata* (London, 1707) (detail) (© The British Library).

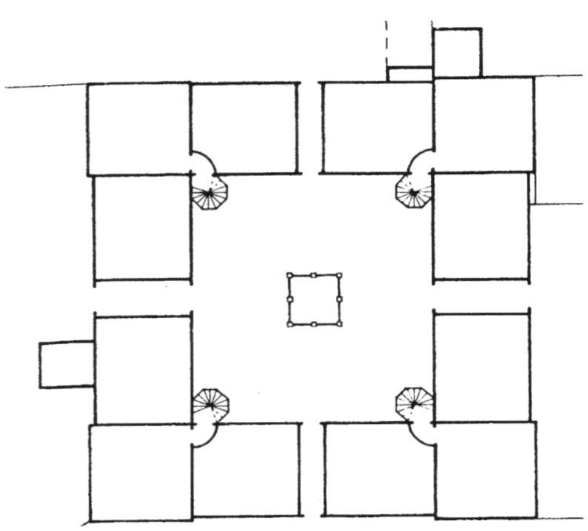

84 (*below*) Plan of the stables of 1669–71 at the King's House, Lyndhurst, Hampshire (from *The History of the King's Works* v, fig. 11).

as a large, regular quadrangle with stalls for 388 horses and room for 42 coaches (fig. 88).[49] However, Charles II did commission expensive stables costing nearly £3,000 for his hunting lodge at Lyndhurst, Hampshire, in 1669–71 (fig. 84).[50] These do not survive but their plan shows that, as with Wren's scheme for the Royal Mews, they maintained the royal tradition of the stable quadrangle and, like the stables at Hampton Court, were served by spiral staircases in the corners of the courtyard.

At Clarendon House, the stables were placed in one of the wings flanking the house. This was to prove a characteristic pattern for post-Restoration stables, with house and stable being treated as an architectural unity. The same was true of two other prominent houses built in Piccadilly in the 1660s, Berkeley House and Burlington House. Though this model had been anticipated by some Jacobean

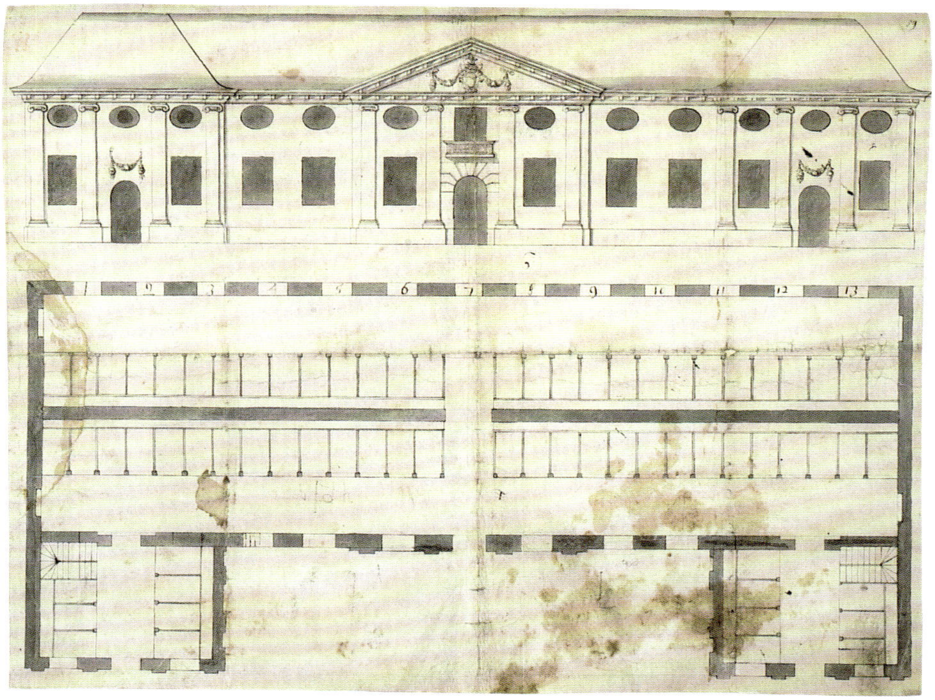

85 Robert Hooke, plan and elevation of the stables at Somerset House, London, *c.*1669 (British Library, Add. MS 5238 f. 89).

86 (*below*) Leonard Knyff and Johannes Kip, Somerset House, London, from *Britannia Illustrata* (London 1707) (detail) (© The British Library).

houses such as Blickling House, Norfolk, it became particularly important in the second half of the seventeenth century, as can be seen with most of the new houses illustrated in *Britannia Illustrata*, including Eaton Hall, Cheshire (1675), New Park, Surrey (1692), Ragley Hall, Warwickshire (circa 1677), Stanstead Park, Sussex (1686), and Uppark, Sussex (circa 1690) (fig. 89).[51]

Alternatively, stables were placed with the other offices to one side of the house. This remained the usual position when the stable was an addition, as at Hampton Court, Herefordshire (fig. 90), but was also common with new houses such as Denham Place, Buckinghamshire (fig. 92). Most such stable courts lacked any overall architectural unity and included a variety of other buildings such as barns, coach houses, laundries, brewhouses and possibly even farm buildings, but some larger houses had regular and substantial stableyards. At Chatsworth House,

87 Office of Sir Christopher Wren, design for the elevation of stables for Charles II, n.d. (The Warden and Fellows of All Souls College, Oxford, II: 100).

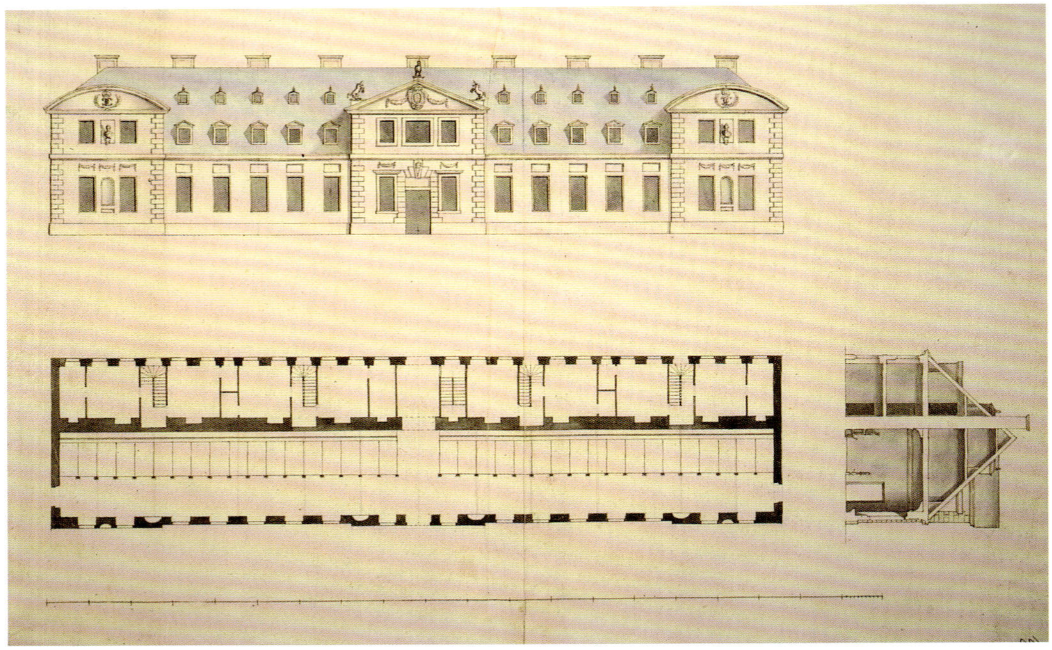

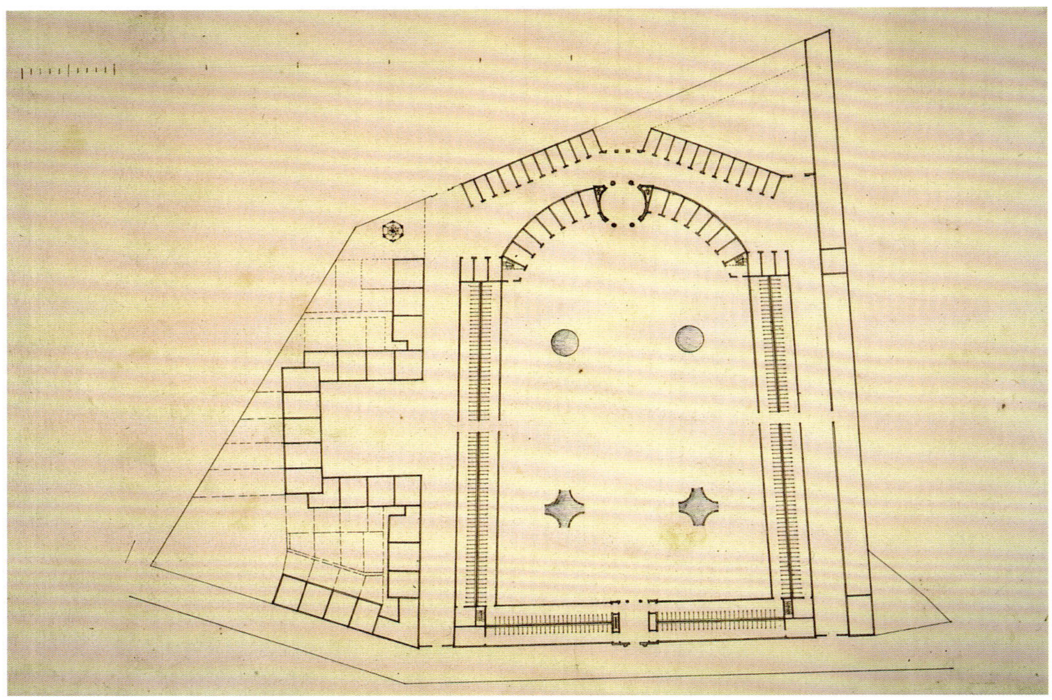

88 Office of Sir Christopher Wren, design for the Royal Mews, Charing Cross, London (The Warden and Fellows of All Souls College, Oxford, II: 105).

Derbyshire, the huddle of stables visible in an anonymous painting of the 1680s was clearly felt to be unsuitable for the remodelled house and was replaced by a regular three-sided courtyard to the left of the main entrance.[52] Chatsworth's stable court was only a single storey high and not especially impressive, despite the pediment in the centre of each range. Similar courts at Milton House, Northamptonshire, of 1690 and Worksop Manor, Nottinghamshire, of circa 1700–04, which both survive, are much grander.

89 Leonard Knyff and Johannes Kip, Stanstead Park, Sussex, 1686, from *Britannia Illustrata* (London, 1707) (detail) (© The British Library).

90 (*below*) Leonard Knyff, *Southeast Prospect of Hampton Court, Herefordshire* (detail), *c.*1699 (Yale Center for British Art, Paul Mellon Collection; photo: Bridgeman Art Library, London). The stables lie to the right of the entrance court.

91 Jan Siberechts, *Cheveley Park, Cambridgeshire* (detail showing the stables), 1681 (Belvoir Castle, Leicestershire; photo: Bridgeman Art Library).

The stables at Holland House were exceptional in integrating stables and coach houses. Usually in the seventeenth century these were treated as independent structures, but, where they were integrated, the resulting building had a much greater architectural impact. At Cheveley Park, Cambridgeshire, built sometime in the ten years before 1681, a typical post-Restoration stable is flanked by a pair of wings housing four coaches (fig. 91). At Badminton House, Gloucestershire, the stables built for the second Duke of Beaufort between 1706 and 1712 ran around three sides of a quadrangle, with the central range pierced by an archway that was carefully balanced on each side by coach houses.[53] This anticipates the detached stable quadrangles that were to become popular from the 1720s.

Most such stables had little or no external decoration. Two exceptions were the stables designed by the Hurlbutts, master builders of Warwick. One was the (now-demolished) stable built in 1667 at Warwick Castle in Warwickshire, which was described shortly afterwards by Thomas Baskerville as 'very fair stables after the new mode'.[54] The other is the surviving stable at Tredegar House in Gwent, probably built for Sir William Morgan who died in 1680 (fig. 95). In both cases, the façades were decorated with Ionic pilasters that stop short of the cornice, interspersed with *oeil-de-boeuf* windows. At first sight this suggests the ill-digested Classicism known as Artisan Mannerism (an impression emphasised at Tredegar by the poorly proportioned pediment which may be a later addition). But closer study of the Tredegar stables shows that the design carefully parallels that of the house, with projecting wings flanking a grand central doorcase with Corinthian columns and

92 (*facing page*) Anon., *View Towards the West Front of Denham Place, Buckinghamshire* (detail), *c.*1705. The stables are to the right of the house (Yale Center for British Art, Paul Mellon Collection; photo: Bridgeman Art Library).

93, 94 and 95 (*above and facing page*) Roger or William Hurlbutt, exterior of the stables of *c.*1670 at Tredegar House, Gwent, before 1680 (William C. Rolf, 1999).

a broken pediment, and transom windows modelled on those on the first floor of the house. This suggests that the fact that the pilasters do not support the cornice, as might be expected, was deliberate, particularly as the pilasters are topped by pineapples. The explanation probably lies in the use of Doric or Ionic columns as heelposts at the foot of the stalls of contemporary stables, usually with some crowning feature, usually a ball or an acorn. A pineapple – the symbol of welcome – might be used for such a feature, and their appearance on the Ionic pilasters at Tredegar suggests that the architect was announcing the purpose of a building in its exterior.[55]

William Talman's stables at Dyrham Park, Gloucestershire, begun in 1698, also break the conventional rules of Classical architecture. These have no entablature, no entasis on the columns, and no capitals; only abaci and the simplest of bases. It has been suggested that this is the result of ham-fisted execution by the builder Edward Wilcox, but this seems unlikely. Wilcox was an experienced mason who had worked for Wren, both at St James's Church in Piccadilly and at Kensington Palace. At Dyrham, similarly, he was working on a building of the highest sophistication for a powerful courtier who was closely involved in the building's construction. It is scarcely conceivable that he would have erected a building with so many apparent solecisms through ignorance. Talman may well have known of another building where the entablature is omitted and the abacus run together, Borromini's cloister of San Carlo alle Quattro Fontane in Rome, from his exceptional architectural library, 'the most valuable Collecion of Books, Prints, Drawings &c, as in any one person's hands in Europe'.[56] Alternatively, he would probably have known the stables at Thorpe Hall, Northamptonshire, of 1654, which he could have

96 (*above left*) Exterior of the stables of *c.*1670 at Llanvihangel Court, Gwent, *c.*1670 (William C. Rolf, 2000).

97 (*above right*) Interior of the stables of *c.*1670 at Llanvihangel Court, Gwent, *c.*1670 (William C. Rolf, 2000).

visited when advising Earl Fitzwilliam on the proposed siting of the offices at Milton House, Northamptonshire, in 1688, only a few miles away. The central doorcase on the Thorpe Hall stables has the same primitive order as Dyrham, with no entablature, no capital except the abacus, no entasis, and only the simplest of bases. Given its date and advanced style, Talman may well have thought that Thorpe Hall was by Inigo Jones, an architect in whom he had a great interest, as shown by his ownership of Inigo Jones and John Webb's drawings. Far from being an example of builder's error, the Dyrham stables can be seen as an attempt by Talman to create a primitive style of architecture suitable for stables by taking the simplest of the orders and stripping it to its fundamental elements.

Several stable interiors survive from the second half of the seventeenth century. A fine set of robust heelposts survives in the elegant stables probably built in the 1670s at Llanvihangel Court in Gwent (figs 96 and 97).[57] A rather more refined interpretation can be found in the contemporary stables built at Arbury Hall in Warwickshire, in about 1674 (fig. 98). Here, the rather crude chess-piece-like forms of the heelposts at Dunster Castle and Llanvihangel Court have been refined into an elegant baluster. The heelpost no longer serves as the support for a hanging bale but as the anchor for a fixed partition, something that becomes increasingly common from the middle of the century.

Powis Castle, Powys, had been badly damaged during a siege in the Civil War and in 1668 a grand new gateway was created, flanked by single-storey ranges and housing stables (fig. 99). Compared with the elaboration of the arcaded interiors of the stables at Whitmore Hall or Peover Hall, or the vigorous moulding of the heelposts at Dunster Castle, the Powis Castle interiors seem modest and restrained, but their architectural quality is clear. As with the early seventeenth-century examples, the rack is still designed as a single timber construction, built out from the rear wall of the stable. That at Powis Castle is designed as an arcade with a full cornice running the length of the rack. Instead of hanging bales between the horses, there are solid timber partitions with an elegant curve that sweeps down from the horse's head (where it is raised to prevent neighbouring horses disturbing each other) and then rises slightly where it meets the square heelpost. This would become a characteristic visual rhythm of stable interiors.

The stables at both Powis Castle and Arbury Hall show the calm sophistication of the mid-seventeenth-century stable interior, but neither could compete with the stables built by the extravagant Marquess of Wharton in the last quarter of the seventeenth century at Winchendon House in Buckinghamshire. These not only had carved pillars, cornices, mangers and racks, but also a partially gilt stuccoed ceiling.[58]

Apart from Gerbier's works, the late seventeenth and eighteenth centuries provide little of interest in the way of stable literature to compare with the series of innovative books on stables in the late sixteenth and early seventeenth centuries. The most prolific source of advice on stables during these years is the series of short works and dictionaries published regularly with names such as *The Gentleman's New Jockey*, *The Experienced Farrier*, *The Farrier and Horseman's Complete Dictionary*, or even *The Gentleman's Accomplished Jockey with Compleat Horseman and Approved Farrier*. Despite the large numbers of titles, these are of little value as sources of information about stables as the first devised a set pattern of advice that

98 Interior of the stables of *c.*1674 at Arbury Hall, Warwickshire (*Country Life*, 1913).

became a formula that was copied almost verbatim until 1796, when it appeared in *Hunter's Compleat Dictionary of Farriery and Horsemanship*. Thus the theories of the early seventeenth century were repeated without change until late into the eighteenth century. Such essential conservatism probably fits well with the opinions of the more old-fashioned members of the stable community, especially grooms and small farmers.

An alternative source of information about stables can be found in the numerous French books on horsemanship. England was heavily influenced by French horsemanship throughout the seventeenth and eighteenth centuries as young Englishmen often studied horsemanship at French academies during their Grand Tour. French ideas about stable design must have been known both from practical experience and through the study of French books. These included books on horsemanship, such as de Solleysel's *Le Parfait Mareschal*, published in 1675 and translated into English as *The Compleat Horseman* in 1696 by Sir William Hope, and de Garsault's *Le Nouveau Parfait Maréschal* of 1745, and conventional architectural books with sections on stable design, such as Savot's *L'Architecture françoise* of 1624, Briseux's *Architecture moderne* of 1728 and J. F. Blondel's *De la Distribution des maisons de plaisance* of 1737 and his *Cours d'architecture* of 1771–77. However, none of these books shows the French to have been more advanced in stable design than the English and in certain respects they were less so than de Grey who wrote in 1639. It is thus unlikely that they greatly influenced English stable design.

There is little sign of significant changes in the practicalities of stable design in the second half of the seventeenth and the early eighteenth centuries. Gentlemen at least considered that their stables needed few improvements. Roger North, writing of stables in his manuscript *Of Building* towards the end of the seventeenth century, declared that 'these are ordinarily so well contrived because men of best geniuses and estates delight in horses and to be provident for them, that the less is to be observed'.[59] Consequently he gives little useful information about stable design. Other stables may not have been so well designed. William Gibson, writing in his *True Method of Dieting Horses* in 1721, noted that only the stables of the nobility and gentry were 'excellent patterns', more ordinary stables being 'but indifferently contrived'. But a complacent stagnation seems to have settled over the practicalities of stable design, a stagnation confirmed by the standard formulae of the books on farriery and the lack of any new books challenging the consensus. Had the thinking behind stable design developed, then it is probable that these developments would in time have been reflected in the farrier's guides. As it is, these disappear at just the moment when the theory of stable design, and indeed of horsemanship, changed markedly, towards the end of the eighteenth century.

99 Interior of the stables of 1668 at Powis Castle, Powys (William C. Rolf, 2000).

5

HOUSING THE URBAN HORSE

FOR CENTURIES THE HORSE WAS an essential part of urban, but above all metropolitan, life and its housing was a critical, though largely unsung, factor in a city's effective organisation. The equestrian needs of the city were not the same as those of the country, as was reflected in the number and type of horses housed in stables, but the principal difference arose from the pressure of space, with all the constraints that ensued.

In any discussion of urban stabling, London must predominate. It was by far the largest city in Britain and housed a disproportionate number of urban horse owners. As the seat of government and primary social centre, London was home for much of the year to the richest members of society, those most likely to own horses and carriages. The sheer size of London, the patterns of social conduct that required those of a certain dignity to travel by coach and the important part in daily social life that riding and driving in Hyde Park played in the nineteenth century made ownership of private transport essential for many. Other cities were not only smaller, but held proportionately fewer horses and carriages. Consequently, in no other city did provision for stabling play so large a role in town planning.

Before the second half of the sixteenth century, stabling was not an important feature of even the grandest London house because the river, or foot, was the principal means of transport. The large retinue of horses that a great noble required in the country would have been expensive to maintain and of little value in London. With the introduction of the coach in the middle of the sixteenth century and the decline of river transport as London spread north and west, the horse became the primary means of city transport for the wealthy. Until the introduction of the omnibus and hackney cab in the nineteenth century, the hack was a popular form of transport, but the coach and coach horse were the dominant element in the urban stable. By 1739, there were reckoned to be 2,484 private coaches in London, compressed into the relatively small part of the city considered fashionable.[1]

The impact of the introduction of the coach and its accompanying horses into a densely established pattern of urban settlement must have caused problems. These were also faced in Paris where coaches were very rare in 1550, but where the provision of a carriage and house with a carriage gateway had become one of the principle conditions of marriage contracts by 1619.[2] Before the introduction of regularly planned new housing, the siting of stables was determined more by the caprice of earlier property boundaries than by practical considerations. The most convenient arrangement was for the stable to have a separate access from the street, but all too often the shape of the available plot prevented this. In Ralph Treswell's survey of about 1607–14 of houses owned by the Clothworkers' Company, Lady Lucy's house in Christ Church had a stableyard with a 'stable and coche howse' at the back of the house, entered separately from Fleet Street (fig. 101); a smaller house in West Smithfield had a stable but could only be approached through the house; and Sir Edward Darcy's house had no room for a stable or coach house at all.[3]

100 George Scharf, *Torrington Mews behind Scharf's House* (detail of fig. 115; larger than actual size).

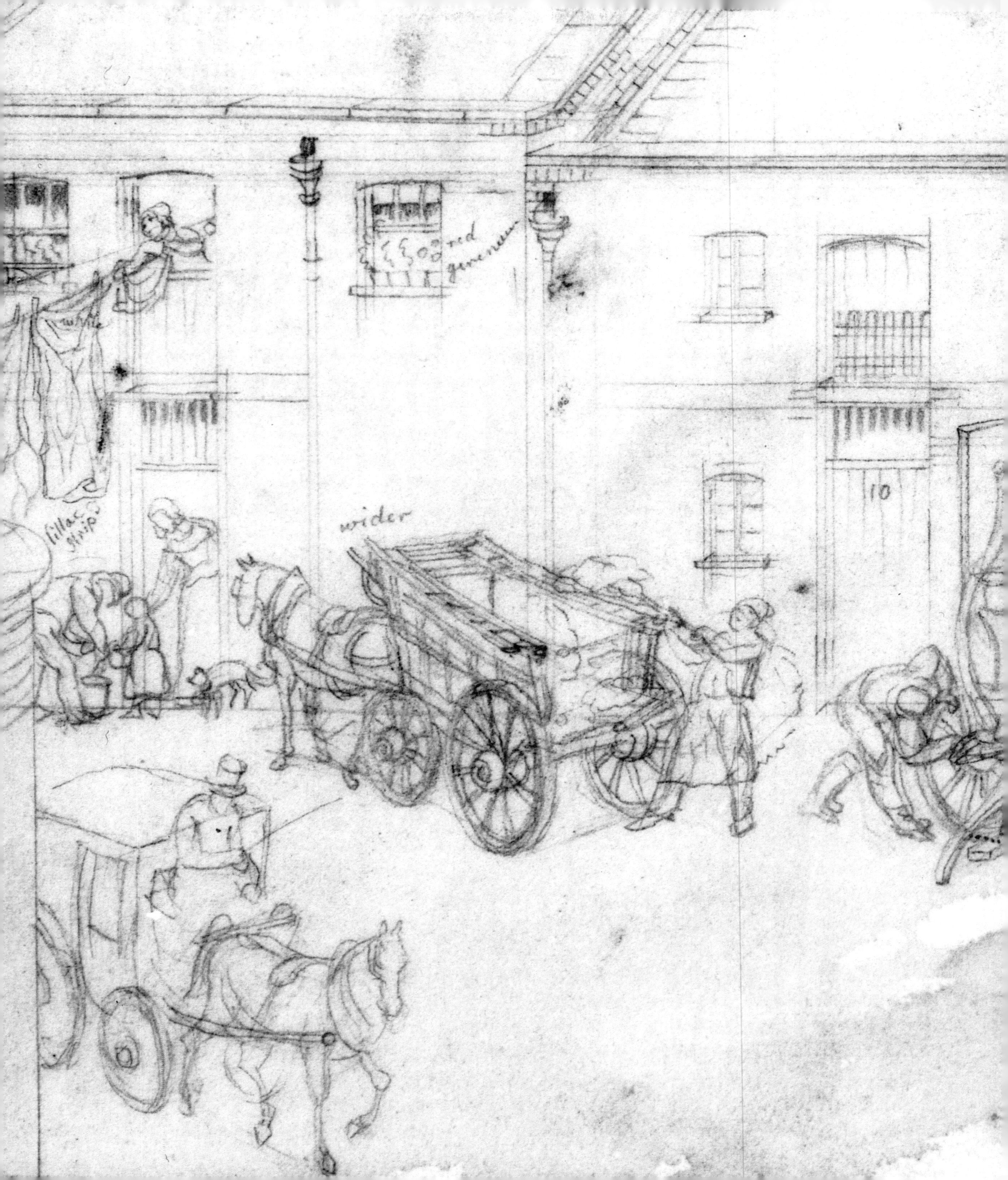
white
red
wider
10

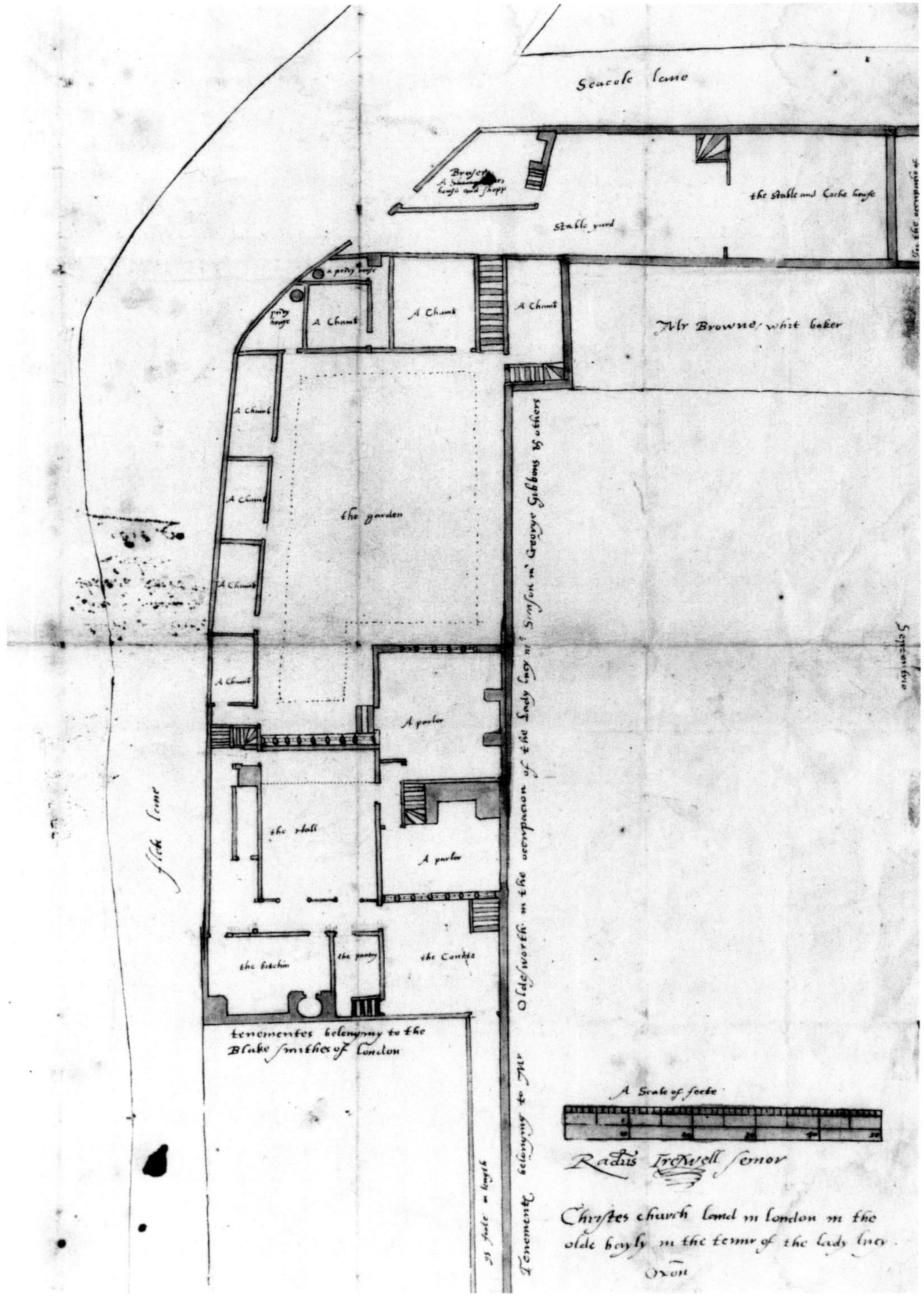

101 Ralph Treswell, survey drawing of Lady Lucy's House, Old Bailey, London, *c.*1607–14 (Christ Church, Oxford, MS Estates 45).

This was equally true of the great Tudor and Jacobean houses along the Strand, the layout of which was determined more by the arrangement of the bishops' palaces they replaced than by the preference of their sixteenth- and seventeenth-century builders. Thus Somerset House had a separate stable yard with its own access to the Strand and Essex House had a separate stable yard but a single entrance to the street, while at Arundel House the entrance to the house was effectively through the stable yard (fig. 103).[4] Alternatively, the stables were detached from the house. At Cecil House and Bedford House, both in the Strand, the only link between the house and its stables was a long passage.[5] At Northampton (subsequently Northumberland) House, the demands of the garden meant that there was no room for adjoining stables, and these were positioned across the street. This can never have been convenient, but it was not until the Duke of Northumberland's remodelling of the house in 1749 that land was acquired beside the house to create a contiguous stable yard.[6]

There was a decline in popularity of the City and the Strand as fashionable places to live in the middle years of the seventeenth century, and many of the great houses along the Strand were demolished to make way for speculative housing by Nicholas Barbon and his associates. Fashionable members of society moved west to find new sites on the outskirts of the city where they could build houses unconstrained by existing building. Behind this change of fashion lay reasons of practicality: the desire to escape the congestion of the more built-up parts of the city and to substitute compact modern houses for old rambling palaces and town houses. The introduction of stables and coach houses must have been a significant element in the move. Critical to its success was the invention of a new building type to house the horse and coach, the mews, which was to prove one of the defining characteristics of Stuart, Georgian and Victorian London.

The mews was a small building for horses and coaches at the rear of a terraced house, connected to it by a garden or yard but approached from the street by a separate alley. As a building type, the mews is a particularly British and overwhelmingly London phenomenon.[7] It is so familiar, especially in Mayfair, Belgravia and South Kensington, that its existence is largely taken for granted, but like so much that is considered Georgian, its roots lie in the middle years of the seventeenth century.[8]

As Fynes Moryson noted in 1617, it was then sixty or seventy years since coaches had been introduced to London, whose streets, he now complained, were almost blocked by them.[9] The introduction of the coach was as significant a social development for the upper classes as the arrival of the motor car three centuries later, and it brought with it serious practical problems.[10] Coaches were bulky and very expensive objects. The horses that pulled them (two, four or even six) were also very costly. Both needed careful housing – indeed, the demands of this new, precious breed of horse had led to a general revolution in stable design in the late sixteenth and early seventeenth centuries – and together they significantly increased the scale of stable accommodation required by an aristocratic London house.

This was not a problem for those who occupied an *hôtel*, an independent town house with its own forecourt and stabling. In great houses such as Arundel House on the Strand, a coach house and new stables could be built in an outer court (fig. 78). With the new Classically inspired *hôtels* that were built after the Restoration such as Clarendon House, Burlington House and Berkeley House, all on Piccadilly, coach house and stables could be placed in the office wings. But these remained exceptions (fig. 102). In London, unlike Paris, most of those who owned a coach chose not to live in an *hôtel* but in a terraced house. Thus in 1721, St James's

102 Leonard Knyff and Johannes Kip, Burlington House, London, *c.*1665, from *Britannia Illustrata* (London, 1707) (detail) (© The British Library). The stables and coach houses lie to the right of the house.

Square, which was composed entirely of terraced houses, held six dukes, seven earls, a countess, a baron and a baronet.[11] The question that the designers of the terraced house faced was where conveniently to house the coach and its horses that figures of such eminence required, given that such houses had no forecourt, and that there were no gaps between the houses where stables might otherwise have been placed. The solution was the mews, without which the terraced house could never have supplanted the *hôtel* as the common London residence of the English upper classes.

It has long been accepted that the introduction of the uniform terrace of houses as the standard London building type happened in the 1630s. These years saw a building boom which began in 1630, following the end of the long and expensive war with France and Spain that had been raging since 1624, and which transformed the area to the west of the City of London around Lincoln's Inn Fields and Covent Garden.[12] These years also saw the introduction of the mews.[13]

In 1631, the Earl of Bedford was granted a licence to build 'howses and buildings fitt for the habitacons of Gentlemen and men of ability' on land he owned to the north of the Strand in Covent Garden, on condition that it was laid out to a regular plan.[14] To design the scheme, as Dianne Duggan has confirmed, the earl employed Inigo Jones, surveyor-general of the King's Works, but, perhaps more significantly in this case, also the key figure in the commission established in 1625 to control new building in London.[15]

Covent Garden is celebrated as the first example of a formal square of terraced houses in London, but attention tends to focus on the Piazza and on its church (fig. 103). The significance of Hart Street, now Floral Street, a small lane laid out at the same time to the north of the Piazza and King Street (its continuation to the west), and of Maiden Lane, to the south of Henrietta Street, which mirrored King Street, has been overlooked. These were London's first mews.

Hart Street ran immediately south of the northern range of the brick wall that enclosed Covent Garden. It provided a service road for the new houses being built along the north side of the Piazza and King Street, each of which sat on a long, narrow plot of land running back to Hart Street.[16] Its purpose is explained by John Strype in 1720 who described Hart Street as 'not over well built or inhabited; the South Side being for the greatest part taken up by Coach-houses and Stables belonging to the Houses in Covent Garden'.[17] Maiden Lane served a similar purpose for Henrietta Street.[18] Although neither street is included in the first undated survey plan of Covent Garden illustrated by Duggan, they are both shown in the second plan.[19]

None of the original mews buildings in Hart Street or Maiden Lane survives, but a detailed picture of these early mews is provided by the three houses completed in 1635 at the north-east corner of the Piazza. These were built by the Earl of Bedford as models for the rest of the development and then leased to Edward Sydenham and Sir Edmund Verney.[20]

Although this area has been completely redeveloped and is now occupied by the Royal Opera House, the building accounts, the inventory attached to the lease, and ground plans made after the Restoration clearly establish the original plan. The building accounts describe the stables of both houses in great detail, showing that they had four pairs of coach-house doors, as well as grooms' quarters, haylofts, stabling and a well. They also include a reference in the paviour's account to 'The backe streete at the end of Sr Edmund Varneys and Mr Sidnams stables . . . 113 yards 3 foote', that is to Hart Street.[21] The linking garden or yard between the stables in Hart Street and their houses is shown in the map of Covent Garden of about 1680–85 in Crowle's extra-illustrated copy of Thomas Pennant's *Some Account of London* of 1790 in the British Museum Department of Prints and Drawings (fig. 104). The drawing also makes clear the layout of houses and mews in the rest of Hart Street and in Maiden Lane.

The germ of the idea for the mews at Covent Garden was probably the Place Royale, now the Place des Vosges, in Paris, which is accepted as one of the key

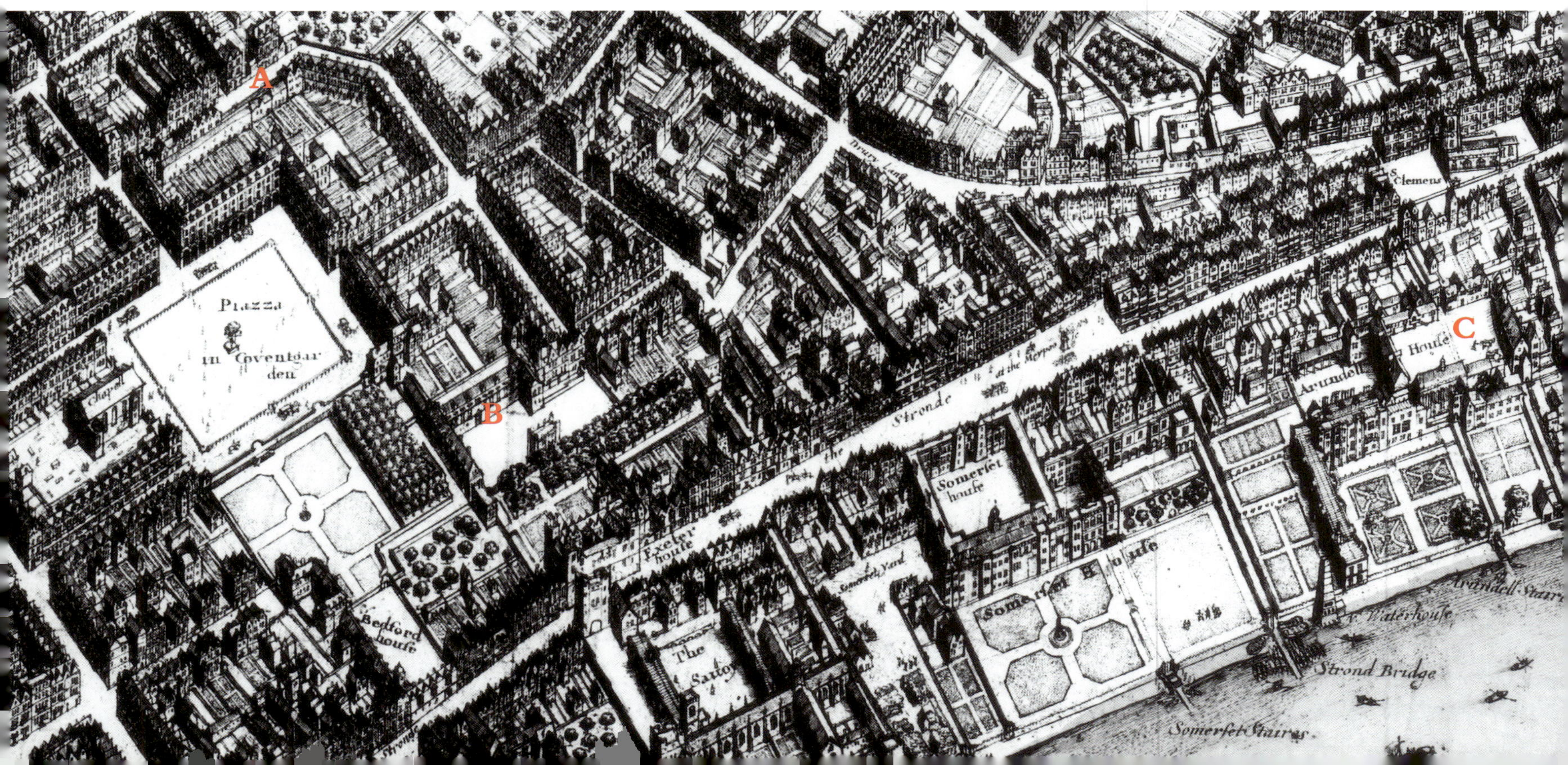

103 Covent Garden and the Strand, London, from Wenceslaus Hollar, *West Central London*, *c.*1658 (detail) (British Museum). A: Hart or Floral Street, the mews for Covent Garden; B: the stables of Bedford House; C: the stable courtyard of Arundel House.

104 Detail from a map of Covent Garden, London, *c.*1680–85, from Crowle's extra-illustrated copy of Thomas Pennant, *Some Account of London* (London, 1790) (British Museum).

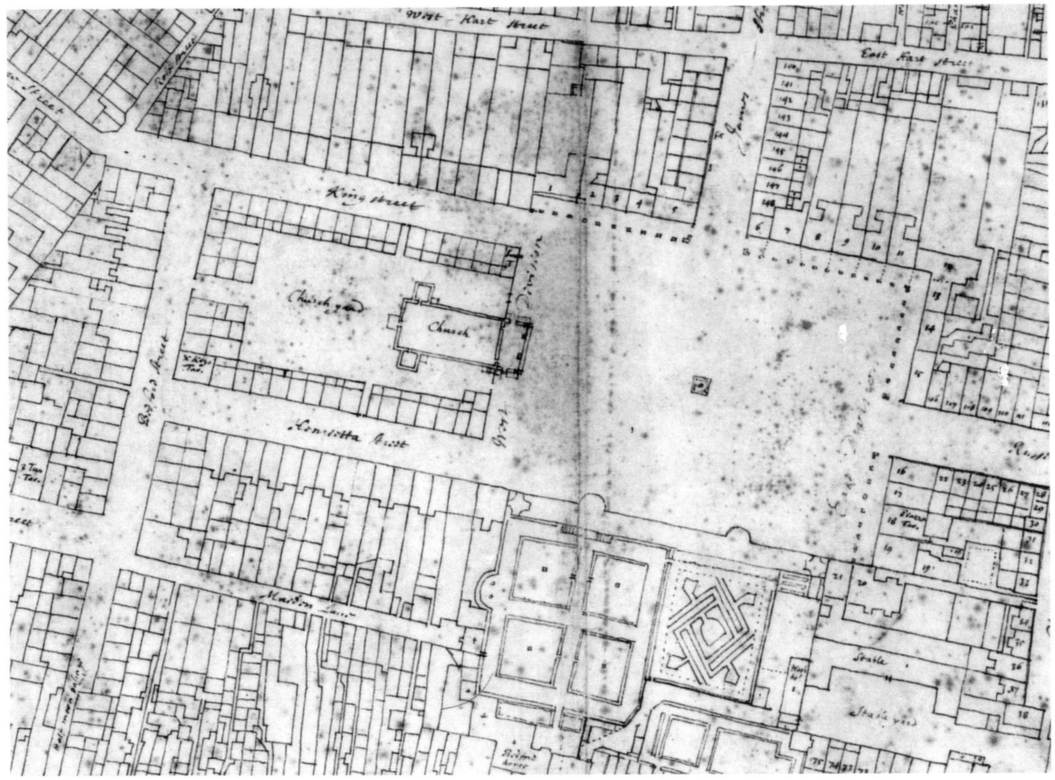

sources used by Jones when designing his new urban quarter. This was begun around 1605 and would have been seen by Jones when he visited Paris in 1609.

The Place Royale is a grand arcaded square of terraced houses, each with an archway under the arcade which led through to a courtyard at the rear of the house. Coaches could drive under the arch, deposit their passengers at the foot of the stairs to the principal rooms on the first floor, and then pass on to the stables and courtyard. The houses in the Place Royale thus combined the idea of a terrace of houses with the traditional approach to a Parisian *hôtel*, which allowed the new arrival to be deposited in his coach at the foot of the stairs. But it was not a successful compromise. The arches, which had to fit under the single-storey arcade, were small and cramped, lacking the grandeur of a gateway to an *hôtel*. At the same time, the attractiveness of the arcade as a place for promenade was compromised by the need for coach access. However, at least some of the stable yards could also be approached from minor streets at the rear, the rue Esgout Sainte-Catherine and the rue du Foin. The Plan Gomboust of 1652 and the Plan de Turgot show that the buildings on these streets were not houses but insignificant one- or two-storey buildings that would appear to be offices or stables.

When Jones came to design Covent Garden he refined the model provided by the Place Royale. Presumably seeing the disadvantages of trying to combine the terrace with traditional *hôtel* coach access, he abandoned the archway onto the street, leaving the arcade of the Piazza solely for pedestrians. Instead, he regularised the role of the rear service street, Hart Street, making this the sole access to the stables and coach house.

The advantage of this arrangement was that the house and the stables remained attached but distinct, only the genteel element being visible from the street. With the stables and coach houses hidden down the back alley, the noise, sight and smell of horses and their dung was kept as far away as possible from the reception rooms. Compared to the Place Royale model, there was the added advantage that any space between the house and the stable building could be given over to a garden or courtyard and need not be compromised by a carriageway.

The disadvantage was that when the coach was called for, it had to go out into the street and round to the front door, where it would have to wait on a possibly crowded public street instead of in the privacy of an enclosed courtyard. Nor was it easy to fit coach, horse and all their necessary offices into the confined space of a mews house. But given the problems of tight urban development, the mews was an ingeniously practical solution that quickly became a standard feature of high-class urban design.

It was certainly a common feature of the better houses in the other key urban scheme of the 1630s, Lincoln's Inn Fields and Great Queen Street. Numbers 1 to 6 Great Queen Street, the houses at the east end on the north side, built around 1639, were designed with mews. Hollar's bird's-eye view of about 1658 shows them linked by gardens to one- or two-storey buildings on Parker's Lane behind (fig. 105). The building contracts explain that these lesser buildings were in joint ownership with the houses on Great Queen Street, and documents discussing their demolition in 1735 confirm that they were stables and coach houses.[22] Some of the houses on the west side of Lincoln's Inn Fields and along Yeates Street, its brief northern extension, for which William Newton received a licence in 1638, also had mews. A man can be seen holding a horse in a yard, clearly a mews, in Hollar's

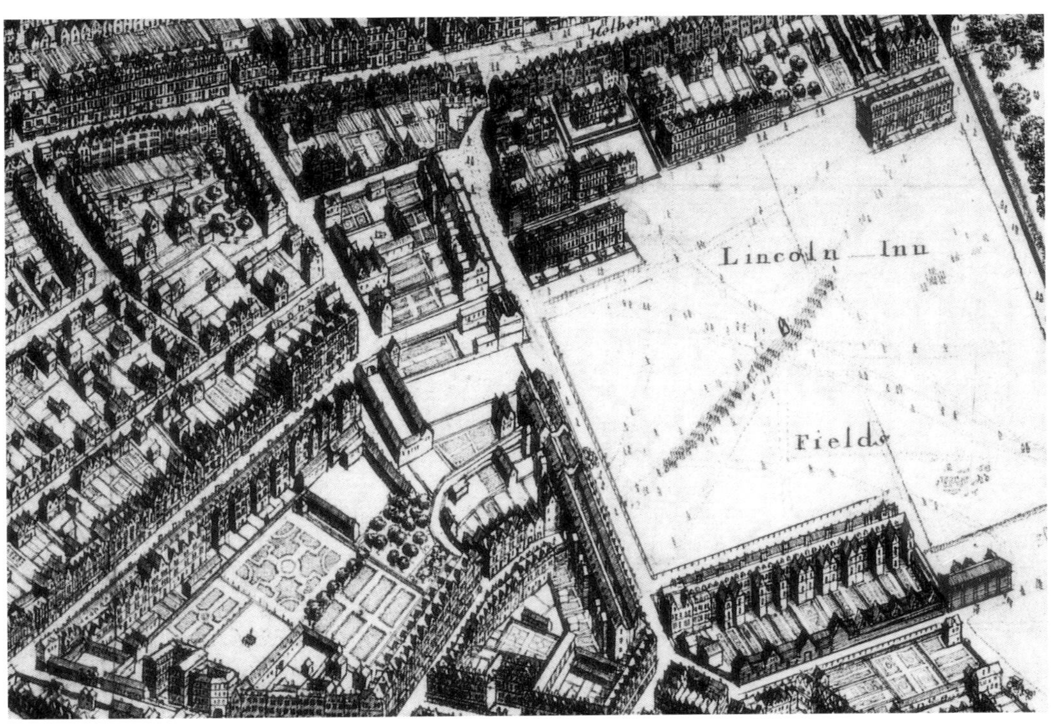

105 The mews behind Lincoln's Inn Fields and Great Queen Street, London, from Wenceslaus Hollar, *West Central London*, c.1658 (detail) (British Museum). Great Queen Street runs down to the bottom left of the picture.

view and when Samuel Pepys visited Thomas Povey, who occupied one of these houses, in 1663 one of the things that most struck him was 'His stable, where was some most delicate horses, and the very racks painted, and mangers, with a neat leaden painted cistern and the walls done with Dutch tiles like my chimnies.'[23]

Hollar's view shows the west end of the south side of Lincoln's Inn Fields as a row of terraced houses with gardens and a regular row of low buildings behind opening onto Portugal Street that have every appearance of being mews. On the north side of the Fields, houses were served by an alley called Whetstone Park. Proof that this was also a mews comes from the description of 13 Lincoln's Inn Fields in 1737 (the original seventeenth-century house was not rebuilt until 1753) as having 'coach house and stables standing behind and fronting to a place called Whetstone Park'.[24]

When large-scale residential development recommenced in London around the Restoration the pattern established in the 1630s was maintained. This began with Hatton Garden off Holborn in 1659, followed by Bloomsbury Square in 1661 and St James's Square, south of Piccadilly, also in the 1660s. All were laid out on a regular grid pattern and included mews. The houses in Hatton Garden were built as residences for gentlemen, merchants and aristocrats, who required coach houses and stables. The development was laid along a central street, Hatton Street, with two parallel lesser streets, Leather Lane and Great Kirby Street, marking the edge of the development and two further streets at right angles. Hollar's post-Fire survey of 1666 shows that the development of Hatton Garden was well under way by that date, but the arrangement of streets is clearer in William Morgan's map *London &c Actually Survey'd* of 1682 and in Abraham Arlidge's survey of Hatton Garden of 1694. This shows a courtyard, approached off Leather Lane, behind numbers 251–61, with a series of smaller buildings in joint ownership with the houses they backed onto. They must be mews houses.[25]

Similarly, at Bloomsbury Square the houses on the east and west sides of the square had stableyards behind them, approached from backstreets (fig. 106).[26] The same arrangement could be found at St James's Square, where Singer Street ran behind the north side of the square, with a further four yards on the east and west sides originally known as East and West Stable Yards, Robin Wood's Yard and Cleveland Yard.[27]

Placing the mews at the rear of the house was not universal at this date. Before regular mews became common, those who wanted to house coaches and coach horses but did not have the necessary space had to rent accommodation in neighbouring yards. This can be seen at Drury House on Drury Lane, where both the original stable yard and the courtyard to Drury House were full of coach houses and stables that were let out to others in the early seventeenth century.[28] As London grew, this arrangement was regularised. Hollar's bird's-eye view of central London shows a long yard with low two-storey buildings on each side on the south side of Great Queen Street, at the junction of Little Queen Street (fig. 105). Rocque's survey describes this development, which Hollar's view shows must have been built as a single development, as a stable yard. It was probably a mews yard. A similar development could also be found off Great Ormond Street. When this was developed in the first decades of the eighteenth century, one of the particular attractions of the grand houses on the north side of the street was the unrestricted view

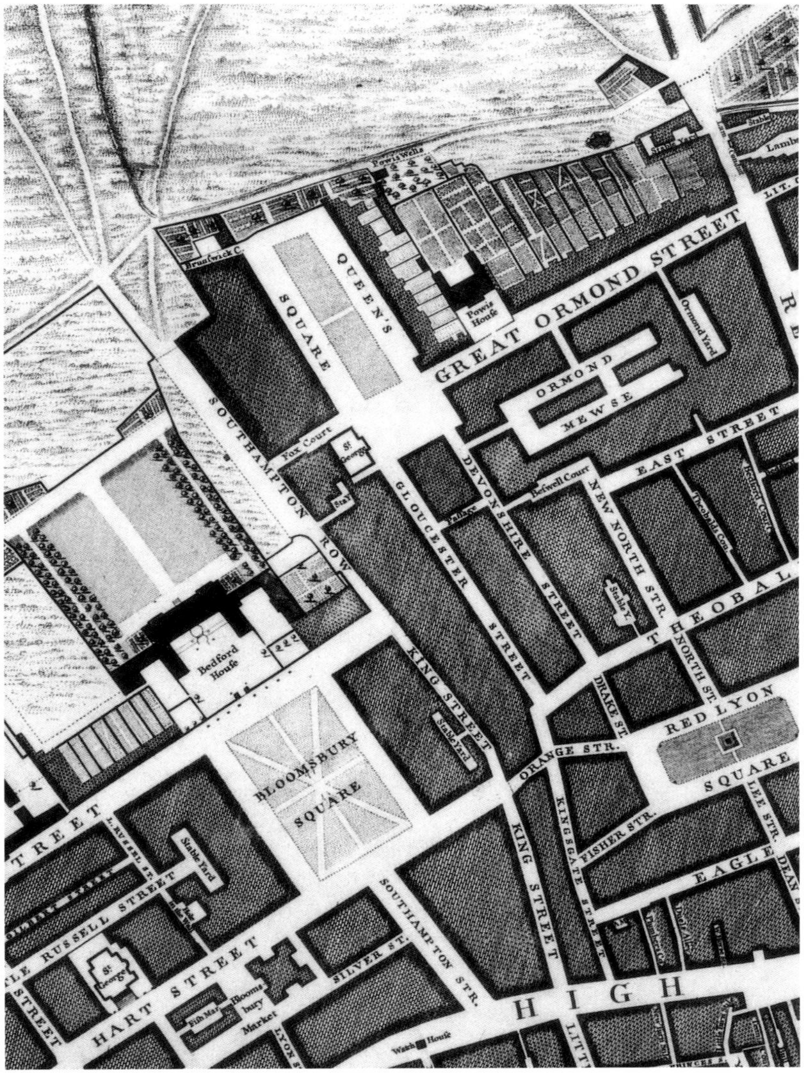

106 Bloomsbury Square, London, detail from John Rocque, *A Plan of the Cities of London and Westminster and Borough of Southwark*, 1747 (British Museum).

north over the fields. Instead of closing this with mews, Ormond Mews was created, a separate mews yard on the south side of Great Ormond Street, with specific mews for the different houses (fig. 106).[29] A similar arrangement could be found in Mayfair, where a mass of small and large stables, together with coach houses, could be found in Davies Mews, for those in need of further stable accommodation.[30]

The development of the mews reached perfection with Grosvenor Square and its surrounding streets in the 1720s, which were laid out around the central square with a grid of major streets, behind which ran mews serving the larger streets on either side (figs 107 and 109). All the building plots were long and narrow to provide a house, garden or yard and mews house. The Grosvenor estate building agreements stipulated that future lessees be granted the right to use the stable yard and horse pond, watering place, pump, water and dung place, and pay the ground landlord costs proportionate to the footage to the yard.[31] This basic pattern was to be the model for all superior building for over a century on the Portland and Portman estates, as well as in Bloomsbury, Belgravia, Pimlico, Kensington and Bayswater.

107 (*right*) Grosvenor Square, London, from John Rocque, *A Plan of the Cities of London and Westminster and Borough of Southwark*, 1747 (detail) (British Museum).

108 (*below*) Philips Vingboons, plan of Joan Huydecoper's house of 1639, Singel 548, Amsterdam, from Philips Vingboons, *Afbeelsels der voornaemste gebouwen* (Amsterdam, 1648) I, p. 30 (British Museum). The mews is at the top.

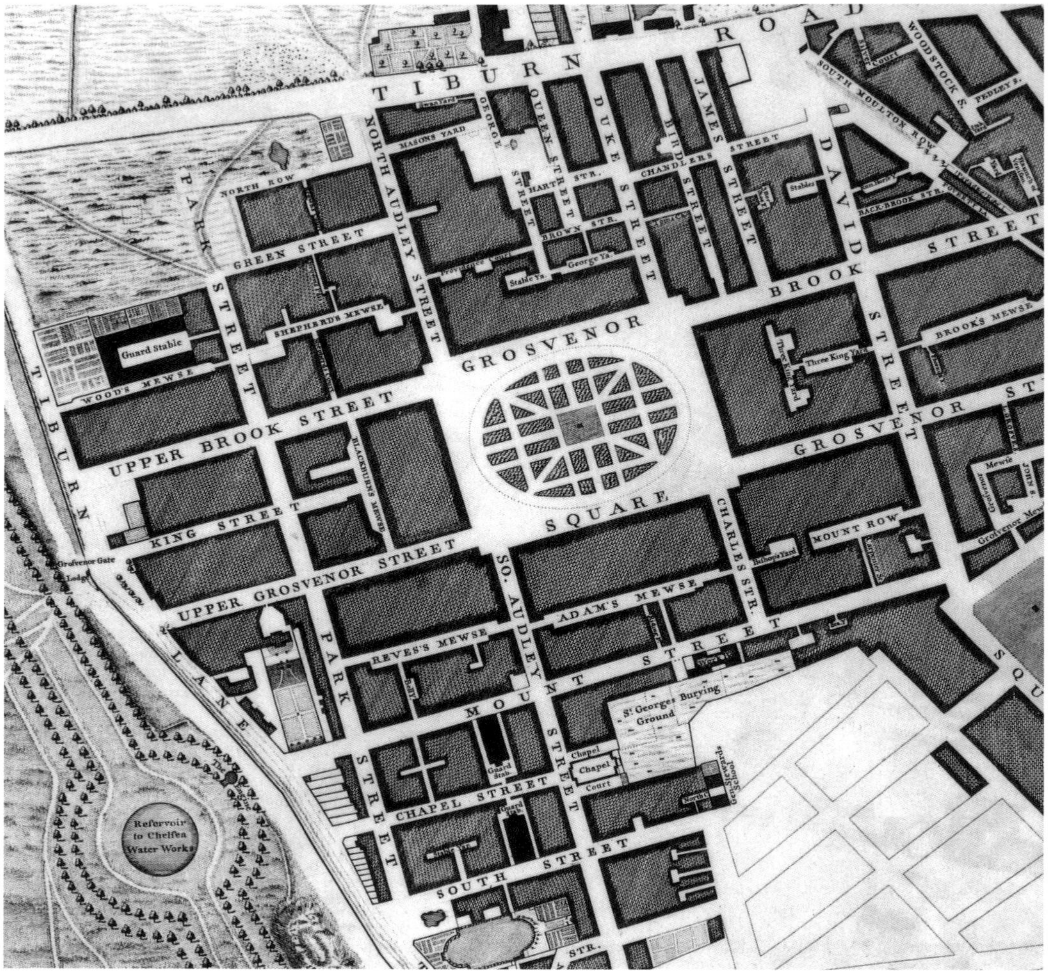

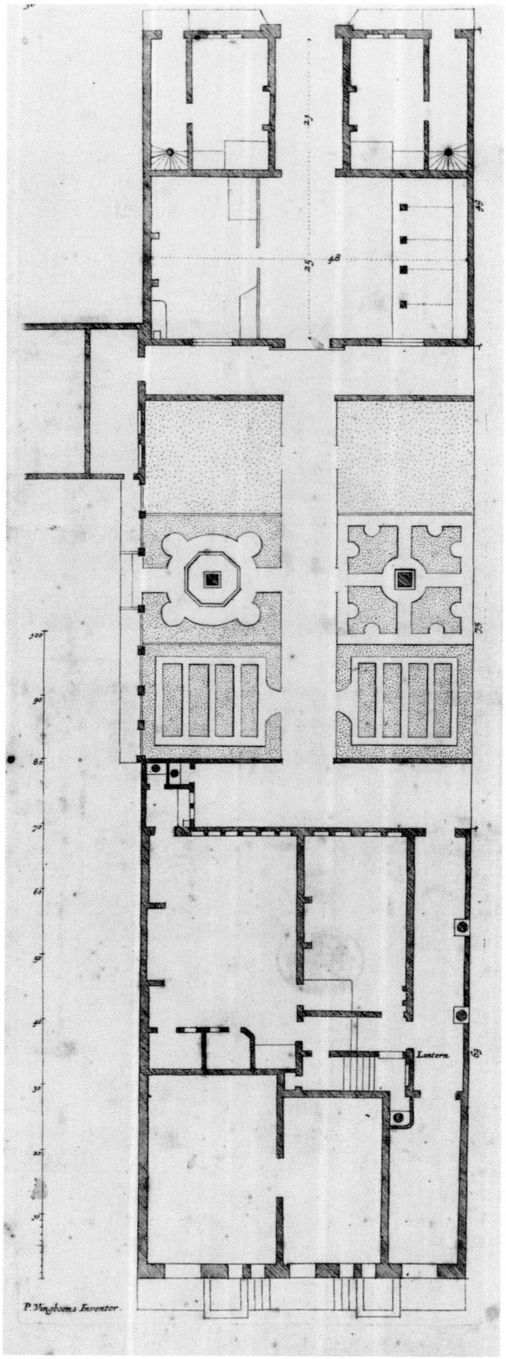

The mews was a distinctively English, and above all London, response to the problem of housing horses and coaches in dense urban developments, though it can also be found in Amsterdam. The Dutch Classicist Philips Vingboons's publication of his own work *Afbeelsels der voornaemste gebouwen*, first published in 1648, includes a house, Singel 548, built for Joan Huydecoper and dated 1639 on the chimneystacks, with stables at the end of the plot, set back behind a garden and entered from the rear (fig. 108). Vingboons's second volume, published in 1674, includes two more Amsterdam houses with mews, one built for Joseph Deutz at Herengracht 450 in 1669–71, and one built for Gillis Marcelis at Singel 460 in 1661.[32]

Because the plots between the canals were narrow and no service street ran behind these houses in Singel and Herengrach, both Deutz's and Marcelis's mews opened onto principal, not service, streets (Keizergracht and Herengracht). This was an extravagant use of prime frontage. It was presumably to overcome this problem that a mews street was included in the second phase of the expansion of Amsterdam, begun in 1662. This expansion, intended to provide sites for Amsterdam's rich merchants, continued the two principal streets of Keizergracht and Prinsengracht laid out in the first phase of the expansion in 1610. In that phase,

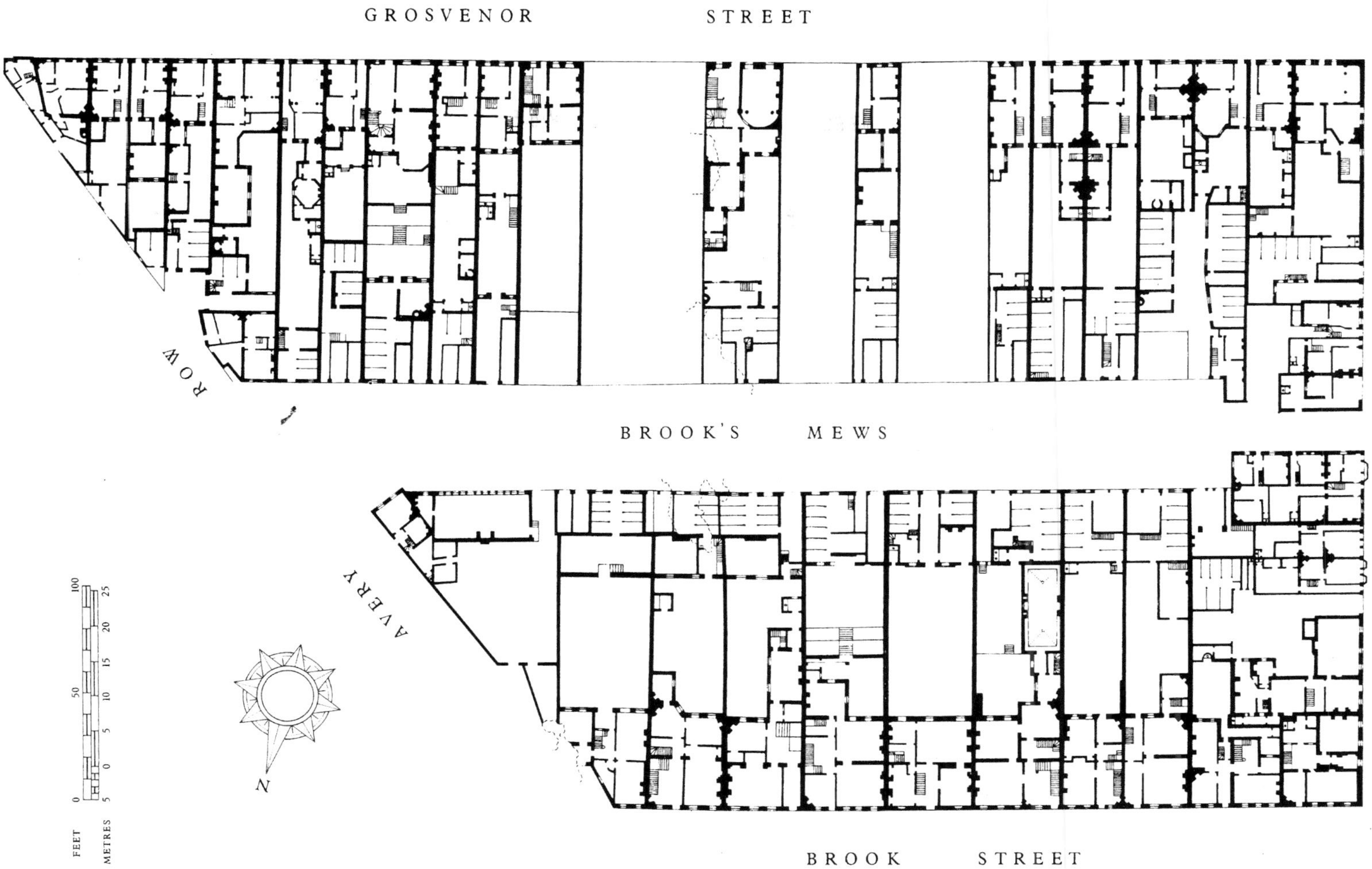

109 Plan of Brook Mews, from *The Survey of London*, XXXIX (London, 1977), fig. 4.

houses between the two streets lay back to back, separated only by gardens. In the second phase, a smaller service street, Kerkstraat, was introduced between them. At least some of the buildings on Kerkstraat served as mews for houses on Keizergracht. This can be seen in Gerrard de Broen's 1783 map of Amsterdam where about eleven houses in two blocks in Keizergracht, between Leydsegracht and Spiegelstraat are shown in common ownership with low buildings without gables, apparently mews, in Kerkgracht.

The dating of the Amsterdam mews so soon after the London mews, is intriguing. We know that at this date connections between London and Amsterdam were close and that the Dutch were particularly interested in the new English Classical style. Constantijn Huygens, patron of the pioneering Dutch Classicist Jacob van Campen, had been struck by the work of Inigo Jones when he visited England. Van Campen's first building, the Coymanshuis in Amsterdam of 1625, was clearly influenced by the Banqueting House in Whitehall. Joan Huydecoper, patron of Vingboons's first mews house, married Maria Coymans, sister of the two Coymans brothers who built the Coymanshuis, in 1624. Given the close connections with London and the general admiration for Inigo Jones's work, it seems logical to deduce that the Amsterdam mews were directly inspired by those in London.

110 Sutton Nichols (engraver), Southampton House, Bloomsbury, London, from the south, 1746. The stables lie to the right of the house.

The alternative to the mews, which was standard in Paris and other French cities, was the *hôtel*. Perhaps the first true *hôtel* in London was Southampton (later Bedford) House in Bloomsbury, begun in the 1630s though not completed until after the Restoration (fig. 110). This was set back behind a courtyard with stables in a wing to the side. Sarah, Duchess of Marlborough, with her keen eye for the practicality of stable layout, praised the convenience of Southampton House with its separate stableyard, adjacent to the forecourt, but with its own access to the street: 'There is more convenience than in any house that I ever saw; and the two courts that are placed on one side of the house for the stables and offices are better placed than I ever saw in town or country.'[33]

Southampton House's pattern was followed by the wave of fashionable *hôtels* built immediately in the 1660s, particularly those on Piccadilly – Clarendon House, Burlington House and Berkeley House – and by Montagu House on Great Russell Street in 1675. More *hôtels* were built in the eighteenth century including Bingley House on Cavendish Square, begun in 1722, Devonshire House on Piccadilly, built by William Kent in 1734–35, Chesterfield House on Park Lane, built by Isaac Ware in 1748–49, Fetherstonhaugh (subsequently Dover) House on Whitehall, built by James Paine in 1754–58, and Melbourne House, now Albany, on Piccadilly, as rebuilt by Sir William Chambers in 1771–72.[34]

On smaller sites, the wings tended to be pressed against the boundary of the plot with no room for a stable yard. The forecourt thus became the effective stable yard and the scene of all the activities, such as mucking out, necessary to service the horse. Kip and Knyff's engraving shows that there was sufficient space at Burlington House for a separate stable yard onto which the coach houses opened (fig.

102), but this was not the case with the five eighteenth-century houses just listed.[35] Where there was sufficient space the stables were always placed at a distance rather than in wings, as at both Buckingham House, built in 1702–5 and Marlborough House, built in 1709–11.[36] This became the standard pattern for later *hôtels*. Carlton House and Grosvenor House had detached stables, as did nineteenth-century *hôtels* such as Stafford House, Londonderry House and Bridgwater House, which generally had no forecourts.[37]

Ware appreciated this disadvantage at Chesterfield House and sought to lessen it by providing a separate pedestrian access for the stables (fig. 111). But that did not make up for the lack of a separate yard and he remained unsatisfied, pointing out in his *Complete Body of Architecture* the advantage of a stable yard in the designs for a house that had been intended to be built opposite Chesterfield House.[38] Where subsequent owners could, they removed the stables from the forecourt. Part of the Duke of York's improvements to Fetherstonhaugh House in 1787 was to build new stables on the other side of Whitehall, and at Burlington House Lord George Cavendish moved the stables out of their wing overlooking the forecourt and built a new stableyard with its own entrance on Piccadilly in 1815–18.[39] At Harcourt (originally Bingley) House, the Duke of Portland employed Samuel Ware to move the stables out of one of the forecourt wings and build a new stable block at the back of the garden, which his son subsequently linked to the house with a tunnel.[40]

Detached stables had the great advantage of being more salubrious and allowing the maximum space to be used by the house itself, providing, of course, that a suitable site could be found close by. Such stables were the grandest of all London stables. Some, as at Stafford House, formed a miniature quadrangle, while others, as with the stables designed by James Brookes for the Marquess of Londonderry's Londonderry House in 1881, formed their own mews yard.

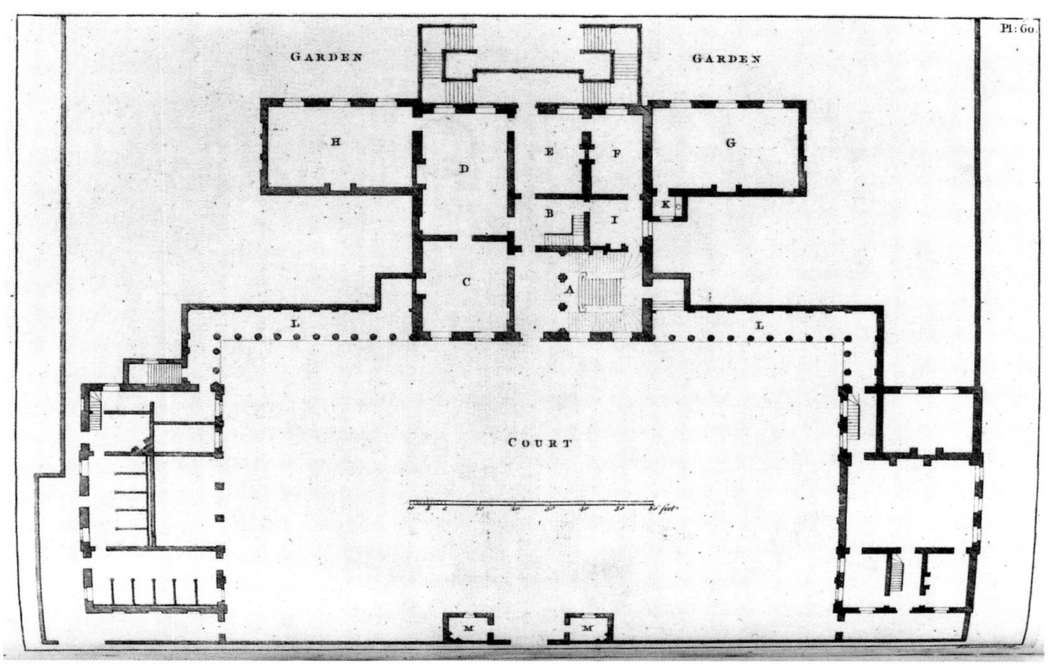

111 Isaac Ware, plan of Chesterfield House, Mayfair, London, 1748–9, from *Complete Body of Architecture* (1758), pl. 60 (© The British Library).

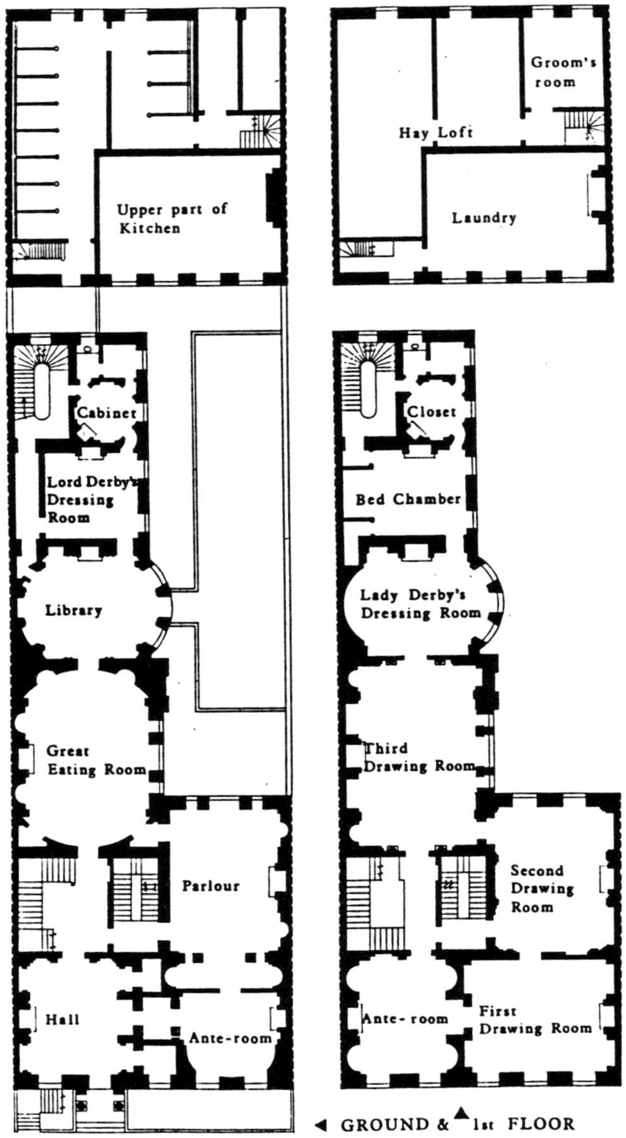

112 Plan of 26 Grosvenor Square, from *The Survey of London*, XXXIX (London, 1977).

Architecturally, seventeenth-century forecourt stables differed little from the common design for late seventeenth-century country houses such as Ragley Hall, Warwickshire, or Eaton Hall, Warwickshire, for unlike their Parisian counterparts there was no tradition of impressive entrance gateways. The one exception was Burlington House, which was enclosed by James Gibbs with a semi-circular colonnade in about 1715, an idea derived from a French source, the Hôtel de Souvre in Paris, and given an impressive gateway by Colen Campbell.[41]

It might be thought that *hôtel* stables would provide more accommodation than a mews house, but this was not necessarily true as two houses occupied by the fourth Earl of Chesterfield show. In 1733 his mews house at 45 Grosvenor Square had accommodation for twelve horses and three coaches, but when he moved into Chesterfield House in 1748–49 he had room for the same number of coaches but one less horse.[42] The mews behind Ormonde House, the largest house in St James's

Square, housed five coaches, twenty horses and fourteen hounds during the time of the first or second dukes of Ormonde.[43] A completely detached stable could house more than that – the Duke of Buckingham's early eighteenth-century stables at Buckingham House had six coach houses and forty stalls – but the advantage of the *hôtel* did not usually include more extensive stabling.[44]

But mews houses were undoubtedly cramped. Provision had to be made for one or more coaches, as well as stalls for the horses, hayloft and grooms' quarters. What may be the earliest surviving design for a mews house is in an album of late seventeenth-century architectural drawings associated with Robert Hooke in the British Library (fig. 113).[45] This shows a utilitarian two-storey building with a double-doored coach house on one side and a stable on the other. Above is a first-floor door giving direct access to the hayloft. Later mews houses followed this basic arrangement and utilitarian character, with half the ground floor given over to the coach house and half to the stalls, and the first floor divided between a hayloft, with its own access, and a room for the coachman. The main difference between this early design and later mews is that there were only two stalls, placed parallel with the party wall, and room for a single coach. Later coach houses were deeper, allowing room for two coaches, and had four or more stalls placed at right angles to the party wall. Where large numbers of horses had to be accommodated there must have been a real problem of ventilation for the further stalls.

The schedule of fixtures agreed for the building of 45 Grosvenor Square in 1733 provides an interesting list of the requirements of a larger mews house:

> Standings for twelve horses with turn'd Columns, Bailes, Rings, Chains, Staples, Racks and Mangers compleat, Two Corn Binns, outside Shutters to the Window, three Coach houses divided, a Hay loft over the Stables with a Window and Shutters to the same, three Windows with Shutters belonging to the Coachmen and Grooms, Rooms over the Coachhouses, a Crane to the Hay loft Door, a

113 Robert Hooke, plan and elevation of a London mews house, *c.*1670s (British Library, Add. MSS 5238 fol. 4).

114 George Scharf, *1 Torrington Square*, 1850 (British Museum, Scharf Book 4, p. 14).

> Lead water trough to convey the wast water into the Stable yard, The water laid into the Cistern in the Stables.[46]

Mews were seldom illustrated, but a good example of a typical and busy mews is Torrington Mews, behind Torrington Square, drawn by George Scharf from his back window in 1850 (figs 114 and 115).

As mews houses were hidden away and not visible from the street, few builders wasted any money on even the simplest decoration. Any architectural distinction was restricted to two areas, the garden façades of the mews of some of the grander house and the arches with which nineteenth-century builders often closed the ends of the mews. Though most houses with mews had little more than an area separating them, some larger houses, particularly those that faced the bigger squares, such as Grosvenor, St James's or Portman Squares, had their mews sufficiently far back to allow for a garden in between. In these cases, the garden front of the mews was often treated architecturally, either as an ornamental temple front, as at 19 and 49 Grosvenor Square, or perhaps with blank arcading, as at Home House in Portman Square. Arches at the street entrance were more common, though they

115 George Scharf, *Torrington Mews behind Scharf's House*, 1850 (British Museum, Scharf Book 4, p. 16).

were never universal and their use was restricted to more important streets, with their decoration being commensurate with the significance of the street they opened onto. The idea was developed by John Nash at Regent's Park and became a particular feature of the Grosvenor Estate in Belgravia and of mews in South Kensington.[47]

The terraced house with its mews was merely the largest in the range of Georgian houses of similar design but varying scale, the great majority of which did not have a mews. These fashionable estates are the exception to the great mass of Georgian houses across the country as the ratio of mews to houses was an indicator of the social standing of any estate. In London, the concentration of wealth was such that in large areas the ratio could approach 1 : 1. In the rest of the country however, there was far less call for houses with mews. An exception was Edinburgh, which had generous provision for mews, although not all of it was taken up. James Craig's New Town, laid out in 1767, was arranged so that every house fronting one of the main streets or squares could have a mews house. But Kirkwood's 1819 plan of Edinburgh shows many of these sites empty, even though all of the houses that they would have served had been built. The plan also shows a definite decrease in the number of mews houses towards the west end of the New Town, which was built later. [48] Nevertheless, all the subsequent developments of the New Town up to and including the Earl of Moray's from 1824 to 1827 were very generously provided with mews. Building leases for these developments included strict restrictions on the use of these sites for anything except mews houses or laundries.[49]

No other city can match London's or Edinburgh's provision. Liverpool, Bristol and Bath had few mews, although some of the grander streets in Bath, including the Royal Crescent, had them. In the early 1720s, Daniel Defoe noted of Bath that 'very few People care to have Coaches here, it being a place where they have but little room to keep them, and less to make use of them', and little had changed by 1793 when Edward Clarke, commenting on the paucity of coaches stated that 'At Bath there are few streets in which a coach may enter; its parades, its by-courts and allies, in which many genteel families reside, are so formed that nothing of this sort, except chairs, can be admitted.'[50] Much depended on the need for a carriage socially. In Bath the promenade was largely on foot, but in Brighton the popularity of driving grew with the town. This helps explain the rows of mews intended for Brunswick Town, although the initial optimistic provision of mews was never realised. Far fewer houses required mews houses than had been expected.[51]

As standards of both equestrian and human sanitation improved during the nineteenth century, mews came to be looked upon with increasing disfavour. William Gordon was dismissive in 1893: 'the stabling in a London mews has not the best of reputations and its accommodation compares unfavourably with that obtainable in country town'.[52] Byng Giraud, writing two years earlier, went into greater detail, pointing out lack of space as the main problem.[53] The capacity of a mews house could only be increased by making it deeper, but this led to greater problems of lighting and ventilation. If the only access was from the mews yard, the deeper stables would suffer from less effective ventilation. Conditions that were fine for the first horse near the door would be less savoury for those further back. And where a loose box was provided in the nineteenth century, it was always placed at the rear of the stable where any restorative value it might possess was negated by the relatively unhealthy atmosphere. The authorities were also concerned about the human occupants of mews. Stable staff, including the coachman and his family, and the unmarried grooms, nearly always slept in the mews, above the horses. Mews were considered notoriously unhealthy, especially for children. To counter these concerns, the Grosvenor Estate redeveloped many of the early eighteenth-century mews during the late nineteenth century, in particular Adams Row, Balfour Mews, Bourdon Street, Mount Row and Three Kings' Yard. These were rebuilt in a fashionable 'Queen Anne Revival' style with individual Dutch gables that gave them a dignity lacking in most mews.

The worst mews were those that no longer served the house for which they had been built because its occupants no longer required the service of a coach. Most landlords wanted their estates to be as fashionable as possible and this was reflected in their provision of mews. Not all estates, however, achieved this standing, and with time others lost it. As early as 1809, George Payne, one of the speculative builders on the Foundling Hospital Estate, complained of the difficulty he found in letting the coach houses and stables of the new houses he was building north of Compton Street. He sought, and was permitted, to build fourth-rate houses on

116 George Scharf, *1 Torrington Square*, 1850 (detail of fig. 114).

the same ground. In 1818, it was reported on the same estate that 'the mews are more generally occupied by poor families carrying on little trades, and by profligate and destitute persons, than as stables'.[54] In 1849, Peter Cunningham reported that few of the residents of the Bedford Estate kept carriages and that the poor state of the mews was a major worry for the estate. To cope with this problem the Bedford Estate operated a policy of demolishing mews in the second half of the nineteenth century and turning the sites into gardens.[55]

The Bloomsbury and Foundling Estates were the victims of the flow of fashionable society westwards through Belgravia and Bayswater to Kensington, taking with it the mews-occupying owners of carriages. Even in the fashionable areas to the west, however, a mews was not always seen as an essential adjunct to a large house. In Tyburnia, the area of building north of Hyde Park largely complete by 1845, virtually every house had its own mews, but further west, the houses stretching towards Ladbroke Grove and beyond were increasingly built without them.[56] Sometimes this trend seems to have taken the builders by surprise. Bridstow Place in Notting Hill was clearly intended as a mews for Hereford Road but was instead built up with a series of semi-detached workmen's cottages, unconnected to the houses behind which they stand. Elsewhere in Bayswater and Kensington most streets were laid out without any intention of mews. In North Kensington, Holland Park, built in 1859–72, is one of the few examples of a street with a regular mews behind and here the houses were exceptional in scale (fig. 117). In Belsize Park, Daniel Tidey's original 1853 scheme included a stable or coach house for every pair of villas. By 1864, only one block of about a dozen stables had been built for more than one hundred families and the land earmarked for horses was used instead for villas.[57]

The last major development incorporating the mews as standard was the redevelopment of Hans Town in Knightsbridge in the 1870s. Thereafter even in the most fashionable areas of Kensington there were increasing requests for the build-

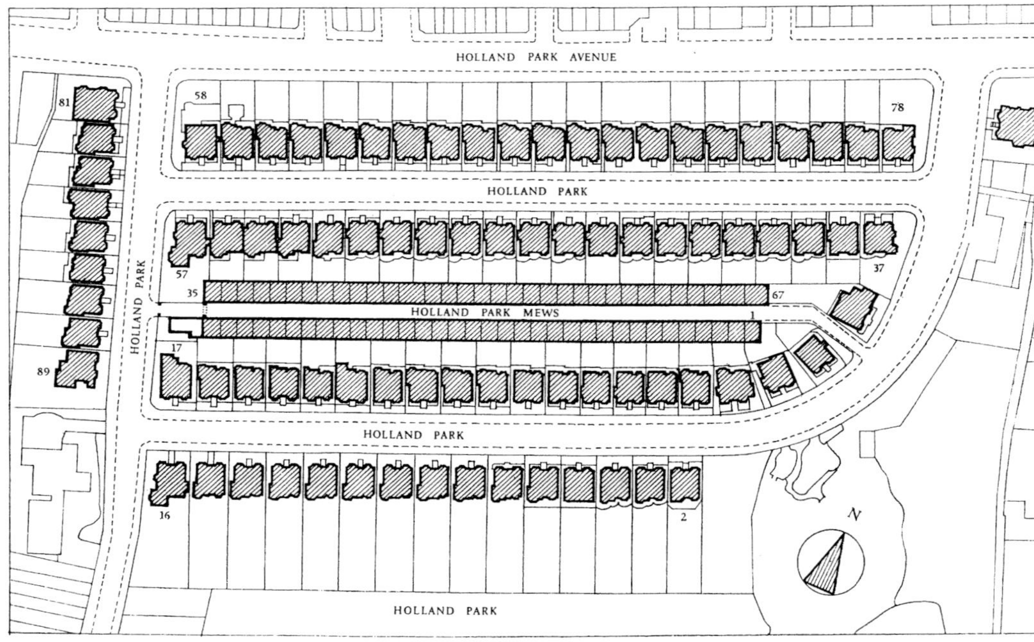

117 Plan of Holland Park and its mews, 1859–72, from *Survey of London* XXXVII (London, 1975) fig. 19.

ing of mews to be avoided. In 1888, the 1851 Commissioners were petitioned against the building of mews behind numbers 170–85 Queen's Gate, a petition they were forced to grant when they admitted that houses were selling elsewhere without them. By 1898, the estate agents were complaining that too many stables had been built for the demand.[58]

As Thompson notes, the rate of growth of private horse owning slowed down markedly towards the end of the century. Private horsekeeping was becoming a headache in London from at least the 1860s because of the unpleasantness of driving in congested streets and the expense of finding stabling.[59] The Victorian distaste for the unsanitory conditions of the mews and the proximity of the horse, especially in summer, must also have played its part. Many, including the aristocracy, turned instead to jobmasters to provide their horses, for, as Sidney wrote in 1875, the value of space was such that few families could obtain enough stabling to house the horses necessary for the round of 'morning concerts, garden parties, shopping, dinner parties, operas and a succession of balls and receptions' of the season, especially if the family had daughters. 'The conveniences of the system have caused its adoption by many noble families who a few years ago would disdain the idea of hiring horses.'[60]

The vast boom in the number of large houses built must have led to the rehousing of many members of the middle classes who had never kept carriages and who had no desire to do so, even though they now lived in the size of house that traditionally was expected to have a mews. The much-improved provision of public transport, omnibuses, hackney cabs and underground trains meant that private transport was no longer essential, despite the increased scale of London. If a carriage was needed, a jobmaster could provide one with far less trouble. At Belsize Park there was a convenient colony of jobmasters just outside the estate, on the far side of Finchley Road.[61] The extraordinary growth of the jobmaster that Sidney noted in the twenty years before 1875 coincided with the years that saw the provision of mews cease to be standard for large houses.

Horse ownership was no longer a prerequisite of respectable society, and once this had been accepted there were many reasons that militated against the mews. The high price of building land was the most obvious; the desire to exclude the undesirable elements always associated with the mews was another; while the layout of Notting Hill showed that many owners preferred private or communal gardens behind their houses to a carriage. But it was the report in 1898 that a 'motor car house' had been built in Queen's Gate mews for Captain G. D. Sampson of Hyde Park Gate that heralded the ultimate demise of the mews, or, it would be fair to say, the imminent end of its traditional equestrian connection.[62]

6

THE PALLADIAN STABLE

Two sets of stables built in the first decades of the eighteenth century reflect the dramatic change that swept over the country house early in that century. At Dunham Massey, Cheshire, Kip's engraving of 1697 shows handsome early seventeenth-century stables and a collection of timber-framed barns set without much apparent architectural thought some distance from the house (fig. 119). Over the next couple of decades these were swept away to be replaced, as John Harris's painting of about 1750 shows, by a pair of surviving buildings carefully placed in parallel to one side of the house (fig. 120). One is the stable, a substantial, thirteen-bay building of brick with stone dressings, with projecting wings, a pedimented centrepiece and *oeil-de-boeuf* windows lighting the first floor (fig. 121). The other, no less substantial, is the coach house, with six individual coach houses, again flanked by wings, set in an arcade on either side of a pedimented archway that supports a cupola dated 1721, when the complex was probably completed (fig. 122). Inside, the stalls survive intact, with their simple racks and mangers, robust partitions, elongated Ionic heelposts supporting the roof and even the original cobbled floor (fig. 123).

All is solid and well built, but, as is so often the case at Dunham Massey, distinctly conservative. But for Kip's engraving and the date on the cupola, it could easily be assumed that the stables and coach house had been built at least fifty, perhaps one hundred years earlier. Indeed, the Dunham Massey stables give us a good idea of what Lord Cottington's famous stables at Fonthill, Wiltshire, built in the 1630s, must have been like.

A decade later, in the 1730s, Sir Verney Cave decided to modernise Stanford Hall, the late seventeenth-century house in Northamptonshire that his grandfather had built. He turned again to the original architect and builder, Francis Smith of Warwick. The east front was to be recast with Palladian-inspired windows. There was to be a new great staircase and wings were to be built flanking the forecourt to house stables and offices. Like the Earl of Warrington at Dunham Massey, Sir Verney's thinking was still stuck in the seventeenth century when such wings were standard on smart new country houses, but Sir Verney died in 1734 and his brother Thomas, who succeeded him, scrapped the scheme. Instead, he commissioned Francis Smith to design what was now fashionable in stable design, a detached quadrangle. The scheme was approved in 1736 and completed in 1740.[1]

The history of the new stables at Stanford Hall encapsulates the most important development in the eighteenth-century stable: the replacement of the relatively small stable that just housed horses, one among a range of buildings lying close to the house, with a detached stable built around a quadrangle. The substantial detached quadrangle, usually elegantly tricked out in neo-Palladian dress, perhaps with a cupola rising above the park, forms one of the most familiar images of the Georgian country house. Too numerous to count, such stables demonstrate again the great importance attached to the horse by eighteenth-century aristocracy and

118 Anon., the interior of the stables of 1768 at Seaton Delaval, Northumberland (William C. Rolf, 1998).

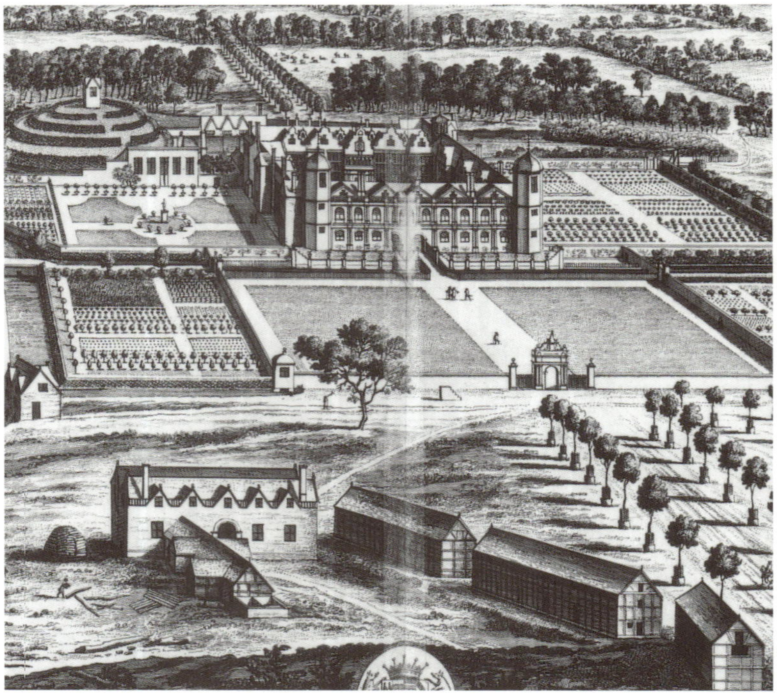

119 Leonard Knyff and Johannes Kip, Dunham Massey Hall, Cheshire, in 1697, from *Britannia Illustrata* (London, 1707) (detail). The seventeenth-century stables are in the foreground (© The British Library).

120 John Harris, *Dunham Massey Hall, Cheshire, from the South-East* (detail), 1751. The stables lie to the left of the house (National Trust Photographic Library/Angelo Hornak).

123 *(facing page)* The interior of the stables of *c.*1721 at Dunham Massey Hall, Cheshire (William C. Rolf, 2000).

121 The exterior of the *c.*1721 stables at Dunham Massey Hall, Cheshire (William C. Rolf, 2000).

122 The exterior of the 1721 coach house at Dunham Massey Hall, Cheshire (William C. Rolf, 2000).

124 (*above left*) William Kent, the stables of *c.*1732 at Houghton Hall, Norfolk (William C. Rolf, 1999).

125 (*above right*) The courtyard of the stables at Houghton Hall, Norfolk (William C. Rolf, 1999).

gentry alike. They can also, perhaps, be seen as symbolic of the newly dominant position of landowners in a country where the absolutist tendencies of the monarchy had been replaced by a constitutional monarchy.

Houghton Hall, Norfolk, built by Sir Robert Walpole in the 1720s, was the most prestigious, high-profile country house of its day, executed with no expense spared for the king's prime minister, the man who dominated British politics from 1721 until his resignation in 1742. Walpole commissioned a house in the fashionable neo-Palladian style, handsomely built in ashlar with sumptuous interiors. It was a statement of his position and authority, set at the heart of his Norfolk powerbase. Here, each summer, influential politicians and local squires met for an orgy of politicking, entertaining and hunting. 'Hunting congresses' was how Lord Hervey described these jamborees, 'dedicated to fox-hunters, hospitality, noise, dirt and business'.[2] When Sir Thomas Robinson stayed at Houghton in 1731, the party hunted six days a week.[3]

Walpole began his transformation of the small, old family manor house into a veritable palace with new stables capable of coping with such demands. These were covered in November 1721 and their plan was published in the third volume of *Vitruvius Britannicus* in 1725 (fig. 126). They did not survive long. Walpole soon decided that they were too close to the house and obstructed one of the main views. They were also badly built, and as a result were demolished in 1732 and replaced by a new block identical in size and layout, designed by William Kent (fig. 128).[4]

Both of the Houghton stables were regular quadrangles, no longer tucked away to one side of the house or placed subserviently in front of it, but standing detached,

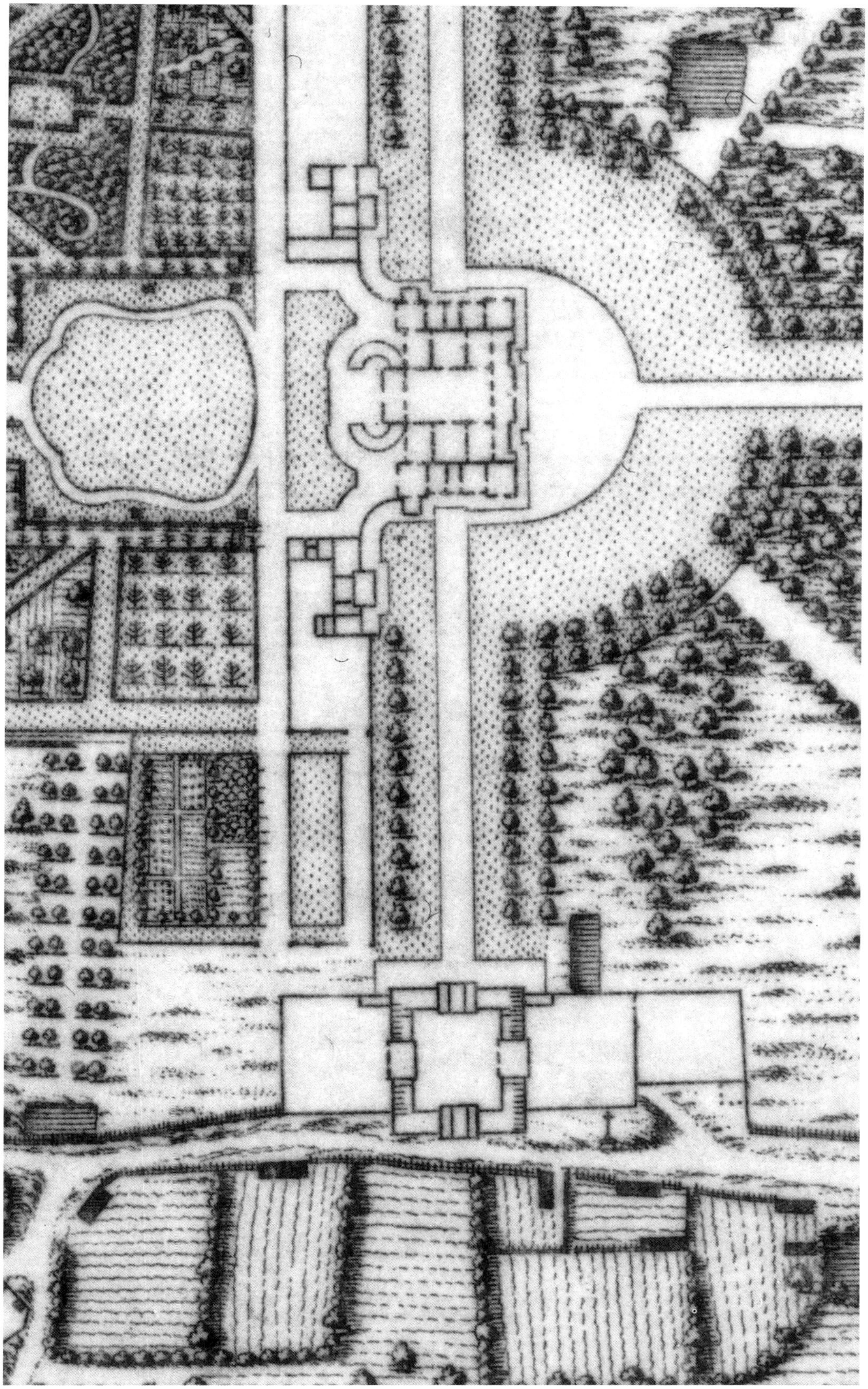

126 Survey plan of Houghton Hall, Norfolk, and its park, from Colen Campbell, *Vitruvius Britannicus* III (London, 1725) (detail). The original quadrangular stables, designed by Colen Campbell, lie below the house (© The British Library).

127 William Kent, the interior of the stables of *c.*1732 at Houghton Hall, Norfolk (William C. Rolf, 1999).

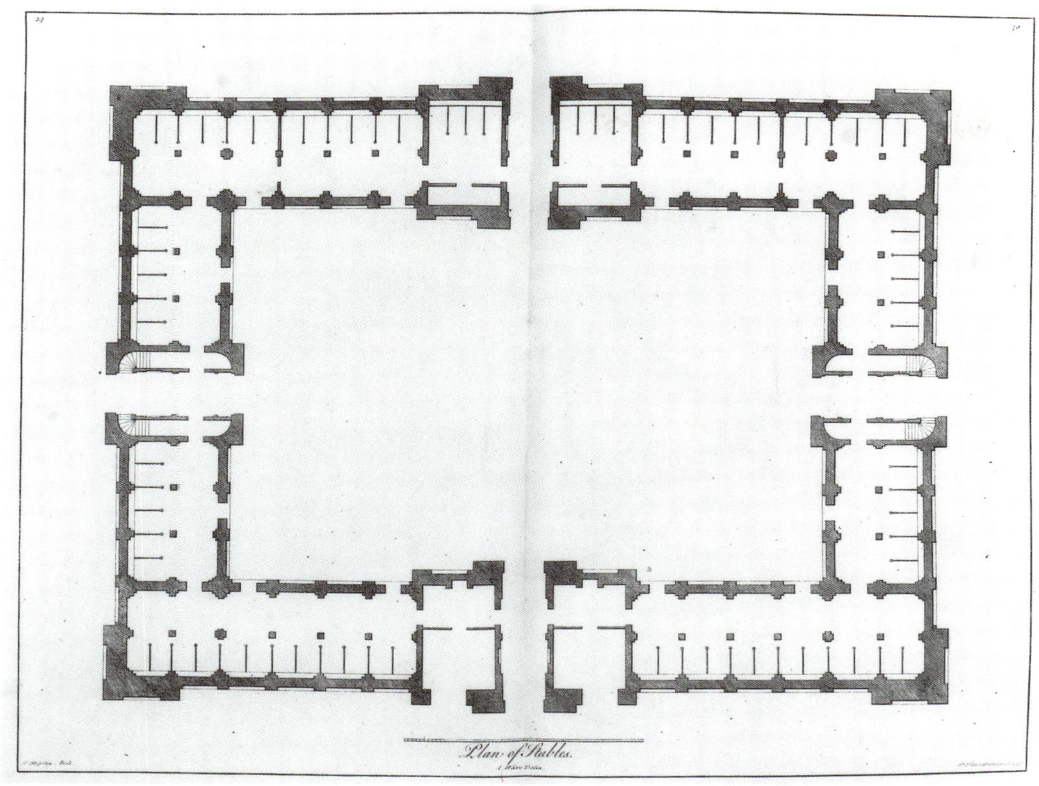

128 Plan of William Kent's stables of *c.*1732 at Houghton Hall, Norfolk, from Isaac Ware, *Plans, Elevations and Sections of Houghton in Norfolk* (London, 1735), p. 33 (© The British Library).

as objects calling for due attention in the landscape, architectural statements in their own right (fig. 124). They were the largest and grandest private stables built at that time in England. No illustration is known to survive of the first stables, but Kent's stables were as innovative in their architecture as they were in their planning. Each of the four facades had blank arcading with Diocletian windows, broken by a central pedimented entrance block, with octagonal cupolas marking the four corners. They were no less impressive within: tall, brick-vaulted stables with giant Ionic columns, topped with balls at the head and feet of the horses, the quality of which is obvious even today (fig. 127). This was as much a statement of Walpole's power and sophistication as the grandeur and architectural novelty of the house. To the hunting squires of Norfolk, the stables were probably the more eloquent statement.

In many ways, Walpole's choice of quadrangular stables is surprising as it is not a Palladian idea. Palladio's villas, amply illustrated in his *Quattro Libri dell'Architettura*, were designed with the house at the centre of spreading, symmetrical ranges of offices, stables and farm buildings. It was this Palladian model that had been followed in fashionable country houses in the second half of the seventeenth century, such as Ragley Hall, Warwickshire, or Castle Howard, Yorkshire, at the beginning of the eighteenth century.

But just at the moment when neo-Palladianism became the dominant architectural movement in Britain, this essentially Palladian approach to the layout of the country house was superseded. Where neo-Palladian houses had wings, as at Houghton Hall, Newby Park, Yorkshire, and Holkham Hall, Norfolk, these were used for offices or apartments, not for stables or farm buildings. Other iconic neo-

129 Tie-up ring in the interior of the stables at Houghton Hall, Norfolk (William C. Rolf, 1999).

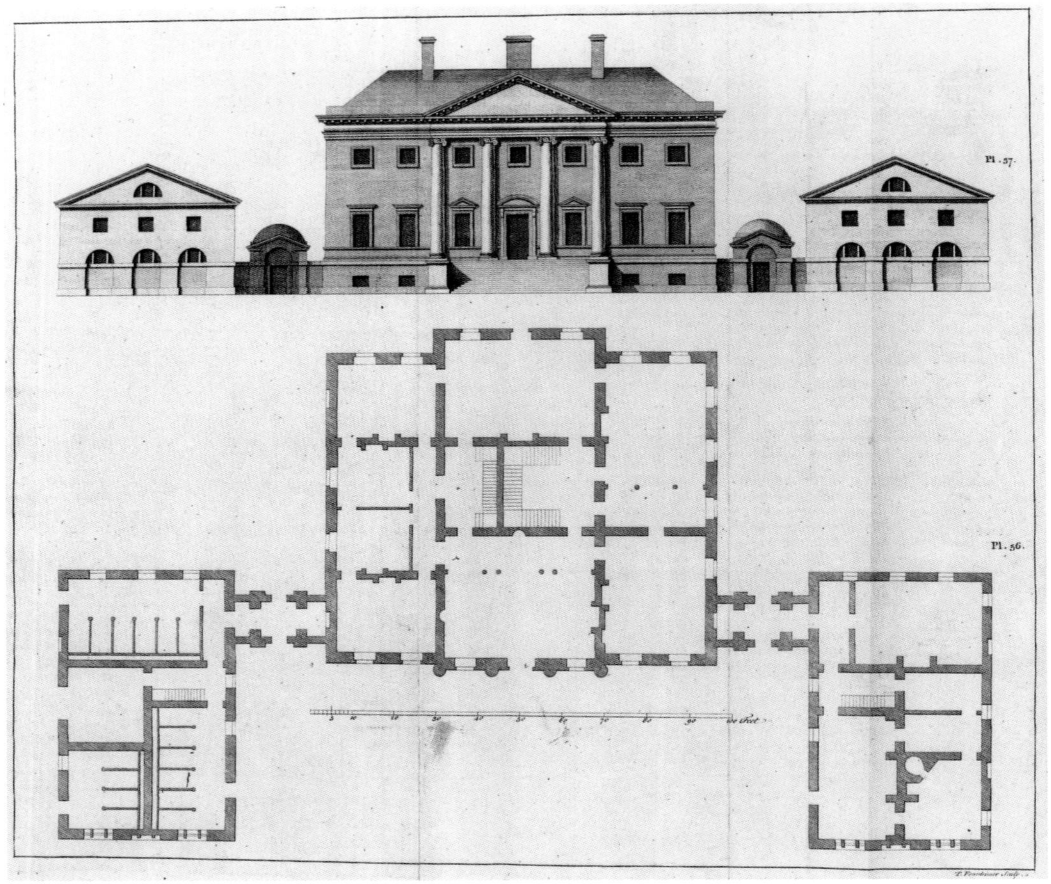

130 Isaac Ware, plan and elevation of house for James Murray of Broughton at Kellie, from *A Complete Body of Architecture* (London, 1758) pl. 57 (© The British Library). The stables are in the pavilion on the left.

Palladian houses were designed with the house appearing isolated in the landscape, as at Wanstead, Essex, Stourhead, Wiltshire, Mereworth Castle, Kent, Marble Hill, Middlesex, Whitton Place, Middlesex, and Combe Bank, Kent. Their stables were placed to one side and played no part in the architectural composition.

Some, rather old-fashioned, architects continued to place stables in wings, particularly James Gibbs, as his *Book of Architecture* in 1728 shows. This was supported by Isaac Ware's *Complete Book of Architecture* of 1756, where almost all the houses illustrated have wings, and all the wings, stables (fig. 130). Ware strongly disapproved of the fashion for distancing the stable: 'We shall shew that these offices are far from being under a necessity to be hid, to be inconvenient, or to be placed improperly. Beauty and use may be consulted together; and together of a plain square house of this extent, it will be possible, at a small advance in the charge, to add wings to the centre, and connect them by passages.'[5] But Ware was not a successful country-house architect and the plans he illustrates were obsolete, except in Ireland where a very different aesthetic prevailed and Georgian houses were distinguished, long, spread-out sequences of stables, offices and farm buildings stretching out from the house in a much more authentically Palladian fashion.

In choosing to place their stables in detached quadrangles, the neo-Palladians deliberately turned away from what might have been seen as correct Palladian design. Indeed, Sir Thomas Robinson, a keen amateur neo-Palladian whose own house, Rokeby Park, Yorkshire, of about 1725 correctly included stables in attached wings, 'argued strenuously' for the new stables at Houghton Hall to be built as wings to the house. But Sir Robert Walpole was determined to see that they were built separately and in a quadrangle.[6]

Regular, detached stable quadrangles had been rare before the eighteenth century and nearly all had been closely associated with the monarchy: Henry VIII's stables at Greenwich Palace and Hampton Court; James I's at Theobalds; the stables built for Charles I's favourite country house, Wilton; Charles II's hunting stables at Lyndhurst; Sir Christopher Wren's unexecuted design for rebuilding the Royal Mews as a regular quadrangle. The one exception was the stable at Petworth, singled out as the greatest of all seventeenth-century stables. But from the completion of Houghton onwards, the stable quadrangle was to be one of the symbols of the newly dominant landowning classes. If quadrangular stables had traditionally been a demonstration of royal power, the stables at Houghton and its successors were a telling illustration of where power lay in Georgian Britain: no longer with the monarch but with the landowning classes, with Parliament, and specifically with the Prime Minister.

The rise of the Georgian stable quadrangle is so noticeable that some commentators have tried to explain it as the result of a major change in horse ownership. These writers have suggested that new stables were needed in the first half of the eighteenth century to accommodate improvements in the design of stable interiors brought on by advances in the breeding of hunters, themselves caused by changes in foxhunting.[7] Their argument is attractive because important early examples, such as the stables at Houghton Hall and Althorp Hall, Northamptonshire, were undoubtedly built for hunting. However, the argument does not stand up to scrutiny.

It is true that foxes began to replace hares and deer as the chief quarry of hunting men in the early eighteenth century, but in 1733, the author of *An Essay on Hunting by a Country Squire* felt he must 'give his vote for the innocent hare above all other game' and Walpole still spent as many days hunting hares as foxes.[8] Hunting hares is a relatively slow business, as *The Spectator*'s account of Sir Roger de Coverley's hunt published in 1711 made clear.[9] It did not demand horses with the speed of modern hunters. Nor at the time did foxhunting. Although there is evidence of long runs, these tend not to be fast runs as meets took place very early in the morning when foxes, with an undigested meal inside them, were loath to run, and the 'flying leap' had not yet been invented.[10] The great hunting writer 'Nimrod' (C. J. Apperley) described how the early eighteenth-century hunter was 'able, from the comparatively slow rate at which hounds travelled, to continue a chase for some hours' duration'.[11]

All this was to change in the second half of the eighteenth century, when the development of the thoroughbred for hunting and racing did lead to major changes in stable technology, and consequently in stable design. But the first half of the eighteenth century was a period of relative stagnation in the design of stable interiors and, by the time these changes become evident, the Georgian stable quadrangle was well established.

Nor can the introduction of the stable quadrangle be explained by the need to incorporate some new element, as all the different parts that made up these new stables had been present for many years. Coach houses, of which there might be any number from two to eight, usually occupied between a quarter and a third of an eighteenth-century stable block. To this should be added the stabling occupied by coach horses, on average about a third, and the extra harness room they needed. Thus the introduction of the coach was ultimately responsible for about half the size of the stable block. But coaches had been common for a century, and for most of that time little architectural effort had been made to integrate stabling and coach house, as the separate coach house and stables at Dunham Massey graphically illustrate. Similarly, the need for hay and straw had remained constant, but where many seventeenth-century stable yards had a separate barn, in the eighteenth century hay and straw were generally kept in a loft above the horses or coaches. Thus for all their noble appearance, the grand quadrangles of the eighteenth century accommodated no more than the muddled, often unimpressive yards of the previous century. What had changed was the decision to integrate all the different elements of a stable yard into a single composition and to treat the result architecturally.

One reason for this may have been perceived practicality. Just as kitchens, which in seventeenth-century houses had frequently been in the basement, were often moved in the eighteenth century to an attached wing to minimise smells and the risk of fire, so it is possible that stables were felt to be less objectionable, especially in summer, if placed at a certain distance. There were also practical advantages to a stable quadrangle, particularly advantages of control that would have appealed to the ordered Georgian mind. A single entrance ensured security and, with all the activity centred around the courtyard, the head groom or the owner could easily supervise what was going on. All the elements of the stable were conveniently to hand and the central courtyard provided a sheltered space for stableyard activities such as harnessing coaches, collecting dung and litter from the stables, and washing and grooming horses. But the principal reasons for the change, apart from the symbolic significance of the quadrangle, seem to have been aesthetics and propriety.

Nothing is more apparent when comparing illustrations in early and late eighteenth-century county histories, such as John Harris's *History of Kent* of 1719 and Edward Hasted's *Historical and Topographical Survey of the County of Kent* of 1778, or Robert Atkyns's *The Ancient and Present State of Glostershire* of 1712 and Samuel Rudder's *A New History of Gloucestershire* of 1771–80, than the transformation of the relationship between the house and its outbuildings. In the earlier book, the house is shown surrounded by a huddled mass of different buildings and formal gardens. The scene is one of considerable activity, with horses being brought out of the stables, coaches appearing at the gates, bowls being played in the garden and the harvest gathered in the fields. In the later book, the house sits alone in the landscape with grass running up to a regular façade, its surroundings empty except for trees, dozing deer, perhaps an ornamental temple and a langorous couple surveying the shrubbery.

Rendcomb, Gloucestershire, is a typical example. Knyff's view in Atkyns's *Ancient and Present State* showed the house attached to handsome stableyards set among formal gardens (fig. 131). In Rudder's *A New History*, both stables and gardens have been swept away and the house lies isolated (fig. 132). Some way to the right is a rectangular castellated building, presumably the stables, though there is no evidence of horses. House and stables stand as if unconnected.[12]

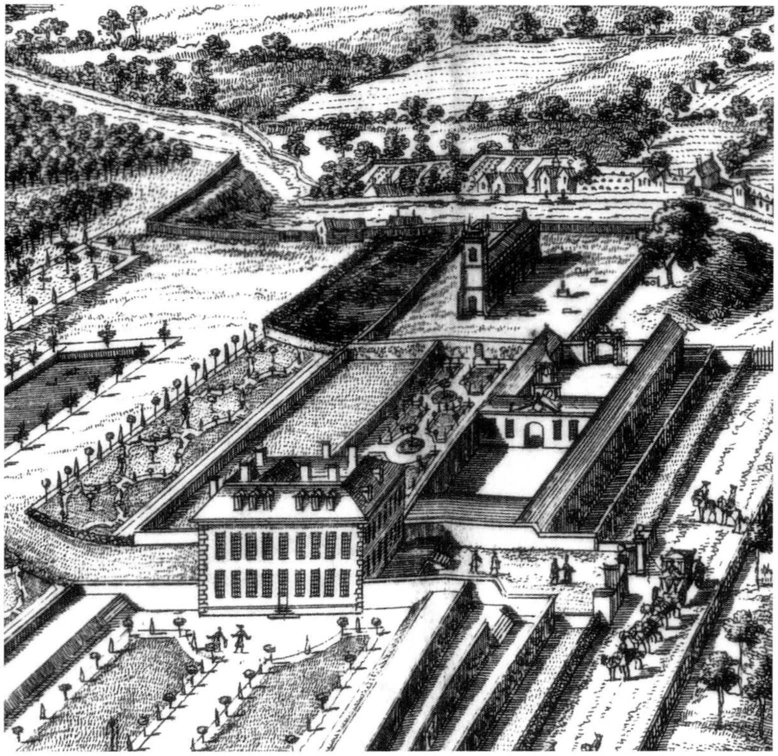

131 (*above left*) Rendcomb Park, Gloucestershire. The stable court lies to the right of the house. Robert Atkyns, *The Ancient and Present State of Glostershire* (London, 1712) (detail) (Society of Antiquaries, London).

132 (*above right*) Rendcomb Park, Gloucestershire. The stables are now hidden from view. From S. Rudder, *A New History of Gloucestershire* (Cirencester, 1779) (detail) (Society of Antiquaries, London).

The change is part cosmetic – a matter of altered viewpoints and careful planting – but it marks a very different attitude, and often a radical rebuilding. Surviving basecourts with stable ranges were swept away, as at Bishopthorpe Palace, Yorkshire.[13] So were many stables set alongside the house, as at Burton Constable, Yorkshire, where the Tudor stables recorded in an early painting (fig. 20) were replaced in about 1757 with a handsome quadrangle, placed at a slight distance from the house, by Timothy Lightoler.[14] Even relatively new stables placed in symmetrical wings flanking the forecourt were demolished in favour of more distant stables, as at Ragley Hall, Warwickshire, where James Wyatt built an extensive stable complex slightly down the hill and out of sight in 1779–97, demolishing the forecourt wings shown in *Britannia Illustrata*.[15] As late as 1809–11, when Humphry Repton improved the setting of the early eighteenth-century Harlestone House, Northamptonshire, one of the most important things he did was remove the stables from the entrance of the house and place them at a distance in a quadrangle.[16]

English neo-Palladian architects seem to have treated the country house as a series of cubic units, with all the different elements essentially self-contained but contributing to an overall harmony. Sir John Summerson talked of the 'curiously pedantic feeling for the separateness of each component in a design'.[17] The closest contemporary theoretical explanation of this comes in Robert Morris's *An Essay in Defence of Ancient Architecture* of 1728. In it, he declares that 'the Correspondency and Agreement of every Part to the Unity of the Whole, are but as so many distinct separate Beauties, connected together in one harmonious Body'.[18] Morris's thinking seems to derive from Vitruvius's definition of Order, the first of his funda-

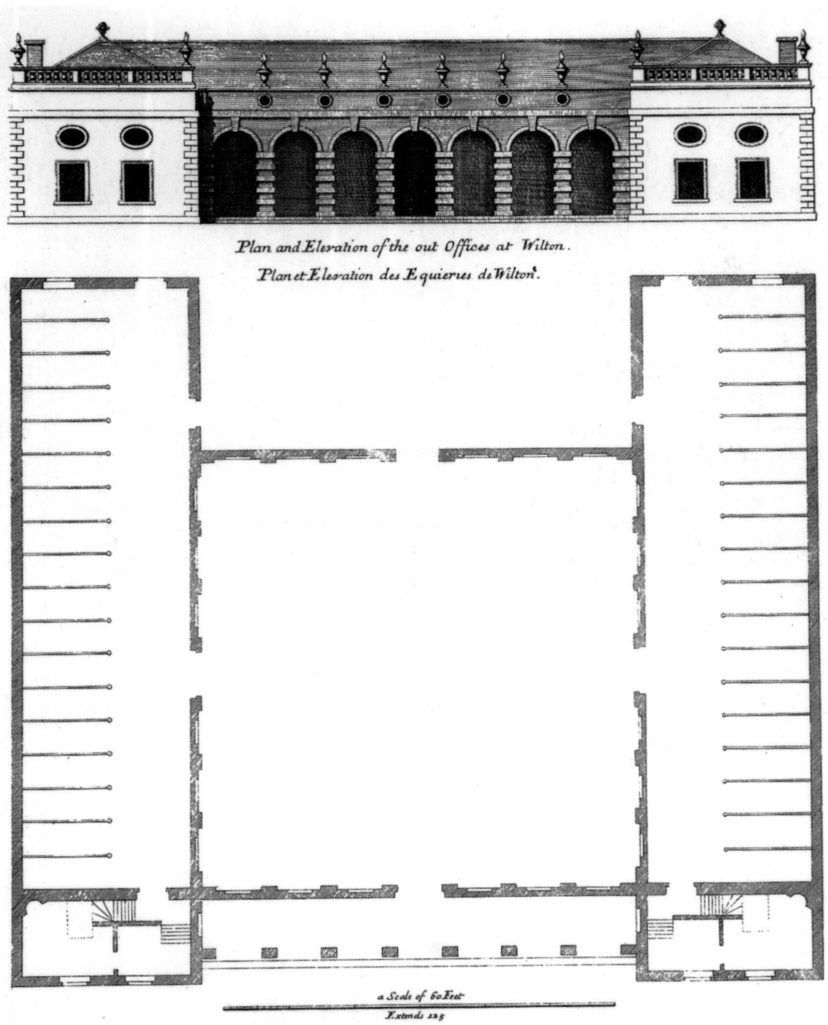

133 The stables at Wilton House, Wiltshire, by Isaac de Caus, 1630s, from Colen Campbell, *Vitruvius Britannicus* II (London, 1717), pl. 66 (author).

mental principles of architecture: 'Order gives due measure to the members of a work considered separately, and symmetrical agreement to the proportions of the whole. It is an adjustment according to quantity. By this I mean the selection of modules from the members of the work itself and, starting from these individual parts of members, constructing the whole work to correspond.'[19] Thus, unlike Sir John Vanbrugh's Castle Howard, Yorkshire, the neo-Palladian country house does not spread out in an expansive sequence of buildings. 'Vanbrugh's houses, as originally sited,' declared Sir John Summerson, 'gathered the whole drama of a landscape into their own complex finality of silhouette and recession. The Burlingtonian house is less a command than a statement; it becomes an object not controlling the landscape but to be seen within it.'[20] The treatment of the country house and its stables, and perhaps other features in the landscape such as an ornamental bridge or temples, as separate elements, follows from this.

In his *Lectures on Architecture*, Robert Morris seems to have recognised the problems caused by placing the stables near the house. He proposed that the ideal villa should be surrounded by gardens on three sides, with the kitchen slightly apart on the fourth side and the stables beyond, 'remote from the house'.[21] This is the closest rationalisation of the neo-Palladian ideal of the stable that exists. Confirmation of the popularity of Morris's views comes from a very different source, the letters of Sarah, Duchess of Marlborough, the scourge of Vanbrugh and a keen Palladian. 'I think the stables stand in a mighty proper place,' she declared in 1732 of her grand-daughter Lady John Russell's stables at Cheam, Surrey, which lay a certain distance from the house.[22] She made a similar comment after visiting Thoresby Park, Nottinghamshire, three months later: 'There's an extreme fine building for stables at a very proper distance which you see and looks extreme handsome.'[23] The building, which replaced earlier stables near the house, had been built seven years earlier.

If the neo-Palladians ignored the way Palladio had treated stables it may have been because what they wanted to achieve had apparently been sanctified by Inigo Jones. The only stables separately illustrated in the first three volumes of *Vitruvius Britannicus* are the quadrangular stables at Wilton, which were believed in the eighteenth century to be designed by Jones (fig. 133). These were drawn for the book by its author, Colen Campbell, probably also the architect of the first stables at Houghton Hall, who subsequently used them as the model for his stables at Studley Royal, Yorkshire, in about 1729 (fig. 134).[24] Campbell took the idea of the quadrangular stable with an arcaded principal elevation flanked by dominant corner pavilions, but refined it. He dropped the *oeil-de-boeuf* windows, which have no Palladian precedent, and emphasised the pavilions. The result is one of the most

elegant of all neo-Palladian stables, but despite its impeccable lineage it had no progeny, perhaps because an open arcade was considered an unnecessary expense for a stable.

The Studley Royal stables were completed after Campbell's death by Roger Morris, who went on to design a third handsome set of early neo-Palladian stables, those at Althorp, Northamptonshire, built in 1732–33 during the fifth Earl of Sunderland's brief ownership before becoming Duke of Marlborough (fig. 135).[25] These, like the stables at Houghton Hall, were designed for hunting, as Wootton's great painted entrance hall to the house with its scenes of the chase and portraits of the earl's favourite horses, makes clear. Here, Morris took a different approach to the question of designing an appropriate form for the neo-Palladian stable block. His model was essentially that of the neo-Palladian house, in particular Houghton Hall, but given a particularly robust treatment appropriate for a stable. Like Houghton Hall, the Althorp stables have a portico flanked by projecting pavilions, but the portico, modelled directly on the portico of Inigo Jones's church of St Paul's, Covent Garden, uses a massive Tuscan order, and the pavilions have simple pyramidal roofs.

Sunderland's mother-in-law, Sarah, Duchess of Marlborough, was appalled by their extravagance:

> whatever the expence may be I am sure Althorp is much the worse for it, & that all which was wanting to make that place the most agreeable habitation that I saw was to make sash windows in the house, and a plain useful stable for thirty or fourty hors's at the most, which is full as many as any man would keep that is not madd . . . but I believe by all accounts I have that there is now built stables for a hundred horse or near it, what can be so ridiculous as this?[26]

One must take any statement by the duchess about architecture with caution, and it would appear that she was misinformed about the number of horses to be housed at Althorp, which was closer to the thirty she recommended than the one hundred

134 Colen Campbell, exterior of the stables of *c.*1729 at Studley Royal, Yorkshire (William C. Rolf, 2000). The stables are modelled on those at Wilton.

135 Roger Morris, exterior of the stables of 1732–33 at Althorp House, Northamptonshire (William C. Rolf, 2000).

she disparaged. Nevertheless, it was too expensive a model for most owners and the only apparent imitator was the stables built, perhaps to the designs of the amateur architect Sanderson Miller, by William Hiorne for the Earl of Aylesford at Packington Hall, Warwickshire, in 1756–58.[27]

The stables at Houghton, Studley Royal and Althorp all show the first generation of neo-Palladian architects searching to find an appropriate response to the question of the detached stable quadrangle. It was William Kent, architect of the second of the stables at Houghton, who devised what was to become the standard neo-Palladian stable formula, the Diocletian window set in blank arcading.

This is not a feature to be found in Palladio's or Jones's work, but was evolved by Lord Burlington and William Kent. It first appears on Burlington-designed school and almshouses at Sevenoaks in Kent of 1727. Though not executed, this was published in *The Designs of Inigo Jones* that year. Kent first used the Diocletian window and blank arcading for a stable at the Royal Mews at Charing Cross, designed, presumably with Burlington's approbation, in 1731 (fig. 136). As the mews was the first stage in what Burlington hoped would be the rebuilding of Whitehall to a neo-Palladian template, he and Kent would have been anxious to mark them out from the common run of stables and to make it a true expression of neo-Palladian architecture. So for the elevations, Kent broke, apparently for the first time, with the ordinary domestic windows, until then usual in stables, and turned instead to the Diocletian window, associated above all with the baths of Ancient Rome. Kent also added a great central archway, inspired ultimately by the ideal of the triumphal arch, and a pair of octagonal arcaded cupolas. Had Kent's complete design for the mews, which envisaged a giant quadrangle, been completed, it would have been one of the most impressive of all English neo-Palladian buildings. As it was, its influence was profound.

The combination of the Diocletian window in blank arcading was particularly felicitous. It introduced Classical elements with echoes of Roman civic grandeur; it satisfied the ideals of architectural hierarchy by choosing a simple arcade and non-domestic windows appropriate for a stable; and at the same time it had a practical basis. Diocletian windows high in the wall provided a soft, diffuse light suitable for stables where direct sunshine needed to be avoided. The result was a specifically equestrian neo-Palladian design, perhaps the only architectural language created deliberately for the horse, though it was subsequently adapted for some industrial and other buildings.

Kent's Royal Mews was soon followed by his rebuilding of the stables at Houghton Hall in 1733–35, which combined the Diocletian arch in blank arcading with handsome, Palladian-inspired gateways and octagonal turrets at the four corners. The result was published in considerable detail in Isaac Ware's *Plans, Elevations and Sections of Houghton in Norfolk* of 1735. The Royal Mews and the Houghton stables were the most prestigious stables of the 1730s, one a royal building in the centre of London, the other built for the Prime Minister and known by readers of Ware's work as well as by the many visitors to Houghton. The Diocletian window set in blank arcading that they pioneered soon became a neo-Palladian orthodoxy of stable design.

136 The south front of the Royal Mews of 1731–33 at Charing Cross, London, by William Kent, demolished in 1830. Engraving by P. Foudrinier (by courtesy of the Trustees of Sir John Soane's Museum).

137 James Paine, courtyard of the stables of 1756–63 at Chatsworth House, Derbyshire (William C. Rolf, 1999).

138 James Paine, detail of the exterior of the stables of 1756–63 at Chatsworth House, Derbyshire (William C. Rolf, 1999).

The influence of the royal mews hangs heavily over what were probably the grandest private stables ever built in Britain, those designed by James Paine for the Duke of Devonshire at Chatsworth House, Derbyshire, in 1756–63 (figs 137 and 138). Paine's pride in these can be seen in the fact that they were the first building to be illustrated in his *Plans, Elevations and Sections of Noblemen and Gentlemen's Houses* of 1767.[28] Their massively rusticated elevations, which recall the equestrian work of Giulio Romano in Mantua, both at the Cortile delle Cavalerizzo and the Palazzo del Te, would have been considered a suitable architectural language for the horse. Chatsworth's only rivals as the grandest of eighteenth-century stables are those at Goodwood House, Sussex, designed by Sir William Chambers for the Duke of Richmond in 1757 and entered through a massive Doric triumphal arch (fig. 139). Sadly, the courtyard of the Goodwood stables has been radically altered, so it is no longer possible to appreciate the full elegance of Chambers's work.

Palladian precedents were sought not only for the exteriors of stables but also for their interiors. Few stable interiors were known to survive from the time of Inigo Jones, but one that did was that of the Holland House stables. These were

139 Sir William Chambers, the portico of the stables of 1757 at Goodwood House, Sussex (William C. Rolf, 2000).

published in Isaac Ware's *The Designs of Inigo Jones and Others* in 1731 (fig. 77), and were closely copied by the anonymous architect who remodelled the interior of one of the stables at Osterley Park, Middlesex, probably in the 1730s or 1740s (fig. 140).[29]

Ware's engraving was also used by William Kent at the Royal Mews (fig. 141). English parsimony meant that no royal stable ever rivalled the splendours of Versailles or Chantilly, but the Royal Mews was a conscious attempt to emulate them. Work had already begun when the original design was changed for a stable of the same dimensions and capacity, but at £14,118 costing more than double the original estimate of £6,500. The principal change was the creation of the arcaded and vaulted double-range great stable recorded by Richardson. The arcade was based closely, but not exactly, on the engraving of the Holland House stables. Not everyone cared for the new mews. According to the Duchess of Marlborough, when the Prince of Wales visited it, he thought the arcading absurd: 'What the devil are these?' he declared. 'The beauty of a stable is to see a great many horses, but in this you only see two or three as you go between the pillars.'[30]

The best Continental stables had long been vaulted in stone and brick. A groin vault supported by double rows of columns forming a nave and two aisles was popular in Italy and Germany, and had been used by both Raphael and Leonardo.[31] In France, the barrel vault, which had been used by Francesco di Giorgio in one of the most renowned of all Renaissance Italian stables, those built in 1482 for the Duke of Urbino at Urbino and at Capraola, was preferred.[32] The advantage of this was that it avoided cluttering the stable with columns, but it required the vault to have a greater span. Barrel vaults survive in the stables at Vaux-le-Vicomte, designed in 1655, and at Chantilly built in 1719–25.[33] In England, one barrel-vaulted stable, built in about 1695 at Burley-on-the-Hill, Rutland, by the Earl of Nottingham, survives as proof of French influence.[34] An alternative, used both at the Grand and Petits Ecuries at Versailles, designed in 1685 by Mansart, and at Lunéville as illustrated by Boffrand in 1745, was to keep the single span and barrel vault, but to break it up with cross-vaulted bays at intervals.[35]

Blundeville called for vaulting as early as 1565 as a protection against fire.[36] His words were echoed by Samuel Wyatt exactly two centuries later when he proposed vaulting the stables at Kedleston Hall in Derbyshire in 1765: 'as it may prevent accidents by fire and the rooms over will be less annoy'd by the horses'.[37] But vaulting remained exceptional, presumably for reasons of cost. Though the Duke of Newcastle used groin vaulting in his stables at Welbeck Abbey, Nottinghamshire, in 1625 (fig. 79), most vaulted stables in Britain date from the eighteenth century.[38] The Earl of Nottingham, who had built a barrel-vaulted stable at Burley-on-the-Hill in about 1695, used groin vaulting supported by columns when he rebuilt the Duke of Buckingham's stables after a bad fire in 1704, a change that is perhaps evidence of the growing Italian influence in British architecture at the time.[39]

Vaulted stables remained rare in England and were built to impress, as they still do at the Horseguards, designed by William Kent shortly before his death in 1748 and erected by Thomas Vardy in 1750–59 (fig. 210). The Horseguards, the ceremonial entrance to the Palace of Whitehall, combined offices for the headquarters of the British army on the first and second floors with stabling for the monarch's Life Guard on the ground floor, part of which is still occupied by the Household

140 Anon., interior of the stables of *c.*1730–40 at Osterley Park, Middlesex (William C. Rolf, 1999).

Cavalry today. With its square piers supporting brick arches, these stables were clearly modelled on Kent's earlier stables at Houghton.

At Althorp, the extravagance of the exterior with its pair of porticoes is matched by the great stable, which has a double row of columns supporting the stone cross-vaulting (fig. 142). A similar double-vaulted stable was built by Samuel Wyatt, following an idea of Robert Adam's, at Kedleston Hall, Derbyshire, in 1767–69 (figs. 144). Adam had originally considered placing the stables in wings attached to the house, with state rooms above. He was dissuaded, but this idea had already been tried out by Thomas Worsley, an enthusiastic amateur architect who built Hovingham Hall, Yorkshire, for himself in 1751–78, a house that united stables, riding house and domestic accommodation in a single structure (fig. 145). Worsley combined a passion for Classical architecture with a love of horses in a pair of beautiful, columned and vaulted stables ('where a Neapolitan would not be surprised' as he proudly explained to Sir William Hamilton). Domestic convenience came a poor third when he placed his main reception rooms over stables, although he must have realised the impracticality of his design for, by the time he died, one of the two stables had been converted into a vestibule, and new stables had been built further away.[40]

But the Hovingham stables were a model of practicality compared with the stables at Seaton Delaval (figs 118 and 146). Although often assumed to be by

Sir John Vanbrugh, architect of the remarkable house to which they are attached, they were actually built much later, after Sir Francis Delaval had visited Hopetoun House, West Lothian, in 1768. As Sir Francis explained in a letter to his brother Sir John Delaval that November: 'I am putting up the grand stable on a plan we saw at Lord Hopetoun's when we were in Scotland, with stone divisions of the stalls, which I am sure you will like as they are very agreeable to the rest of the building and look very magnificent at Lord Hopetoun's from whence I have got a plan. No man has finer horses than we saw there and they have received no inconvenience from the stone.'[41]

Here, perched above the North Sea in chilly Northumberland, the entire construction – walls, floors and partitions – is of stone, the ceiling is high and the east

141 Interior of the Royal Mews of 1731–33 at Charing Cross, London, by William Kent, demolished 1830. From A. C. Pugin and Thomas Rowlandson, *The Microcosm of London* (London, 1808) (Society of Antiquaries, London).

142 Roger Morris, the interior of the stables of 1732–33 at Althorp House, Northamptonshire (National Monuments Record, 1998).

wall is largely given over to a great Venetian window. In Italy, such a stable would be admirably cool. In England, nothing could be less suited to its position. Why Vanbrugh's conventional plan for the stables, shown in *Vitruvius Britannicus*, was abandoned is, from a practical point of view, incomprehensible. But then Seaton Delaval was always a work of sculpture on the grandest scale and as architecture the stables are superb, with their finely dressed stone wall, great soaring arches and carefully detailed niches for the racks. Just pity the horses that had to live there in winter.

There was a very practical reason for the stone-vaulted stables Ralph Allen built at Prior Park on the outskirts of Bath. Allen was a great stone magnate and Prior Park – including the stables – was a flamboyant advertisement for his quarries and the benefits of stone:

> The Stables and Hay House are Arched, or Vaulted over, and were from the very first, Designed to be Arched with Stone . . . The Sides of the stables are divided into Recesses, Arched over set in every side; and in these the Racks and Mangers are placed for the Horses to Eat out of, as weel as the Bins for holding their daily Food: This makes the Curiosity of the Stables; and they want nothing but Recesses at the ends like those on the Sides, with a little more Magnitude to render them as Beautiful within as stables need be, since they are entirely lined with wrought Free Stone.[42]

143 (*right*) Samuel Wyatt, the exterior of the stables of 1767–69 at Kedleston Hall, Derbyshire (William C. Rolf, 2000).

144 (*facing page*) Samuel Wyatt, the interior of the stables of 1767–69 at Kedleston Hall, Derbyshire (William C. Rolf, 2000).

146 Anon., the interior of the stables of 1768 at Seaton Delaval, Northumberland (William C. Rolf, 1998).

145 (*facing page*) Thomas Worsley, the interior of the former stables of 1759 at Hovingham Hall, Yorkshire (William C. Rolf, 1999).

147 (*below*) Edward Lovett Pearce, interior of the stables of *c.*1729 at Castletown House, County Kildare (*Country Life*).

The source for the stables proves a surprising one. John Wood, the architect, revealed that the model was not an Italian stable but a Welsh one: 'Mr Hanbury's Stables at Ponty Pool in Monmouthshire, which I went purposely to see, furnishing us with the Example.' Wood must be referring to John Hanbury, who began to enlarge Park House at Pontypool in 1701 and who died in 1734. His grand stables probably date from after 1720 when a legacy of £70,000 may have turned his mind to ideas of more conspicuous consumption. The stables have been demolished and no view or plan survives of the interiors.[43]

Vaulting was too expensive for most owners and remained exceptional even in aristocratic stables in England, though in Ireland it was a constant theme of the architect Richard Castle. The earliest vaulted stables in Ireland are at the great house at Castletown, County Kildare, just outside Dublin (fig. 147). This was begun in 1722 to the designs of Alessandro Galilei and constructed under the aegis of Edward Lovett Pearce, but was not fitted up by the death of its owner, Speaker Conolly in 1729. As vaulted stables in Ireland are normally evidence of the hand of Richard Castle, Pearce's assistant, it is possible that he may have designed the Castletown stables. However, it is more likely that these were Pearce's work and the source for

148 Plan of George Steuart's stables of 1785 at Attingham House, Shropshire, from George Richardson, *The New Vitruvius Britannicus* II (London, 1808), p. 30.

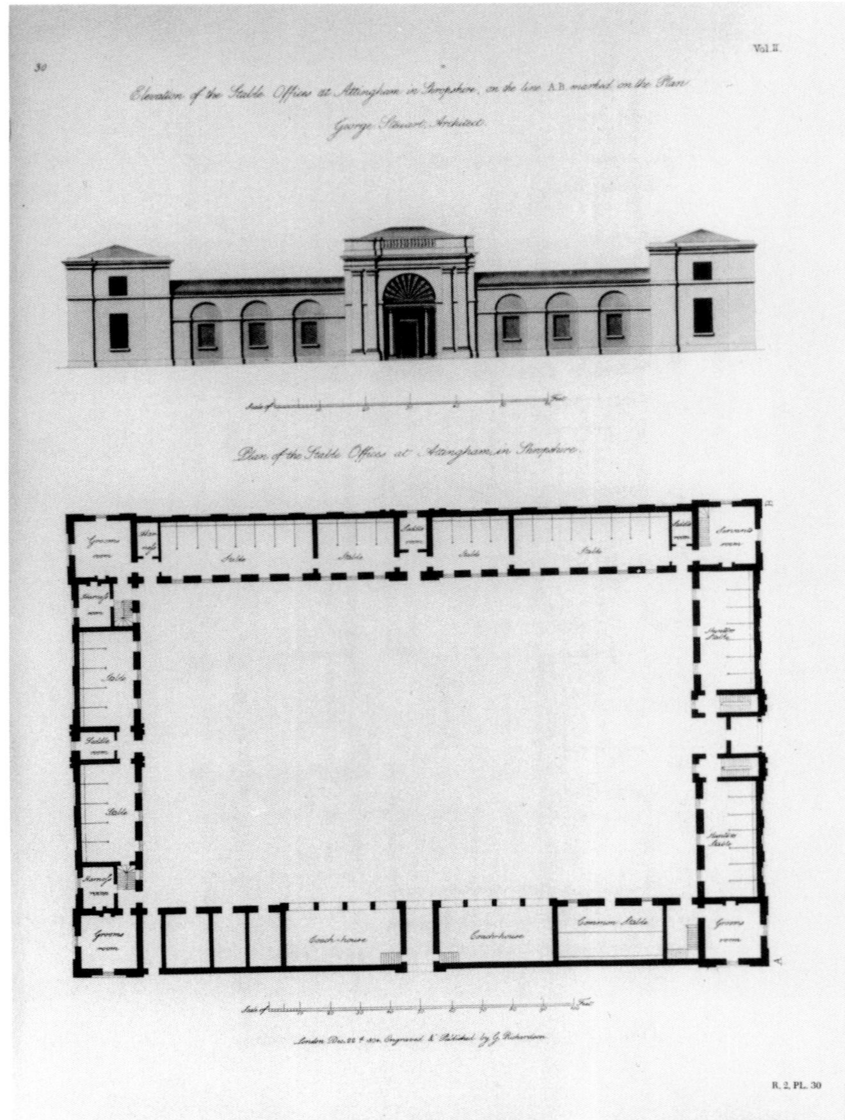

Castle's later obsession. In the 1730s and 1740s Castle designed at least five similar vaulted stables: those at Gill Hall, County Down, Strokestown, County Roscommon, Carton, County Kildare, Ardbraccan, County Meath, and Russborough, County Wicklow.

Vaulting was always exceptional in British stables, but no less attention was paid to creating handsome interiors that showed horses off to their best than had been the case in the seventeenth century. A good example of a late eighteenth-century stable with intact interiors is the stable at Attingham Park, Shropshire, designed by George Steuart for the first Lord Berwick in 1785. The stables at Attingham were among the finest built in the second half of the eighteenth century, and their reputation was such that they were illustrated in the second volume of George Richardson's *New Vitruvius Britannicus* in 1808, the only stable to be singled out in either volume (fig. 148).

149 and 150 (*left and below*) George Steuart, interior of the stables of 1785 at Attingham House, Shropshire (William C. Rolf, 2000).

Steuart's design shows the strength of the neo-Palladian tradition begun at Houghton Hall over half a century earlier. Though the detail has been refined down from the vigorous forms of the 1720s and 1730s and the Diocletian windows have been replaced by more conventional domestic windows, the form is unmistakably still that of Houghton, a substantial quadrangle with blank arcading, a grand central arch and emphatic corner pavilions.

Inside, a fine cornice runs round the interior of the stable, which is divided into partitions, whose heelposts are topped not by large balls, as at Houghton, but by delicate urns (fig. 149). Rack and manger are designed as part of a single timber structure that stretches the full length of the stables, a direct successor to the timber racks at the stables at Whitmore Hall and Peover Hall nearly two centuries earlier (figs 37 and 41).

151 Samuel Wyatt, a stall in the stables of 1767–69 at Kedleston Hall, Derbyshire (William C. Rolf, 2000).

152 Samuel Wyatt, design for the stables at Kedleston Hall, Derbyshire, *c.*1767 (National Trust Photographic Library/John Hammond).

By the late eighteenth century, such substantial timber constructions were no longer standard in stables. The growing availability of iron meant that racks were often designed in metal and set in individual niches in the wall, as at Kedleston, which was designed by Samuel Wyatt, whose connection with the industrial north-west meant that he was often innovative in his use of materials (fig. 151). Metal racks had the obvious advantage of being harder than timber, and less vulnerable to chewing by horses. They also meant that the substantial timber structures traditionally used for racks could be avoided, a saving both of cost and space. The potential presented by metal racks was most successfully exploited by John Carr at stables such as Wentworth Woodhouse, Yorkshire, built for the second Marquess of Rockingham from 1766, probably the largest country-house stable of the eighteenth century (figs 153, 154 and 156). As can also be seen in a contemporary painting of the stables of Ribston Hall, Yorkshire, Carr placed each rack under a semi-circular arch and linked them by a short length of cornice, repeating the pattern he used for the windows of the stable façades and creating an unparalleled unity between interior and exterior of the stable (figs 155).

The stalls at Attingham Hall, like those at Houghton Hall, still have their original clinker brick floor (fig. 149). During the eighteenth century, clinkers or some form of brick became the standard stall flooring (though the grand stables at Kedleston used stone). Planking had taken a long time to fall out of fashion. Even at the very end of the seventeenth century the stables at Winslow Hall in Buckinghamshire, built in 1699–1702, a house where some of the best craftsmen of the day were employed and where the bills were checked by Sir Christopher Wren, had planking in the stables.[44] But it was becoming obsolete and though Gibson mentioned planking in 1721, he said that paving was more com-mon.[45] By 1798, John Lawrence was able to write dismissively of earlier authors who had advocated planking, while adding that 'At Vienna and perhaps most parts of the Continent, the horses, even in the first stables, stand all day upon boards, without litter.'[46] This was the case with Prince Eugene's stables at the Belvedere in Vienna, as shown in an engraving of 1738. Similarly in Germany both Sturm and Delsenbach depict planked stalls.[47]

De Garsault, writing in 1741, talks of copper being used in the mangers of the 'rich and curious' and indeed the mangers in the Royal Mews at Charing Cross were covered with copper, but this was exceptional.[48] Leather was more common and the mangers at Houghton Hall in Norfolk still have the remnants of their leather covering. The mangers in the stables at Kedleston were of finely carved stone, but perhaps the most unusual facing for the manger was slate, as used by Samuel Wyatt at Penrhyn Castle in Carnarvonshire, which was built for the great slate developer Richard Pennant in 1782. Pennant used slate extensively in this stable (as all over the estate). Not only was it faced in slate, but the pilasters that divided the stalls were made of slate as well.[49]

The familiar image of the eighteenth-century stable is of a Classical building, but this was not universal. At Berkeley Castle, Gloucestershire, the impressive medieval seat of the Earl of Berkeley, Thomas Wright was probably responsible for the mid-eighteenth-century stables with their pointed-arch windows and crenellated roofscape, which was clearly intended to be a feature in the park (figs 158 and 159).[50] The delightful stables at Megginch Castle, Perthshire, with a low courtyard, pretty Gothic windows and an almost Chinese dovecote in the centre, built by Robert Drummond in 1806–7, show a similar respect for the character of the castle (fig. 160).

Scotland was the home of perhaps the most unexpected feature of the Palladian stable, the Scottish tradition of over-sized cupolas and steeples on stables that is

153 John Carr, the entrance to the stables of 1766–83 at Wentworth Woodhouse, Yorkshire (William C. Rolf, 1999).

154 John Carr, detail of a downpipe at the stables of 1766–83 at Wentworth Woodhouse, Yorkshire (William C. Rolf, 1999).

155 Anon., *The Interior of the Stables at Ribston Hall, Yorkshire*, by John Carr, *c.*1773 (Country Life).

156 (*right*) John Carr, the interior of the stables of 1766–83 at Wentworth Woodhouse, Yorkshire (*Country Life*).

157 John Carr, exterior of the stables of 1766–83 at Wentworth Woodhouse, Yorkshire (William C. Rolf, 1999).

unknown in England. Among the grandest of all Scottish stables were those built at Hopetoun House in Midlothian by John Adam in 1750–56, with cupolas that appear to float above the stables in a conventional way, but sit on a large tower marking the entrance to the stables and dominating the stableyard. Another tower, this time castellated, rises over the entrance to the stables at the Lees in Berwickshire. A similar, almost tower-like effect is given by the enormous cupola at the Castlemilk stables in Glasgow and at Darnaway Castle in Morayshire, built in 1802–12 by Alexander Laing. There may be a parallel between these over-sized cupolas and the excessively large cupola over the coach house at Wallington Hall in Northumberland, built in 1754, whose authorship has always puzzled architectural historians because it is unlike anything else in England. Because of the Scottish tendency to combine the domestic stable with the steading or farm building, these large cupolas often housed dovecotes or pigeon houses.

This desire to give a dominant accent to the stables reaches its most extreme in the Scottish use of steeples. Unlike the English, the Scots never associated steeples exclusively with churches. They also used them to mark tollbooths and town halls, such as William Adam's now-demolished Town House in Dundee of 1731.[51] The

158 Thomas Wright, detail of the stables at Berkeley Castle, Gloucestershire (William C. Rolf, 1999).

159 Thomas Wright, the stables at Berkeley Castle, Gloucestershire (William C. Rolf, 1999).

160 Robert Drummond, the stables of 1806–7 at Megginch Castle, Perthshire (William C. Rolf, 1999).

161 Sir James Clerk and John Baxter, the stables of 1762–65 at Penicuik House, Midlothian (William C. Rolf, 2000).

Adams seem to have had a particular interest in secular steeples. A drawing dated 1753 in James Adam's sketchbooks shows an arched entrance, possibly to a stable, with a steeple above it,[52] and Robert Adam's unexecuted 1759 design for a stable at Kedleston Hall in Derbyshire also included a steeple.[53] The first stable to be built with a steeple was that at Penicuik House in Midlothian, built from 1762 to 1765 by John Baxter to the designs of the owner Sir James Clerk (fig. 161).[54] As a Mr Adams, presumably John Adam, was paid five guineas in 1761 for a consultation, it is possible that the Penicuik steeple may be an Adam inspiration.[55] Robert Adam later included steeples in his designs for stables at Kirkdale in Kircudbrightshire, Dalquaharran Castle in Ayrshire and Findlater Castle, Banffshire, though none of these was executed.[56]

Perhaps inspired by the stables at Penicuik, steeples became a popular feature on late eighteenth-century Scottish stables. Blackadder Farm in Berwickshire has a

162 Anon., Rosebery Farm, Midlothian, early nineteenth century (Mark Fiennes, 1987).

steeple dated 1785, closely modelled on that at Penicuik, although it has a wooden rather than a stone spire. Other examples, such as that at Rosebery Farm, midlothian (fig. 162), are less elaborate. That at Tarvit House in Fife is a plain octagonal steeple. That at Doune Park in Perthshire, built around 1807–9, was only a little more elaborate. Despite their ecclesiastical character to English eyes, the Gothic steeple at Duns Castle in Berwickshire, built in 1792–94 by John Baxter, is exceptional.[57]

[The art of riding] has so long been neglected and despised, that one would almost be prompted to conclude that a fatality had constantly attended it in this country . . . for a length of time it has been able to boast but a very few persons who have stood forth its avowed friends and protectors . . . but since the accession of his present Majesty, the prospect has brightened, and better times begin to dawn. Since this happy event the *Art* has raised itself a little, and given some signs of recovery; public riding houses have been opened, which are largely encouraged, and frequented by the youth of the nation . . . several private *Maneges* have likewise been erected by the *Princes* of the blood, some of the *Nobility* and *Gentry*; and to crown it all his Majesty has erected one for his own immediate use.[1]

7

THE GEORGIAN RIDING-HOUSE REVIVAL

IN 1771, RICHARD BERENGER was at the centre of the London equestrian world. Gentleman of Horse to George III, translator of Bourgelat's *New System of Horsemanship*, author of *The History and Art of Horsemanship*, friend of all the leading horsemen of his time and adviser on the scheme to establish a riding academy in Edinburgh, his authority is unimpeachable. There may be a slight note of hyperbole in what he says, but the essence is true. Only three new private riding houses are known to have been built between 1660 and 1750 but between 1750 and 1780 there were about fourteen, most of them built by devotees of *haute école*, together with a number of public riding houses. The fashion had begun with a string of military riding houses in London.

In 1728 Benjamin Timbrell contracted to build a riding house for the Earl of Hertford, future seventh Duke of Somerset, behind Grosvenor Square, between Adam's Mews and Mount Street. This was to be 86ft long and 40ft wide and was to be accompanied by stabling for thirty horses, for which Lord Hertford was to pay an annual rent of £80.[2] As this was at some distance from Lord Hertford's house in Grosvenor Street, and provided an unusual quantity of stabling for a London house, and as Lord Hertford was captain and colonel of the second troop of Horse Guards from 1715 to 1740, it was probably intended for military use. Ratebooks show that it was not built, but a riding house was included in substantial new stables erected under Lord Hertford's colonelcy. These were built by Roger Morris in 1738 between Green Street, Park Street, Wood's Mews and Park Lane.[3]

Given the constraints on space in London, it made sense for training to take place under cover and Morris's riding house was one of a sequence of riding houses built with new cavalry barracks in the city. A riding house had been included with the new stables erected in about 1730 for the first troop of Horse Guards on the south side of Chapel (later Aldford) Street.[4] Another was built in Curzon Street, just to the south of Chesterfield House, in 1751–52, and a fourth was erected for the first troop of Grenadier Guards in Great Portland Street in 1752.[5] A drawing in the Dashwood Collection at West Wycombe reveals the Great Portland Street

163 Thomas Rowlandson, *The Riding School* (detail of fig. 174).

barracks to have had two ranges of stables flanking a riding house that was 40 ft by 86 ft. This was subsequently extended to 120 ft, probably in the 1770s.[6] The Curzon Street barracks lasted only a few years, disappearing from the ratebooks in 1756, presumably when its occupants went off to fight in the Seven Years War. Its valuable site was swiftly redeveloped as housing. The Great Portland Street riding house survived the demolition of its barracks and in 1823 was being used as Hall's Riding School.[7] Its presence is recorded in the name of Riding House Street. In Dublin, the chapel built by Arthur Jones at the Royal Barracks was converted into a riding house in 1752.[8]

Berenger described riding as consisting of two parts, the useful and the ornamental: 'That the latter of these may be dispensed with, is most readily granted; but that it behoves every one who puts himself upon a Horse to have some Knowledge of the first, is most evident.' Nevertheless, it was the revival of ornamental riding, one of the 'superfluous but refin'd Pleasures of Life', that Berenger was celebrating in 1771.[9] Eighteenth-century *haute école* was not aimed at satisfying the need for courtly carousels as it had been under the Tudors and Stuarts. It was a more private activity practised for the pleasure of the individual and his friends, not for public display. But it achieved considerable popularity, interesting people as diverse as Lady Mary Coke and Samuel Johnson. London riding houses, such as that of Angelo's academy in Soho Square or that of Sir Sidney Medows in Half Moon Street (where Medows would entertain his visitors with a constant stream of conversation even when mounted), became fashionable places of resort.

The public riding houses Berenger refers to may have included two built on the edge of London, by William Rastall at Hyde Park Corner in 1738 and by James St Amour on the north side of Oxford Street in 1740.[10] The first of the new generation of private riding houses was probably that built by the third Earl of Bute at Mount Stuart on the Isle of Bute in 1744. He later built a second riding house at his marine villa at Highcliffe, Dorset. Other enthusiasts followed his example in the 1750s and 1760s, including Thomas Worsley at Hovingham Hall, Yorkshire, Sir Henry Calthorpe at Elvetham Hall, Hampshire, George Pitt, first Lord Rivers, at Stratfield Saye, Philip Meadows in Curzon Street, Mayfair, and the tenth Earl of Pembroke at Wilton House, Wiltshire, and Pembroke House, Whitehall.[11] These were the 'few persons who have stood forth as its avowed friends and protectors' mentioned by Berenger. All belonged to an older generation than George III. Sir Sidney Medows was born in 1700 (fig. 164); Thomas Worsley in 1710; the third Earl of Bute in 1713; John Elwes in 1714; Sir Henry Calthorpe in about 1717; George Pitt, later first Lord Rivers, in 1722. Only the tenth Earl of Pembroke, born in 1734, was the future king's near contemporary. All had learnt their skill in foreign academies. The riding master at the academy at Geneva was said to have had three of the best horsemen in Europe in the early 1730s: Medows, Worsley and Elwes.[12] Bute spent two years at Groningen in 1730–32.[13] Pembroke passed 1754 travelling from riding master to riding master, from d'Eisenberg in Pisa to Bourgelat in Lyon, and returned to England fired with a love of *haute école*.[14]

These were years of renewed interest in *haute école* on the Continent, where new developments were codified in the publication of three important treatises: Baron d'Eisenberg's *Description du manége moderne* in 1727, François de la Gueriniere's *Ecole de cavalerie* in 1733 and Claude Bourgelat's *Le Nouveau Newcastle* in 1744. In these

164 Silhouette of Sir Sidney Medows from Strickland Freeman, *The Art of Horsemanship, Altered and Abbreviated According to the Principles of the late Sir Sidney Medows* (London, 1806) (© The British Library).

books, the authors sought to improve on the methods of the seventeenth century and crystallised much of *haute école* as it is practised today. De la Gueriniere was director of the French Royal Manege from 1730 to 1751 and his book is still the equestrian bible of dressage,[15] but in the mid-eighteenth century both d'Eisenberg and Bourgelat, with his academy in Lyon, exercised considerable influence. A comparable growing interest in England can be seen in the publication of John Vanderbank's *Twenty Five Actions of the Manege Horse* in 1729 and the translation in 1743 of Gaspard de Saunier's reinterpretation of Newcastle, *A General System of Horsemanship*.

In England the Royal Mews provided a centre for the revival. Both Berenger and Worsley were equerries, and it was probably another official, Edmund Charles Blomberg, whose skill at *haute école* was such that he had won the *carrousel* that celebrated the marriage of the Duke of Wolfenbuttel in 1715,[16] who instilled a love of *haute école* in Worsley. Blomberg, who was first an equerry and then from 1740 till his death in 1757 was the avenor or senior administrator in the mews, was a neighbour and close friend in Yorkshire (despite their difference in age) and eventually made Worsley his heir.

Lord Bute, Worsley's old schoolmate from Eton, probably inspired the future king with his interest. Bute was appointed Groom of the Stole to the then Prince of Wales in 1756 and had an enormous influence over him. Henry Angelo later describes George III discoursing at length on the pleasures of the Great Horse, and it is known from Mrs Papandiek that he practised daily.[17]

Most of the known enthusiasts had already built themselves riding houses before the accession of George III. The exception was Berenger who for years was unable to leave the security of the Royal Mews for fear of being arrested for debt. Eventually David Garrick came to an arrangement with his creditors.[18] Pembroke had been planning to build a riding house even before he returned to England and had brought back a set of designs from Paris by the French architect Vallin Delamothe. He did not use them, but he did build two riding houses almost immediately, one at Wilton House in Wiltshire and the other at Pembroke House in Whitehall (fig.

165 The distance riding calculator in the riding house of *c.*1755 at Wilton House, Wiltshire (William C. Rolf, 2000).

165).[19] Sir Sidney Medows had his riding house in Half Moon Street in Mayfair where Mrs Thrale compared his horses to Kemble, Bernini and Pasiello: 'I have seen a high-dressed Horse in Sir Sidney Medows Riding House guilty of the same Fault. Affectation is not confined to Man, educated Animals all have a Share.'[20] Worsley had begun the slow process of building his house at Hovingham, Yorkshire, in 1751 (fig. 172), and in 1755 built himself a temporary riding house 37 ft square, where he was able to work four horses in bad weather, although it was not until 1769 that his permanent riding house was complete.[21]

But as Berenger makes clear, it was the king's example that made riding houses fashionable, especially in London. On his accession he bought Buckingham House for his private residence and built himself a riding house there in 1763–66, perhaps with the help of Thomas Worsley, who had been appointed surveyor-general of the Board of Works in 1760 thanks to Lord Bute (figs 166 and 167).[22] The king's two brothers soon followed his example: the Duke of Cumberland is reported as having a riding house but no manege horses to fill it in 1766, and the Duke of Gloucester built one in Hyde Park, opposite Gloucester House, in 1768 (fig. 168).[23]

In the country, Sir Henry Harpur built a riding house at Calke Abbey in Derbyshire in 1760.[24] Not far away, Sir Henry Calthorpe spent his years as a lunatic in his riding house at Elvetham Hall and at Stratfield Saye Lord Rivers 'studied and indeed brought to perfection the noble exercise of Horsemanship'.[25] Humphry Morrice built a riding house for his pampered horses at Grove House in Chiswick in the 1770s and in Yorkshire Robert Adam designed a riding house at Nostell Priory for Sir Rowland Winn in 1772.[26]

The one unifying strand among all these riding houses is the love of horses shown by their builders. Pembroke described himself as 'being, I fear, as horse mad as my father was tennisly so'[27] and was the author of a book that revolutionised cavalry training. Thomas Worsley, whom Horace Walpole dismissed as 'rider of the great horse and architect',[28] wrote: 'I am humble enough to be content with the

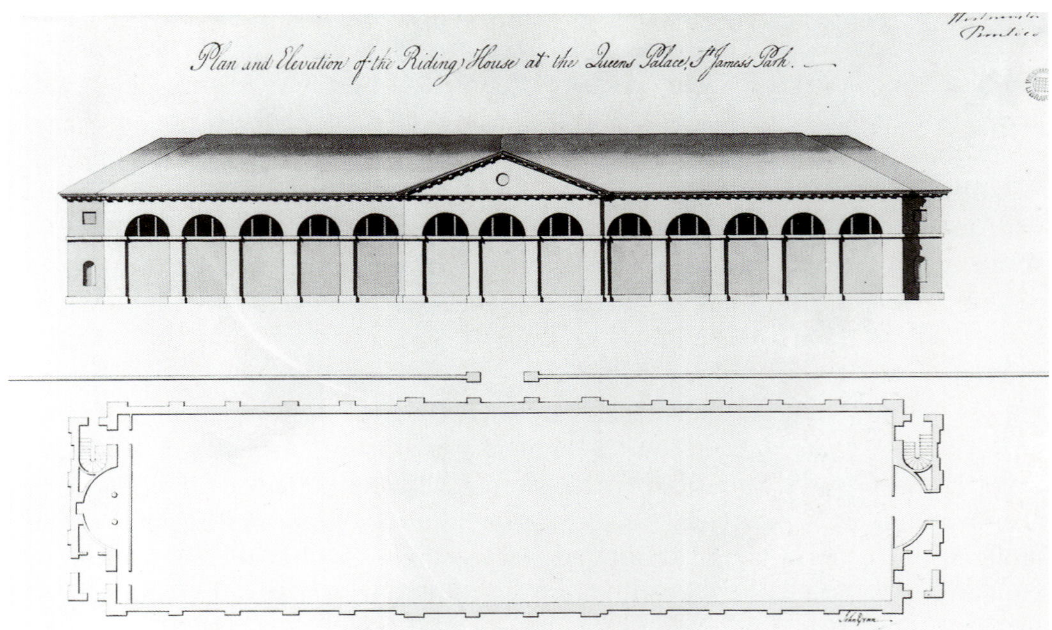

166 John Yenn, survey plan and elevation of the riding house at Buckingham House, London (City of Westminster Archives Centre).

167 Daniel Morier, *James Montagu in the Riding House at Buckingham House*, 1766 (Sotheby's).

adagio, upon some pretty well drest horses of my own breed, which I enjoy daily, without fear. They are my musick and I thank God I have yet an hand & ear for them.'[29] Humphry Morrice's indulgences towards his horses – he had stabling for thirty – went so far that he employed a boy to wave away the flies. 'The honours shown by Mr Morris to his beasts of burden were only inferior to those which Caligula lavish'd on his charger', declared one commentator. Sir Francis Delaval was another whose love of horses led him to want a riding house. He was a great spendthrift and when eventually forced to retrench in 1771 described his ideal of a quiet retirement to his brother: 'I have taken a sweet house with stables for eight horses and coachhouse for two carriages for £120 per year and can have for £25 a year more a piece of ground to build a riding house upon within 20 yards which bounds all my wishes.' But he died before he could build it.[30] These riding houses were built for utility, not show, and were not considered essential by those without a passion for horsemanship, even for the grandest stables such as those at Good-

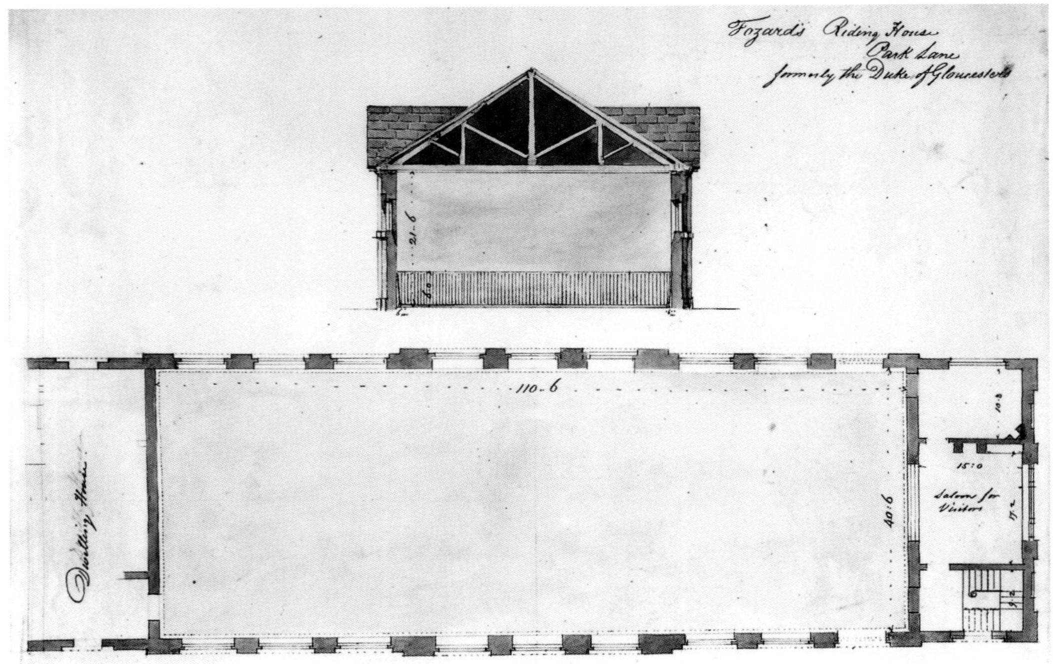

168 Anon., plan and section of Fozard's riding house, Park Lane, built for the Duke of Gloucester in 1768, London, *c.*1823 (West Yorkshire Archive Service).

wood House, Sussex, or Chatsworth House, Derbyshire. Only Sir Rowland Winn's equestrian credentials look weak. Although he had nine horses in his manege in 1772, he did not handle them himself. That he left to his riding master.[31]

Because they were built for utility, most eighteenth-century riding houses, apart from the royal examples, were plain buildings, and few were placed for architectural effect. Some were detached buildings, as at Buckingham House and Wilton House. Others were placed at the back of the stables, as at Calke Abbey. Occasionally they formed one of the ranges of a quadrangle, as at Nostell Priory. Only Thomas Worsley, as passionate about architecture as he was about riding, made an architectural statement with his riding house at Hovingham Hall (fig. 2). Ornamented with stone columns and grisaille paintings, it served as the formal entrance to his house, which, with its stone-vaulted, columned stables, was a veritable palace of the horse.

Design and decoration of the riding house were usually determined strictly by their function. The unknown architect of Prince Henry's riding house at St James's Palace had worked out how this could best be satisfied: with a rectangular building about 40 ft wide and about three times as long, with plenty of indirect light, easy access and a viewing gallery. Eighteenth-century riding houses differ from his design only in details. The one exception, William Newton's unexecuted design for a 'Menagery' (probably a proposal for a riding house at Buckingham House) in the form of a Greek cross was an attempt to impose an ideal architectural form by an architect who clearly had no practical experience of *haute école*.[32]

There is little theoretical writing on the subject until Strickland Freeman's *Art of Horsemanship*, whose first chapter is entitled: 'Description of the riding-house best adapted to the ease of the horse, and to the convenience of the rider'. Although this was published in 1806, what it says holds true for the previous century as the book's subtitle makes clear: *Altered and Abbreviated According to the Principles of the Late Sir Sidney Medows*. Despite this recommendation, Freeman brings a little more theory to a simple subject than is really necessary. This is most obvious in his discussion of the best shape for riding houses. Virtually all riding houses were rectangular, except that at Pembroke House in Whitehall, whose shape was determined by its unusual site on the edge of the Thames. Freeman, however, argued that it was better for one of the ends to be rounded: 'By this, there is less constraint to a colt, when beginning to be *longed*, than what is occasioned by the corners',[33] and built his own riding house at Fawley Court with a curved end (fig. 169). Most of his fellow enthusiasts would have said that the diameter of a circle within which

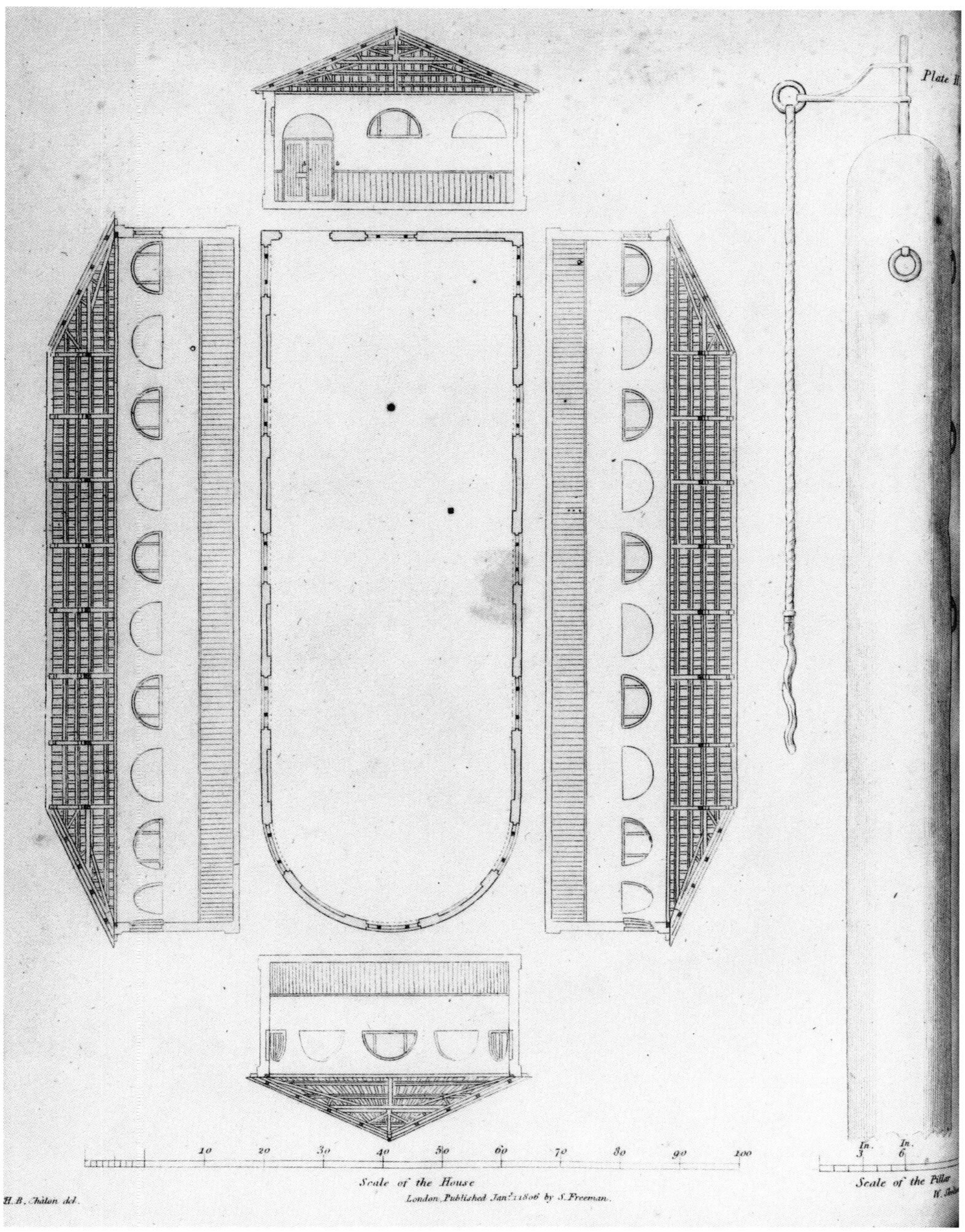

169 Plan of the riding house at Fawley Court, Oxfordshire, and detail of an exercising pole, from Strickland Freeman, *The Art of Horsemanship, Altered and Abbreviated According to the Principles of the late Sir Sidney Medows* (London, 1806) (© The British Library).

the horse could be exercised remained the same whether the end of the riding house was semi-circular or not. Adding a semi-circular end with all the extra expense that such a complicated roof would entail was simple extravagance. Only Thomas Worsley might have sympathised with Freeman. Some of his early designs for the riding house at Hovingham Hall show walls in the centre curving out to form a circle.[34]

Apart from the curved end, Freeman's dimensions of 3:1 for a riding house would have generally been accepted, although figures as low as 2:1 were not uncommon. For private use, Freeman reckoned that a 120ft by 40ft riding house was large enough. He did not object to it being longer, but argued that if it were any wider it would be too fatiguing for the horse when doing exercises that required it to hold its haunches while crossing the riding house. Few eighteenth-century riding houses were more than 40ft wide, the king's riding house at Buckingham House being an exception at 50ft.

Though the basic construction of the riding house was simple, its size meant that roofs were both wider and longer than any other that most country-house builders would otherwise have had to construct without intermediate supports. The number of surviving surveys of riding-house roof structures shows the interest that they held for builders. These were important because no architectural book dealt specifically with the riding-house roof, and Freeman, being a horseman and not an architect, makes no reference to roof design. The architectural drawings of Thomas Worsley, an amateur architect with limited practical experience in building, are most revealing because the need to cover a riding house 35ft wide and more than 75ft long might well have caused him difficulties. Among his collection of architectural drawings are nine for riding houses, including several attributed to Isaac Ware, and the survey of a roof structure very similar to that of the Buckingham House riding house by Chambers. There are also sections showing the roof structure of contemporary warehouses and one inscribed the 'Argentine Theatre at Rome'.[35] It is not surprising that the one detail Lord Bute sent to Worsley of his proposed riding house for comment was a drawing of its roof, and that the one element of the Carlton House riding house to be recorded in Holland's sketchbooks was a cross-section of the roof.[36] When Lord Grantham, later Lord de Grey, another amateur architect, proposed building a riding house early in the nineteenth century he first had survey drawings made of three riding houses in London, detailing their roof structure and noting in particular that 'Horizontal Beams would be of great Utility to this Roof'.[37] Other architects took their solution from pattern books. The roof of the Calke Abbey riding house, for instance, is derived from one in Prince's *British Carpenter*.[38]

The problem of roofing wide structures such as riding houses had been solved by the designer of the St James's Palace riding house in 1607 when he had used a king- and queen-post roof (fig. 55).[39] His example had been followed by Inigo Jones in his second riding-house design and was repeated at many eighteenth-century riding houses, such as those at Hinton House, Somerset, the Duke of Gloucester's riding house in Hyde Park (fig. 168) and at Wilton (These last two have identical roof structures).[40] However, when the 50ft-wide riding house at Buckingham House was designed, it must have been felt that a more elaborate roof structure was needed to span the distance, and it was given a collar beam with scissor braces.[41] The Carlton House riding house, which was 51 ft wide, had an identical roof structure.[42] Collar beams without the scissor braces were commonly used, as at Hovingham Hall and in the nineteenth century at Burton Constable, Kirby Hall and Stoneleigh Abbey. Samuel Wyatt, who was famed for his expertise in carpentry, showed more technical innovation at Fawley Court, using an ingenious and unusual queen-post truss very similar to those he invented for the Greenwich Hospital Chapel in order to cover a wide span in relatively short lengths of timber.[43] Similarly, William Porden later covered the Royal Pavilion riding house with a unique construction based on the Halle au Blé in Paris as was the dome itself. In order to gain as much height as possible, the roof was constructed without tie-beams, and the main timbers of 12 in by 9 in scantlings were built in the form of an arch.[44]

Having built his riding-house roof, the architect had to decide whether or not to give it a ceiling. Many architects took the practical view that in a riding house there was no reason to do so and many remain open to the roof revealing its roof timbers, as at Hinton House, Firle Place and Wilton. But those who were anxious to make the riding house less of a barn and more of a distinguished space in which to display horses, as at Nostell Priory and Hovingham Hall, added ceilings.

Until the nineteenth century all riding houses were lit by windows in the walls. Toplighting would have weakened the roof and been technically very difficult. Riding houses were generally lit by windows similar to those in contemporary use for stables. Riding houses needed plenty of indirect light as direct light would dazzle the horse, so windows were nearly always placed high up the wall. The Diocletian window, commonly used in Palladian stable designs, was well suited to this, and was first suggested for use in a riding house in Kent's abortive 1730s design for the Royal Mews (fig. 170). Freeman recommended 'Dioclesian' windows, but he called for a window that differed slightly from the conventional, with a wider central division and casements carefully designed so that when opened for ventilation they kept out the rain. As he did not want too much light in the riding house, he recommended that alternate windows be blank.[45]

Even riding houses that did not have Diocletian windows generally placed them high up in the wall, whether they were many and relatively small, as at Woburn, or few and relatively large, as at the Royal Pavilion and Hovingham Hall. At Hovingham, Worsley had a problem because his riding house was higher than most and Diocletian windows would not have provided enough light. Instead, he used deeper tripartite windows with a shallow curve. Such windows are usually associated with the Wyatts, but the riding house at Hovingham was completed in 1769 before the Wyatts had built a single building. Another was used for the riding house of the second troop of Horse Guards near Portland Place, built in the 1770s, possibly by the Adams brothers.[46]

Internally, riding houses were plain, the four walls sometimes plastered as at Woburn but generally with the stone or brick left bare and limewashed. No executed riding house was ever as grand as William Newton's unbuilt design for a riding house at Buckingham House, which envisaged a columned and arcaded interior with life-size statues of horses in *haute école* postures in each arch.[47] The Royal Pavilion riding house was unusual in having its walls entirely panelled in wood and Woburn Abbey was exceptional in having a *trompe l'oeil* cornice and frieze with *trompe l'oeil* painted decoration. At Wilton, Lord Pembroke placed versed celebrating the maneged horse:

> First ERICTHONIUS dar'd with dauntless Skill
> To yoke four Steeds and guide the Victor's Wheel
> THESSALIA taught the conduct of the Bit,
> To mount the Steed, and from his pliant feet
> To paw the Ground, to wheel, to turn with Grace
> To tread the Plain with more majestic pace . . .

Architectural treatment was generally reserved for the viewing gallery, an essential part of any riding house, 'pour les dames' as Lord Bute put it.[48] Because *haute école* never became a central feature of court life in England after the reign of Charles I, there were none of the grand *carrousels* with large audiences common at Continental courts. Consequently, there was no need for grand riding houses with galleries for spectators on all four sides, like those of the Spanish Riding School in Vienna or at the Palace of Belem outside Lisbon. In England, the riding house remained private and hence the galleries needed only be large enough to hold a handful of spectators.

The usual place for a gallery was at one end of the riding house, probably facing the main door, although at Nostell Priory the gallery is in the middle of one of the longer walls. At Bolsover Castle (fig. 61) and Wilton House they were given Serlian openings, but elsewhere in the eighteenth century, several of the grander riding houses placed the gallery in a semi-circular apse, perhaps with a coffered ceiling and possibly matched at the other end by another apse. This simple but effective way of giving variety and distinction to the interior of the riding house was first suggested in Kent's design for a riding house for the royal mews at Charing Cross, and survives in the Buckingham House riding house. At the beginning of the nineteenth century at Middleton Park in Oxfordshire, Cundy produced a dramatic scheme for an apsed gallery that was considerably larger than usual, approached from outside through a portico standing at the top of a short flight of steps.[49]

It became common in the eighteenth century to panel the bottom five to six feet of the wall with wood to protect the legs of horsemen from being crushed by horses riding too close to the walls. These were often placed at a slant, although Freeman disapproved of this, saying it did not look so good and only taught the horses to lean on them. Such panelling can still be seen in the Buckingham House riding house, and the grooves where it would fit survive at Hovingham Hall. Lord Bute suggested an alternative at Mount Stuart when he talked of building a 3 ft high and 1 ft wide projection at the bottom of the wall, but such a structural solution would seem to have been rare. In the nineteenth century at Windsor Castle a patent cork and rubber composition called Kamptulicon was used instead of boarding in the riding houses.

Riding houses needed a soft but firm surface for the horses to work on, and tan was one of the most common floorings, but as most have been converted to other uses, few retain their original floor. Freeman did not recommend a tan floor but the sort of clay with the least mixture of sand in it, laid 2 ft deep and well rammed. This would form a tenacious surface when dry on top, on which the horses' dung

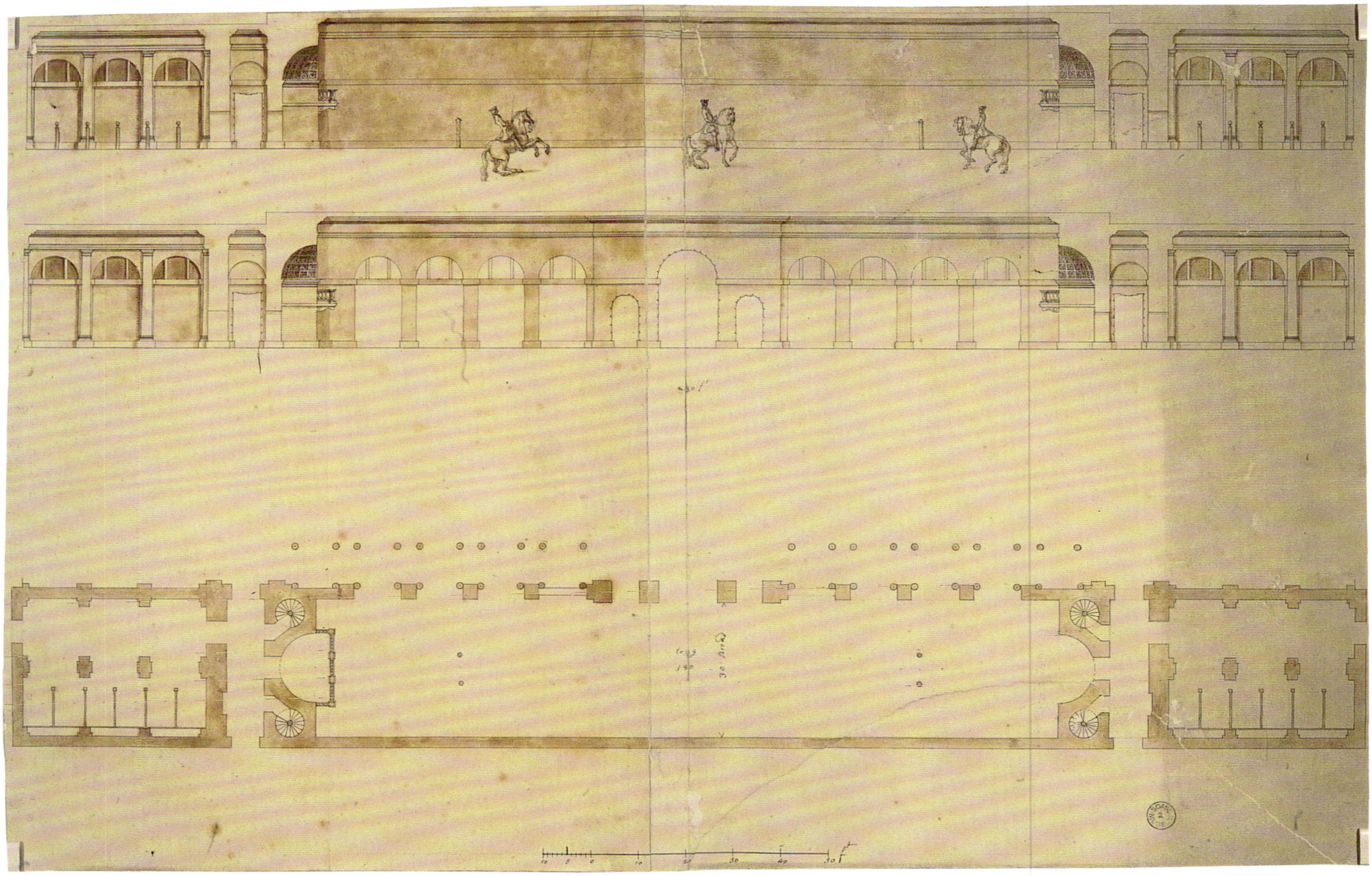

170 William Kent, sections and plan for a proposed riding house at the Royal Mews, Charing Cross, 1732 (by courtesy of the Trustees of Sir John Soane's Museum).

would form a mould sufficient to soften the upper surface and prevent it from becoming slippery.[50]

Because they were so large and were seldom used for display, the exteriors of most riding houses were very plain. They often received no articulation at all beyond that of their windows. Even Wilton, where Pembroke first commissioned an elaborate French design and then a neo-Palladian scheme with attached twin Ionic columns and a pediment, ended with a very simplified version with a basic pediment, no columns, no balustrade and plain side walls.[51] The exceptions were the royal riding houses. The blind arcading and Diocletian windows of Kent's design for the Royal Mews was imitated at both Buckingham House and Gloucester House, whose arcading continued round the whole building (fig. 166). Both riding houses articulated their façades with the central three bays breaking forward under a pediment. The riding house at Woburn Abbey of 1790 was a rare example of an architecturally ambitious non-royal riding house, with rectangular windows interspersed by attached columns, supporting a bold cornice over blank arcading.

Most riding houses would have had an attached open manege for the practice of *haute école* out of doors. Hugo Arnot's description of the Edinburgh Academy refers to 'an area without the building sufficient for an open manege',[52] and as Strickland Freeman explains: 'Having now described every thing that constitutes a drest horse *within* the house, I shall point out as well as I can, a lesson for a drest horse *out* of the house . . . the *scene of action* must be changed to the lawn before my riding house.'[53] Lord Pembroke had a different surface at Wilton: 'I worked her in the Manege . . . I then took her to the open Manege which is chiefly of small gravel.'[54] A design for an open manege with a central space 42 ft 6 in by 127 ft 6 in and two concentric circles 60 ft apart survives among the Pembroke papers, although it is not clear if it was used at Wilton.

An open manege need be no more than the word suggests, but some riders felt that *haute école* needed more of a setting than empty space, the most famous example

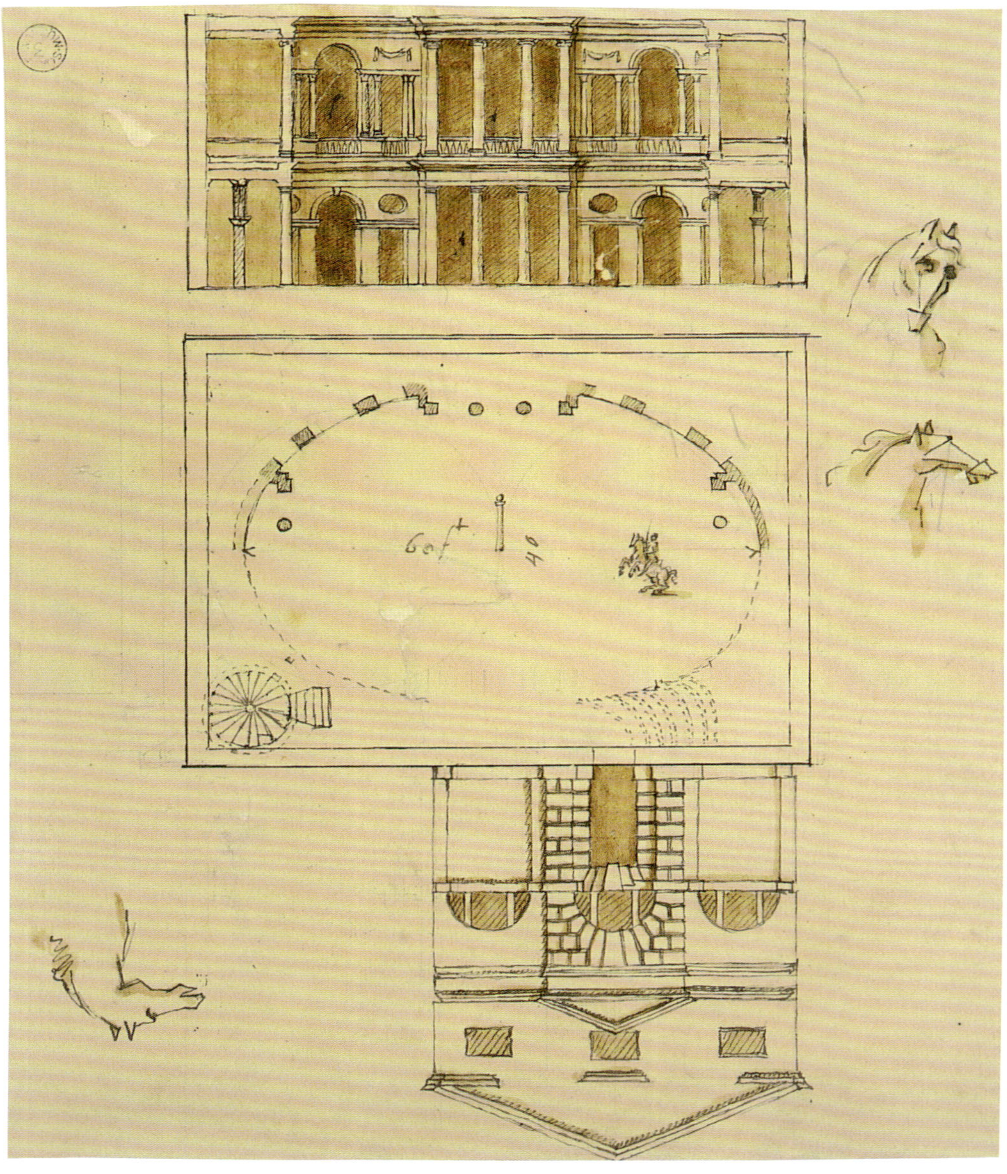

171 William Kent, design for an open manege at the Royal Mews, Charing Cross, London, 1732 (by courtesy of the Trustees of Sir John Soane's Museum).

172 Thomas Worsley, the exterior of Hovingham Hall, Yorkshire, of 1751–66. This was probably intended as the backdrop to an open manege (William C. Rolf, 1999).

being Giulio Romano's Cortile della Cavallerizza in the Ducal Palace at Mantua of 1558. De Pluvinel set the illustrations of his equestrian exercises against the elaborate background of the Louvre.[55] The Duke of Newcastle displayed his horses in front of his houses.[56] The most extreme example of this is the Eisenberg series of *haute école* paintings at Wilton with their fantastic architectural backdrops. These were used as the illustrations for his *Manege Moderne*, published in 1747, but without the backdrops. Such elaborate architectural settings were not restricted to books. In Vienna, the Comte de Par had an open manege with an elaborate architectural setting worthy of a theatre.[57] At Chantilly, the Prince de Condé's great stables of 1719–35 have a circular open manege with walls pierced by arches. Except for an unexecuted design for an elaborate backdrop to an open manege designed by William Kent for the Royal Mews, perhaps inspired by Chantilly, such ambitions generally seem absent in the pragmatic world of British *haute école* (fig. 171).[58] The one exception was Hovingham Hall where Thomas Worsley was not only a skilled horseman but an enthusiastic architect (fig. 172). Its courtyard (one wing of which was never built) with its rusticated facades was probably intended as an architectural backdrop to an open manege.

Haute école was not only practised in private riding houses. As Berenger noted, its fashionable status also boosted the popularity of the riding academy, which had maintained a tenuous existence during the first half of the century. For a time Salomon Durell, who succeeded his uncle Henry Foubert, described as Major and Equerry, Master of the Royal Riding School at his death in 1743, was fashionable, having taught George III to ride. His activities with the future king were recorded by Henry Angelo: 'O mine star!', he would cry out. 'How se royal youth carry himself as Alexandre on Bucephalus. Now, Sare, advance your pace . . . Hold up you august chins – turn in your royal toe.'[59] But Durell's star waned when George III's younger brothers were taught by his rival Domenico Angelo and in later years the academy declined into a simple riding school. By 1782 it was no more than livery stables.[60]

The academy that stole Foubert's royal patronage was run by an Italian, Domenico Angelo Malevolti Tremamondo, more commonly known as Domenico Angelo, who had learnt his art under de la Gueriniere and come to England in 1753. Angelo ran the most successful academy of its day – if one believes his son and biographer Henry. Certainly, his arrival just as *haute école* returned to fashion was well timed. He first entered service with the Earl of Pembroke and later claimed to have been responsible for the new system of cavalry training that Pembroke promoted in his *Method for Breaking Horses and Teaching Soldiers How to Ride* of 1762. He was engaged to teach the future dukes of Cumberland and Gloucester to fence in about 1758 and subsequently to ride in the manege. Angelo's prowess on the horse led George III to declare him 'the most elegant horseman of his day'.

Such royal favour must have been instrumental in the success of the academy he founded in 1763 or 1764. His first pupil was the future Duke of Devonshire. 'Such was the estimation in which the elder Angelo was held', recorded Henry Angelo, 'that the sons of many persons of rank were placed under his tuition and were attended by masters to cultivate their minds and form their manners, previously to their entering society and taking on themselves the responsibilities of manhood.'[61]

To house his academy and live in style, Angelo rented the grand but unfashionable Carlisle House in Soho Square, where he was able to take boarders. With fees of 100 guineas per year, 'his income at the period was great, the household establishment was commensurate and his table elegant'. Garrick and Reynolds were among the frequent visitors to his table and the gallery of the riding house that he built in the back garden was a favourite retreat for fashionable ladies. These included the Duchess of Queensbury (whose husband kept a stud of capital manege horses). Among her protégés was the son of a black slave whom she sent to Angelo's academy, where he became one of his most accomplished riders. His skill gave him airs above his station and the duchess was forced to persuade Angelo to take him on as an assistant. Even this failed to humble him and he was eventually sent to Bengal where he established an academy himself, but was eventually thrown and killed by an Arabian he was riding.[62]

In 1763, at about the same time that Angelo was establishing his academy in Soho Square, the similarly named Domenico Angelo Tremamondo, presumably a member of the same family, who later claimed to have spent some time in Russsia, was appointed riding master of the new academy in Edinburgh. Unlike the London

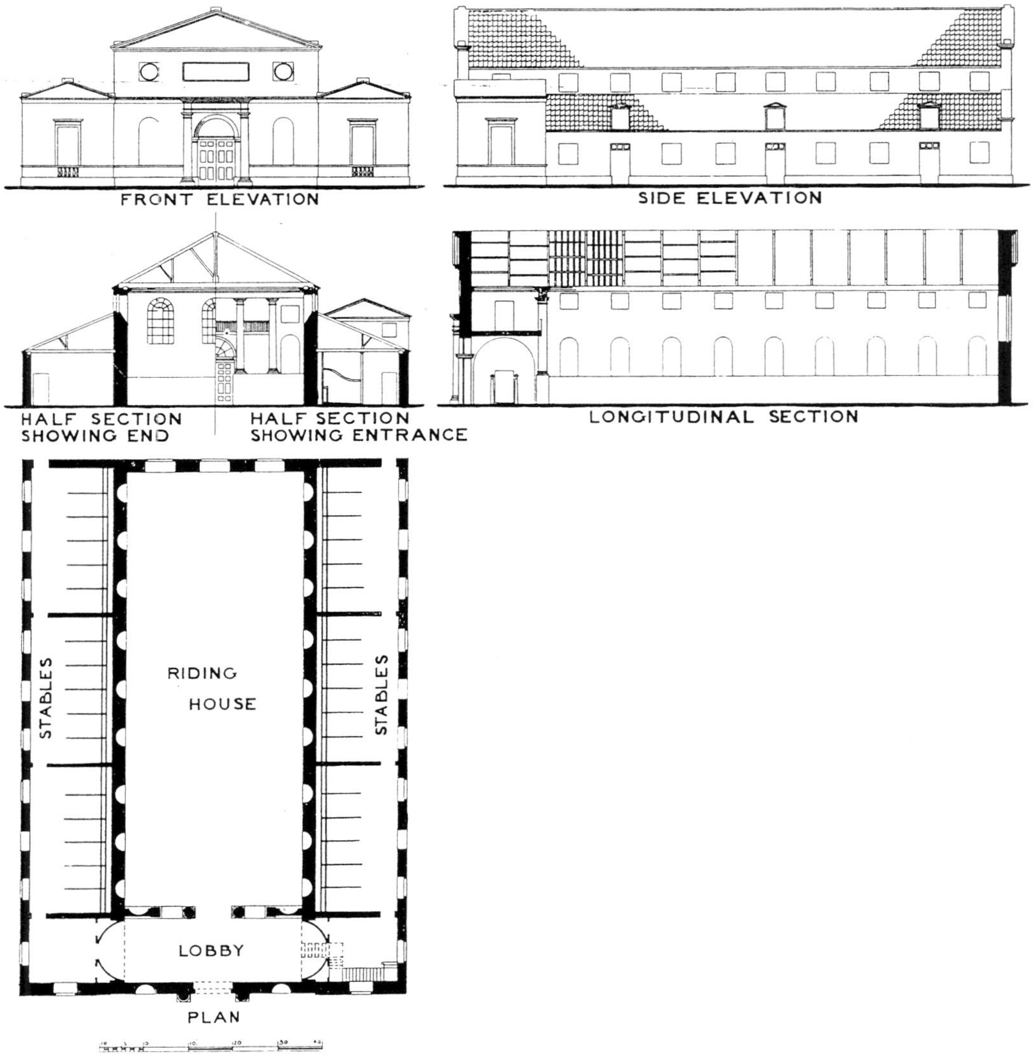

173 Robert Adam, plan, sections and elevations of the proposed Edinburgh Academy, 1763. From A. T. Bolton, *The Architecture of Robert and James Adam* (London, 1922).

academies, this was always a public institution, with the city donating the site on the newly laid out Nicholson Street. Robert Adam provided a plan for a handsome riding house flanked by stables and subscribers raised £2,733 15s towards its construction (fig. 173). A royal charter was granted for 'The Royal Academy for Teaching Exercises' and a warrant for a £200 salary obtained for the master.

Adam's plan was ambitious, with giant Doric columns supporting the gallery and niches along the walls. He also hoped to extend the scheme to form a complete academy with houses for various masters set on a regular plan around the riding house. Adam's grander ambitions were not realised, although his riding house, or something very similar, was built and annual *carrousels* displayed the talents of the pupils. In 1776, the *Weekly Magazine* described the one 'at which gentlemen performed their various equestrian exercises with great dexterity'. Angelo continued to teach at the academy until his death in 1805 but by 1788 it was clearly ailing. In that year, Hugo Arnot complained that 'The advantages accruing from this branch of education are so various that we must express equal applause to the promoters of the institution, and surprise at its not being more generally attended.' By 1830 the academy had folded. The Surgeons' Hall was built on its site.[63]

At least two other academies were established in London and a public academy in Oxford was endowed, though never built. The Oxford academy was suggested by Henry Hyde, Viscount Cornbury, MP for Oxford from 1732 to 1750, who left the royalties from the memoirs of his great-grandfather the first Earl of Clarendon, together with other unpublished papers, to the University to found a riding academy. The earl had vigorously disapproved of foreign academies, declaring in *Concerning Education* that 'Since those places abroad obtain the name and reputation of academies, by their instructing youth in the exercises of riding, dancing and fencing; it be part of the ornament of our learned academies to add those qualities at those hours when more serious exercises should be interrupted.'[64]

Lord Cornbury's plan received the support of the Vice-Chancellor, who believed it was necessary to provide a sufficient variety of innocent amusement in places of education to keep young men free from more pernicious pleasures, and for several months in 1775 Samuel Johnson busied himself about the project. Others were less enthusiastic, declaring that 'A Riding House was given by Hyde/ because his Lordship could not ride', and in the end nothing came of it. Plans to hire a tennis court to act as a riding house were abandoned when none could be found and a project to build a new one for £450 was abandoned when it was found that the fund was insufficient. Eventually in 1868 the money was spent on building the Clarendon Laboratory.[65]

The most interesting of the London academies was set up in 1761 by Thomas Marquois, 'Professor of Artillery and Fortifications', at Norlands, near the Kensington Gravel Pits, on the site of what is now Holland Park Avenue. Marquois listed his facilities and curriculum in an ambitious advertisement that included a plan. As well as Classics, French, writing, mathematics, fortification, navigation, geography, dancing and fencing, riding was offered at two guineas per month. The academy included a riding house 86 ft by 42 ft with an open manege, or 'course where various exercises of horsemanship are occasionally exhibited'. It was not a success and four years later Marquois was forced to auction the lease of the academy, and even the horses and their equipment. The school closed in 1792 when the site was redeveloped.[66]

The other London academy, founded by a Belgian, Lewis Lochée, on the Fulham Road in 1773, was more successful. Lochée described himself in the various books he wrote as 'Master of the Military Academy at Little Chelsea'. His pupils wore uniform, mounted guard and set up camp on Wimbledon Common each summer. A letter to John Grimston of Kilnwick Hall in Yorkshire from Joseph Groves in 1777 explains the thinking behind these military academies, which were favoured by young men planning to join the army. Groves noted disapprovingly that though Lochée's cadets put on the greatest parade, with pupils wearing uniform, standing guard and setting up camp on Wimbledon Common in the summer, they had no riding house. That deficiency was corrected the following year.[67]

The fashion for military-style academies was brief. Angelo left Carlisle House in the 1780s, Lochée's academy closed in 1788/89 and the Edinburgh Royal Academy folded around the same time. Their collapse may have been hastened by the end of the short-lived eighteenth-century revival of *haute école* in Britain. By the time

George III was disabled by porphyria in 1788–89 it was already in decline. In 1796, John Lawrence could write: 'It is well known, that the grand manege has been long out of fashion in this country.' He was echoed by John Adams in 1805: 'How strange then it is, that the art of riding and managing horses should be so much neglected, that very few, indeed, know or think that any such art exists.'[68]

Haute école, though briefly fashionable, owed much to a small group of enthusiasts. With their deaths – Worsley and the Duke of Queensberry in 1778, Berenger in 1782, Morrice in 1785, Calthorpe in 1788, Elwes in 1789, Bute and Medows (who rode daily for two or three hours until within a few days of his death aged 91) in 1792 and Pembroke in 1793 – and the loss of royal patronage, it faded rapidly. Strickland Freeman optimistically dedicated his book to the Prince Regent, but he was never a great horseman and never lent the sport his patronage.

Ironically, just when *haute école* seemed finished, two impressive books on it were published, in 1805 and 1806: John Adams's *The Analysis of Horsemanship*, and Strickland Freeman's *The Art of Horsemanship Altered and Abbreviated According to the Principles of the Late Sir S. Medows*. While Freeman, as his title makes clear, represents the last gasp of the former generation, Adams admits the decline of *haute école* and tries to revive it with a modernised style. Neither seems to have had any effect, although Lawrence, who was so critical of *haute école* in his earlier book, could later write of *The Analysis of Horsemanship*: 'to my most agreeable surprise, I there found moderation and the soundest practical lessons'. In practice, his views had changed little:

> With respect to a comparison of the *manege*, or in the old style, *riding the great Horse*, the English method, my sentiments, published some years since, have sustained no alteration. The grand *manege* is an antique and cumbrous supefluity which ought to be laid aside, or exhibited only in a depository of heavy carriages, and heavy starched apparel. They are all equally incongruous with the rational simplicity of modern taste.[69]

Haute école never recovered its popularity in Britain. M. H. Hayes knew so little about it in 1900 that he described it as 'a system of horse dancing' and had to travel to Paris to see it, while James Fillis, one of the greatest exponents of *haute école*, despite being an Englishman, was forced to make his living on the Continent.

But the end of *haute école* did not mean the end of the need for riding houses. When Richard Berenger wrote in 1771 of the revival of horsemanship, he noted that 'public riding-houses have been opened which are largely encouraged, and frequented by the youth of the nation'.[70] He was not just referring to the newly opened academies but also to riding schools that taught nothing but riding (fig. 174). Most of those visiting riding schools simply wanted to learn to ride, not to master the aristocratic intricacies of *haute école*. As Lawrence put it in 1796: 'The common business of our town riding schools, is to teach young gentlemen and ladies, and to set ill-broken horses upon their haunches.'[71]

The first riding schools were in London, often in riding houses that were no longer needed for their original use. The Duke of Gloucester's riding house in Hyde Park ended up as Fozard's riding school and the riding house built for the second troop of Horse Guards in what is now Riding House Street, off Langham Place, was used for a time as Hall's riding school (figs 168 and 177).[72] Knightsbridge

174 Thomas Rowlandson, *The Riding School*, n.d. (Yale Center for British Art, Paul Mellon Collection; photo: Richard Caspole).

and South Kensington, on the western edge of the more fashionable part of London, were always popular places for riding schools, especially in the nineteenth century. The riding house at Nanning's riding school in Halkin Street off Belgrave Square was one of those surveyed for Lord Grantham in 1823 (fig. 175), while J. A. Preece opened a riding school on the site of Lochée's academy on the Fulham Road, with a new riding house, in 1883.[73]

Outside London, riding schools were established in the fashionable resort towns of Bath, Brighton and Cheltenham. The first riding house in Bath was erected in 1772 in Montpellier Row, behind the New Assembly Rooms, and by 1795 there were three rival riding schools in Bath.[74] In Cheltenham, where there was considerable competition between the rival spas, the Montpellier riding school, built at the same time as that spa, would have been considered part of its amenities. Likewise, in Brighton the planned development of Kemptown in 1825 included a large circular riding house among the regular squares and crescents. The development was not as successful as was hoped and the riding house was never built, but it had clearly been intended as a draw for the new area.[75] In Brighton itself, the Brunswick Market, built in 1828, was enlarged and converted into a riding house for a riding academy, which from 1875 was known as Mr Dupont's Riding Academy.[76]

The function of such establishments was summed up by an advertisement for the original Montpellier Riding School in the *Bath Directory* of 1846: 'Carter's Riding School, Livery and Commission Stables . . . Ladies and Gentlemen taught the Polite Art of Riding by an Experienced Master', and by the description in *The Original Bath Guide* of that in Monmouth Streen: 'extensive and commodious Riding House in Monmouth Street, near the Elephant and Castle, where Ladies and Gentlemen take equestrian exercise when the weather will not permit them to go on the roads or mount the downs'.[77]

Urban riding houses were also used for early circuses, not inappropriately as in England the circus can be seen as the illegitimate, less respectable, but more successful, offspring of *haute école*. Philip Astley, whose talents for trick riding and skills as an impresario stimulated the development of the circus, had been trained in horsemanship by Domenico Angelo when he was serving in the Earl of Pembroke's regiment. Such was Astley's success that by 1779 he was able to build the first dedicated circus building in England, the Amphitheatre Riding House, which despite its name was not a conventional riding house but a circular building.[78] Astley often used existing riding houses on his tours, playing at Bristol and Manchester in this way. At Nottingham he used the Yeomanry's riding house built by William Stretton in 1798.[79]

Riding houses proved inconvenient for the hippodrama, complete with plot, that became popular in the late eighteenth century.[80] To accommodate these, special equestrian theatres, similar in appearance to conventional theatres except for the ring that took much of the space in the stalls, were built in most of the major cities. Examples were built in Liverpool and Newcastle in 1789, in Edinburgh in 1790, in Bristol in 1791, in Manchester in 1793, in Brighton in 1808, in Glasgow by 1810, in Birmingham about 1830 and in Sheffield in 1836.[81] Between tours, or when the circuses fell on hard times, these were sometimes used by riding schools. In 1805, Elizabeth Chivers commented on a visit to Edinburgh that the amphitheatre served as a riding school as well as being used for dancing and tumbling.[82] Similarly, in Newcastle a plan of 1830 refers to a 'Circus and Riding School' near Forth Walk. On the other hand, a separate hall or riding house was added to the rear of Ginnett's Royal Hippodrome in Brighton to serve as a riding school in 1877.[83]

The other continuing use for the riding house was for military training (fig. 176). In Prussia, Frederick the Great had made riding houses central to cavalry

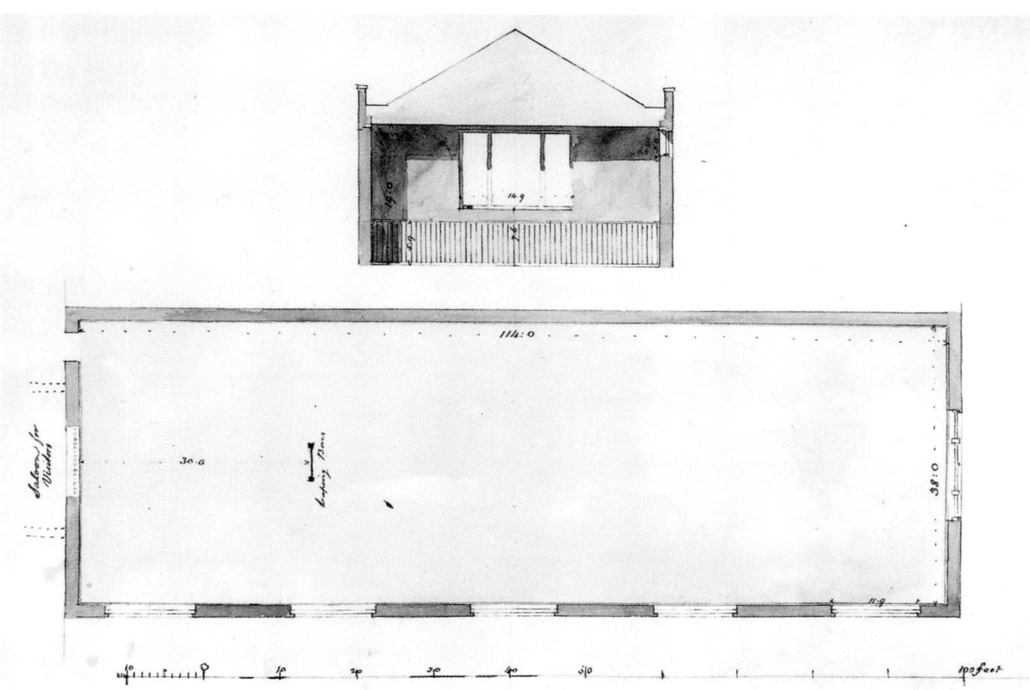

175 Survey plan and section of Nanning's Riding School, Halkin Street, London, *c.*1823 (West Yorkshire Archive Service).

176 Charles Samuel Keene, *An Indoor Military School*, n.d. (Yale Center for British Art, Paul Mellon Collection).

training, insisting on one per regiment. In France, cavalry riding schools with riding houses were organised by the state. Following their example the tenth Earl of Pembroke strongly encouraged the use of riding houses to train military horsemen from 1759, employing Domenico Angelo to teach recruits in his riding house at Wilton House, Wiltshire, and publishing his *Method for Breaking Horse and Teaching Soldiers how to Ride* in 1762. This became the basis of British cavalry training for many decades, but it was left to individual colonels to carry through one of his key recommendations, the use of riding houses. When in 1773 Colonel Blaquiere in Ireland enquired of the commander-in-chief what experience the army had in building riding houses, he learnt that only two riding houses had recently been built in England, one by the King's Own Regiment of Dragoons, the other by the Blues at Nottingham. Both had been built by the commander of the regiment at a cost of just over £400. A set of designs for a riding house and drill hall to be used in dragoon encampments, dated 1773 and made by Christopher Myers, an architect working in Ireland, may be the result of Blaquiere's enquiry. The riding house was to be very plain, 90 ft by 40 ft, and to cost £616.[84]

Unsurprisingly, riding houses remained rare. When Lord Pembroke's protégé Major Floyd was quartered in York for the winter of 1780, he was delighted to have the use of a riding house, the converted former tennis court of Ingram's Palace, for the first time. The following year he was in Manchester, from where he wrote that he had an excellent riding house that had been a great help.[85] Such riding houses were the result of the efforts of individual officers or units, as was the surviving riding house in Oakham, built for the Rutland Fencibles in 1784.[86] But it was not until the Napoleonic Wars that riding houses became commonplace in cavalry barracks outside London.

In response to the threat of French invasion, and of domestic unrest, the early 1790s saw a concentrated campaign of barrack building, which included a string of new cavalry barracks, all, except those intended only for single troops, with a riding house.[87] One of the best examples was Hyde Park Cavalry Barracks

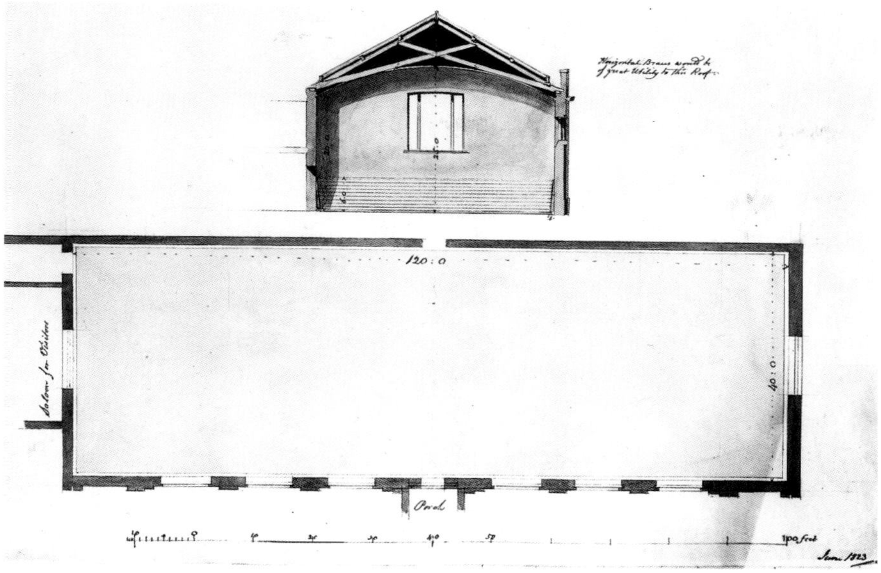

177 Survey plan and section of Hall's riding house, Regent's Park, built in 1752 for the second troop of Horse Guards, 1823 (West Yorkshire Archive Service).

in London, built in 1792–93 for the Life Guards, where a riding house with Diocletian windows was added in 1795.[88] The best surviving examples of this date are the riding houses of the Higher Barracks in Exeter and the Marabout Barracks in Dorchester, both built in 1794.[89] One of the largest examples must have been the London Cavalry riding house, off City Road, shown in Horwood's map of London of 1792–99. Meanwhile at the Royal Artillery Barracks at Woolwich, the largest single barracks in the country housing 3,492 men and 1,718 horses, no fewer than three riding houses, at least one of them designed by James Wyatt, were built between 1786 and 1808.[90]

Most eighteenth-century military riding houses were demolished along with their barracks after the Napoleonic Wars and only three survive from the nineteenth century, despite being a ubiquitous feature of nineteenth-century cavalry barracks.[91] A standard model for military riding houses was drawn up by the Office of the Inspector-General of Cavalry. This recommended that riding houses should be three times as long as they were broad, be high enough for a lance to be carried and have a spectators' gallery for superior officers.[92] Where eighteenth-century riding houses had been little larger than civilian examples – Myers's 1773 riding house was only 90 ft by 40 ft, and that in Riding House Street was 120 ft by 40 ft – the Hyde Park Barracks riding house (considered perfect by Thomas Cundy in 1816) was 180 ft by 40 ft and the St John's Wood riding house, built in 1824–25, was 184 ft by 64 ft. This seems to have become the standard size for the military riding house.[93] The St John's Wood riding house, the largest riding house in the army at the time, is still used for its original purpose. The other surviving nineteenth-century military riding houses are at the Le Cateau Cavalry Barracks in Colchester and the Beaumont Cavalry Barracks, a particularly grand example built at Aldershot in 1854–59 (fig. 214).[94]

8

THE THOROUGHBRED AND THE LATE-GEORGIAN STABLE REVOLUTION

The sixteenth-century revolution in stable design had been triggered by the emergence of a new, expensive breed of horse that needed particular pampering, the coach horse. In the last decades of the eighteenth century the emergence of the thoroughbred, as racehorse and hunter, triggered a second revolution in stable design. Initially at racing studs, then in the stables of keen hunting men, a wave of innovations spread during the 1790s, the 1800s and the 1810s. They became commonplace in the 1820s and 1830s, and were to reach their final form in the mid-Victorian stable.

From the 1750s, the pace of the hunting field began to quicken, particularly under Hugo Meynell, Master of the Quorn Hunt from 1753 to 1800, the first master to introduce quick hunting. It is during these years that the 'flying jump' emerged and in the end even Meynell began to find the pace too hot, declaring that 'he had not had a day's happiness' since the 'splittercokation pace' became the thing. This new pace and the demands of the flying jump required a new fleetness in hunters.[1]

Nor was it just on the hunting field that the demands placed on horses were changing. On the racetrack, the tradition of long-distance races ridden with heavy weights gave way to shorter courses covered at greater speed. The Eclipse, the most famous racehorse of his day, bridges this gap. Foaled in 1764, he raced in conditions similar to those earlier in the century. The races of his immediate progeny differed little from those of the early twentieth century.[2]

The new demands placed on horses both on the hunting field and the racecourse could be met only with an improved strain of horse, the thoroughbred, Britain's greatest contribution to the equestrian race. Improvements were marked over the second half of the century and by 1800, according to George Burrows, the English thoroughbred was the perfect hunting instrument.[3] The result was a rapid rise in horse prices. Before 1770, 200 guineas was hardly dreamed of for a hunter. During Lord Sefton's mastership of the Quorn from 1800 to 1802, 500 to 800 guineas was common.[4]

The change in the manner of hunting was accompanied by a change in the style of hunting. In the first half of the century, most hunting was generally still a modest affair. Meets were informal gatherings of a squire and his friends, like those of Squire Western in *Tom Jones* (published in 1749), and hunt establishments were small. Thus William Somerville (1675–1742), author of the best-known hunting poem of its day, *The Chace*, published in 1735, had a very modest stable: 'four small nags' and 'Old Bill' whom he hunted four times a week – a clear sign that his style of hunting

178 Early nineteenth-century loose box at Osterley Park, Middlesex (William C. Rolf, 1999).

was less punishing than it was to become. Similarly, Squire Draper of Beswick Hall, Yorkshire, the founder of the Holderness Hunt and one of the most renowned early Yorkshire masters of foxhounds, took up foxhunting in 1726 and kept his hounds, horses and fourteen children on £700 a year, which does not argue for an extensive stable.[5] Even a formal pack was run with an establishment that would have been considered absurd by the end of the century. When Sir John Tyrwhitt, Charles Pelham and Robert Vyner signed an agreement in 1713 that was to form the basis of the Brocklesby Hunt the establishment consisted of only three horses and one huntsman.[6]

All this began to change during the second half of the century. As organised hunts replaced informal squires' hunts, hunt establishments grew and so did the number of hunters a gentleman was felt to need. Thus by 1775 the Pytchley Hunt Club, in the heart of England's finest hunting country, needed twelve grooms, two coachmen, four postilions and four men to look after the hacks in the stables.[7] In about 1816, Sir Bellingham Graham gave the Shropshire Hunt an establishment of twenty-six hunters, and by the 1820s the town of Melton Mowbray, Leicestershire, housed three hundred hunters, an average of ten horses to each horseman.[8] One horse was no longer enough. A serious hunting man would need a hack to take him to the meet, his hunter, and a replacement for later in the day. After a hard day's hunting, horses would need resting and so a replacement set would be needed for the next day.

With this rapid improvement in the quality and cost of horses, and in the number of horses that needed housing, came a parallel improvement in stable design. This revolution can be traced in buildings, drawings, books and magazines, with a new wave of writing about stables beginning in the 1780s.

Most eighteenth-century writing on stable design is to be found in books on farriery, horsemanship or more general works, and repeated conventions accepted since the mid-seventeenth century. This changed in 1788 when James Clark, a veterinary surgeon from Edinburgh, published *A Treatise on the Prevention of Disease Incidental to Horses from Bad Management in Regard to Stables*. Although it said little that was new on the specifics of stable design, its scientific approach and the stress it placed on the health of the horse mark it as the first 'modern' book on the subject.

Clark's book was followed by a string of similar works. Four books wholly or partly on stable design were published in the last decade of the eighteenth century, five in the first decade of the nineteenth century, another four in the second.[9] Critical attention at the time was focused on the gentleman's stable, and here the rapid advances were widely praised. William Taplin, writing in *The Gentleman's Stable Directory* in 1791, commented on 'the present improved state of gentlemen's stables in every part of the kingdom, where the mode of management is approaching too near a degree of perfection to admit the aid of instruction, from the pen of theoretic information, or practical experience'.[10] The extent of the new interest in the subject is shown by the fact that by 1796 his book was already in its thirteenth edition.

Though attitudes to gentlemen's stables were positive, this was not always true of other types of stable. 'You inconsiderate and cold-hearted owners of such stables', wrote John Lawrence of urban stables in 1809, 'ye deserve not the services of that

179 Thomas Rowlandson, *Interior of a Stable*, n.d. (Yale Center for British Art, Paul Mellon Collection; photo: Richard Caspole).

generous and noble beast, whose comforts and health you thus sacrifice', before going on: 'But to proceed to the pleasing part of our subject, the comfortable and hospitable British stable, that of the country gentleman.'[11] Lawrence was one of the most balanced writers on stable matters of his day. He praised the best contemporary English stables as far preferable 'in the more essential respects of utility, convenience and comfortable accommodation' to grand Continental stables such as those at Chantilly, the greatest of the French stables. But he admitted that English stables could be improved, as indeed they were over the coming decades.

The most significant innovation of late eighteenth-century stable design was the introduction of the loose box. Later known sometimes as the cage box, this was a pen 10 ft or more square, enclosed on all sides, within which limits the horse has freedom of movement, as opposed to the enforced idleness of the stall. A good early example with elegant tall iron columns supporting the door, which probably dates from the early nineteenth-century, survives in the stables at Osterley Park, Middlesex (fig. 178).[12] The loose box was especially favoured for sick horses and horses in foal, and for horses that took violent exercise in relatively short and concentrated periods, followed by periods of idleness, in particular racehorses and hunters.

It is telling that George Steuart's stables at Attingham Park, Shropshire, designed in 1785 and singled out for their excellence in George Richardson's *New Vitruvius Britannicus* of 1808, had stalls for forty-eight horses, as well as a common stable with no divisions, eight coach houses, three separate saddle rooms, three harness rooms and four grooms' rooms, but no loose boxes (fig. 148). In 1785, even the most sophisticated country-house stable did not have loose boxes. Their rise can be charted in three books by John Lawrence, published in 1798, 1809 and 1829. When Lawrence recommended that all the stables be converted to loose stables in his first book, he clearly considered that the idea would be unfamiliar and strongly opposed by the majority of grooms and horsekeepers. By 1809, he could note that his suggestion

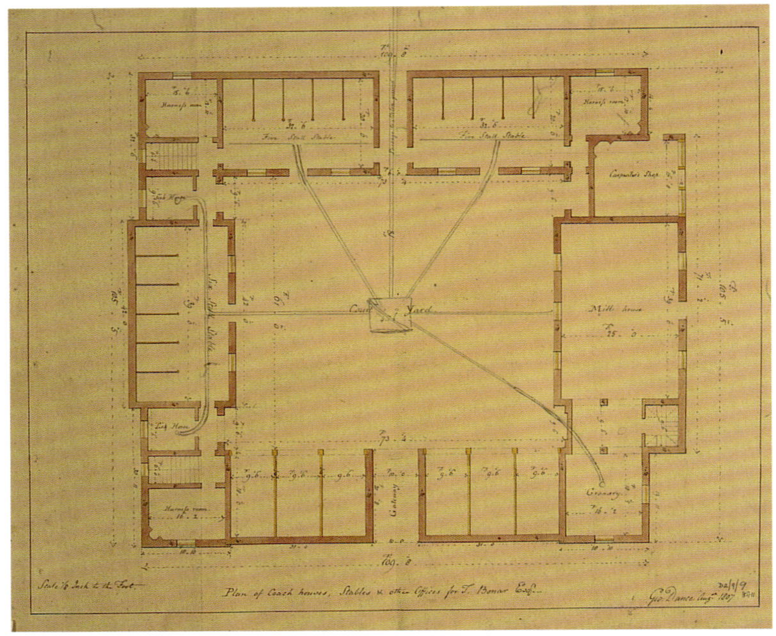

180 George Dance, plan of the stables at Camden Place, Kent, 1807 (by courtesy of the Trustees of Sir John Soane's Museum).

had been put into practice in various stables, 'and I believe proved satisfactory', though grooms still distrusted loose boxes because of the extra trouble they believed they caused.[13] By 1829 he was able to say that the loose box was in considerable use.[14]

Lawrence explained that the lead in the introduction of the loose box came from racing stables, declaring in 1809 that 'in conveniences and comforts for the horse, every system must give place to that of the turf'. One of the principal advantages of racing stables in his eyes was what he called the loose stable or well-littered room or box: 'This may be truly called the grand restorative of the Horse, second only to a run abroad.'[15] The first experiment with such loose boxes that he knew of, 'by a gentleman in Westmorland', must have been almost contemporary with his first book, for in 1829 he observed that this had happened thirty years ago.

The loose box seems to have developed from two sources: from isolated stalls for foaling and for sick horses, and from those in the late eighteenth century who believed that stalls were too restricting for horses. It was obviously sensible for sick horses to be kept separate from the main stable, and in the late eighteenth century their stalls were sometimes replaced by boxes where the horses could move about at will. Such large pens were also useful for foaling. Soane included individual 'loose stables' on a number of his designs for stables, including those at Lees Court, Kent, in 1789 and at Tyringham Hall, Buckinghamshire, and Tendring Hall, Suffolk, in 1793, though these were by no means standard. George Dance's designs for stables at Camden Place, Kent, of 1807, also included two boxes for sick horses (fig. 180).[16]

The first architect who seems to have taken to providing significant numbers of loose boxes instead of the occasional loose stable seems to have been Thomas Cundy, who included them in designs for Normanton Park, Rutland, of 1803 and Tottenham Park, Wiltshire, of 1816. Sir Jeffry Wyatville included loose boxes – which he described as loose stalls – in designs for Endsleigh, Devon, of about 1810, and

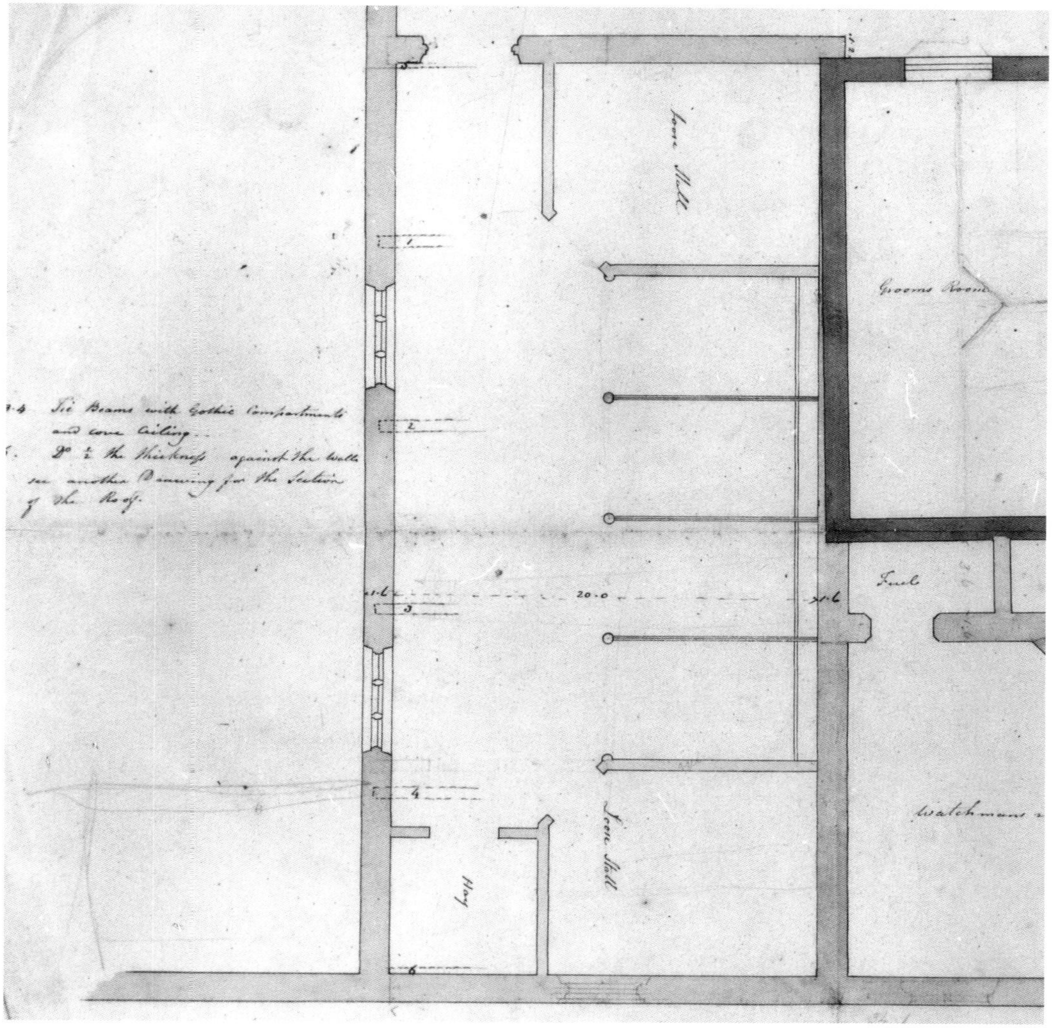

181 Sir Jeffry Wyatville, detail from a plan of stalls and loose boxes at Ashridge Park, Hertfordshire, 1817 (RIBA Library Drawings Collection).

Ashridge Park, Hertfordshire, 1817 (fig. 181).[17] The Ashridge stables, which are particularly handsome, with oak stalls set behind an arcaded screen deliberately reminiscent of Jacobean stables, were designed with a loose box at each end of a row of stalls. This was to become one of the standard ways of arranging loose boxes in the Victorian stable.

At the beginning of the nineteenth century, the loose box remained an innovation that had yet to be commonly accepted. Robert Lugar's 1807 design for what he describes as an ideal hunting stable did not include a single example.[18] But its spread was rapid. In 1818, William Scott, writing in *British Field Sports*, referred directly to John Lawrence when he described the loose box as 'the noblest of all remedies'.[19] By the late 1820s and 1830s the series of stable designs made by William Donthorn show they were common even for smaller country houses in Norfolk.[20] In 1836, J. C. Loudon was able to talk of a loose box 'such as is usually erected for gentlemen keeping hunters'.[21]

The introduction of the loose box was not the only evidence of a revolution in stable design in these years. There was also growing concern about ventilation and drainage. Most theorists had long been agreed on the importance of a well-ventilated stable. Blundeville, Clifford and Morgan writing over two centuries earlier had all believed that well-ventilated stables prevented disease and aided the horse's digestion. Among serious theorists, only Markham had argued against a ventilated stable. English gentlemen, he claimed, had found from experience that it was better to keep a stable 'temperate and proportionable warm'. The 'noisesome smells and vapours' of an unventilated stable could be prevented by burning Storax or Benjamin, sweet-smelling gum resins, twice a day.[22]

The despairing tone of subsequent authors suggests that Markham's beliefs continued to be held by grooms. Thomas Tryon, writing in about 1684, started a litany that would continue well into the nineteenth century when he complained that most stables were kept so close that fresh air was excluded and that few horse-keepers thought once in their lives of the ill they were doing their horses. To prevent this, he wanted large open windows on all sides of a stable, 'that the hot streams may pass away and refreshing breezes freely enter'.[23] Tryon's argument was taken up by eighteenth-century writers, including William Gibson in 1721, John Reeves in 1758 and John Mills in 1776.[24]

Though grooms may have seemed obscurantist in their obsession with blocking every crevice, they were not without reason. Ventilating stables was more complicated than the early theorists acknowledged. Authors implied that it was enough just to open the windows, but fresh air from outside would almost certainly be colder than the air within and thus create draughts that could harm horses. The difficulty lay in arranging a fresh, but not cold, draught-free stable.

Gibson was the first to treat the issue at some length, discussing the problems of draughts and the importance of having a large enough volume of air for the horse to avoid breathing its own exhalation.[25] Though he relied entirely on windows for ventilation, he argued that these should be placed high up in the wall so that any draughts would not affect the horses. At the time that he was writing, stables generally had large windows, like houses. Within ten years these were being challenged by the half-lunette set high up in the wall, designed in the form of a Diocletian window. The Diocletian window was a neo-Palladian motif, but it suited Gibson's requirements admirably. Its introduction is, perhaps, an example of practicality and fashion going hand in hand.

An extreme position was taken by William Ellis in his *Modern Husbandman* of 1750 and later in *Ellis's Husbandry Abridged and Methodised* of 1772. He argued that bad ventilation was so harmful to horses that it was better to keep thick-skinned horses out of doors all the year with only a shelter to protect them, 'for indeed a stable is only a prison for this useful animal'. He even recommended this for hunting and saddle horses, citing an example at Great Gaddesden, Hertfordshire, where such horses were always kept out of doors.[26]

Ellis's radical suggestion was ignored. Indeed, two quotations from the end of the eighteenth century show that the advocates of the well-ventilated stable had failed to persuade grooms of the value of fresh air. 'Upon entering the major part . . . the olefactory and optic nerves are instantaneously assaulted with the volatile effluvia of dung and urine, equal to the exhalation from a stock bottle of harthorn at the shop of any chemist in the neighbourhood', wrote William Taplin in 1788 of London stables; 'every crevice [is] as strongly stopped as if the external air was infectious and its admission must inevitably propagate contagion.'[27] John Lawrence agreed in 1796: 'It is, perhaps, still the fashion to keep our stables too hot . . . neglect of airing stables of all descriptions, is too general . . . Yet when the stable is empty of horses, and enveloped with a hot mist, which makes one's eyes water, I have found the grooms averse to leave even a crack open, whence the foul air might escape.'[28]

James Clark went further in his *A Treatise on the Prevention of Diseases Incidental to Horses from Bad Management in Regard to Stables* of 1788, claiming that close stables were the rule rather than the exception, and were the main cause of equine illness. He clearly struck a chord because fifty years later John Stewart declared that 'previous to his publication people never thought of admitting fresh air into a stable'.[29] Though Clark was in fact only one in a long line of stable theorists attacking ill-ventilated stables, the manner in which he did so marked the beginning of an increasingly scientific attitude towards stable design and heralded a nineteenth-century obsession with stable hygiene. In previous centuries this had only been one among several themes examined by stable writers. Displaying horses to their best advantage was equally important. In the nineteenth century, hygiene, and in particular ventilation and drainage, dominated the literature and preoccupied nineteenth-century stable theorists, as they did those writing on many other types of buildings such as hospitals, schools and housing.

The next forty years saw an increasingly vocal campaign by writers against inadequate ventilation, which coincided with a changing attitude to fresh air in the stable and with the introduction of new methods of ventilation. John Lawrence, writing in 1798, declared that he was 'agreeably surprised this spring, at Epsom, to find the doors wide open at stable-time, and to observe that the horses generally enjoyed a reasonable portion of air'.[30] Not that the battle ever seems to have been entirely won. William Roper, writing in 1844, recorded how he visited a 'great stable' on an intensely hot day, to find, when the stable door was quickly shut behind him, that 'the first respiration, literally and truly speaking, nearly suffocated me'.[31]

One solution was to introduce other forms of ventilation. John Mills, writing in 1776, realised that the hot air of horses' breath rises and could be made to escape through holes in the ceiling, but did not explain how his system would work in practice.[32] His idea was taken up towards the end of the century. Taplin refers to such a ventilator in 1791, again without going into any detail.[33] In 1800, White suggested that adjustable apertures should be placed in the ceiling to let air out.[34] Lugar recommended ceiling ventilators in 1807 and both Richard Lawrence in 1816 and Lawson in 1825 advocated 6 in tubes through the roof.[35]

Country-house architects were not slow in putting these ideas into practice. Soane, whose early stables in the 1780s and 1790s had not included any ventilators, suggested one in his design for a stable at Albury Park, Surrey, in 1805–8.[36] Henry Holland included two copper ventilating spinners in the roof of the stables at Southill House, Bedfordshire, in 1796–1800, and a drawing by Wyatville for the stables at Endsleigh, Devon, of about 1810 is inscribed: 'ventilators to be fixed in the ceiling over each window and door and pipes to convey away foul air out of the sides of the dormer windows'.[37] Such ventilating tubes became common in early nineteenth-century stables. Surviving examples can be found at the Exeter House stables at Newmarket, built in the 1820s. In 1841, Tattersall believed that ventilating tubes through the roof were the generally accepted method of ventilation.[38]

Hand in hand with this greater interest in ventilation went greater concern with drainage, where covered drains in the centre of each stall, linked to a main drain along the back of the stalls, were being advanced as an alternative to the open drains standard in earlier stables. It is telling that Soane, who showed little interest in ventilation or drainage in his earlier stable designs, marks covered drains and a cesspool, as well as a ventilator, on his 1805–8 drawings for stables at Albury Park.[39]

In the late eighteenth century there was also an increase in the width of stalls. Lawrence in 1798, noting complaints formerly made about the narrowness of the stalls, commented that the best stalls no longer suffered from this fault. William Ward, writing in 1776, had recommended that stalls should be 6 ft wide and this became the common width for stalls. George Dance allowed a generous 6½ ft for the stables at Camden Place in Kent in 1807 (fig. 182).[40]

The eighteenth century also saw an increase in the length allowed for a stall. In the seventeenth century, if the rack is discounted, 8 ft was a standard length (although the Theobalds stalls were only 6 ft deep). Robert Smythson's stable at Clifton Hall had 8 ft stalls and so did Pratt's Clarendon House, although his stalls at Kingston Lacy were 7½ ft. By the end of the eighteenth century, 8½ ft to 9 ft was common. It was generally believed that the passageway behind the stalls should be at least as wide as the stall.[41]

Rising standards in stable design can also be seen in the facilities for the breeding of horses. Horsebreeding has always been a favourite occupation of landowners and until the nineteenth century most of the breeding took place on gentlemen's estates, but eighteenth-century country-house stables that included a stud differed little from other country-house stables in design or appearance. The stables of keen eighteenth-century Yorkshire breeders, such as those of John Aislabie at Studley Royal, Thomas Worsley at Hovingham and John Hutton at Marske, were particularly handsome – Worsley noted that the £950 profit he made from selling horses covered the cost of building the new stable and riding house at Hovingham[42] – but nothing marked them out as studs. Indeed, it is striking how often the stables of renowned studs were of no architectural distinction.[43]

The demands of breeding could mean that the stables of some horsebreeders were larger than they might otherwise have been. Horace Walpole, visiting Lord Rockingham's Wentworth Woodhouse, Yorkshire, declared that 'this Lord loves nothing but horses, and the enclosures for them take place of everything'.[44] Certainly the stables, designed by John Carr and built between 1766 and 1783, were the largest of any eighteenth-century country house. But then everything built at Wentworth Woodhouse was on an exceptional scale.

The enclosures that Walpole noted at Wentworth Woodhouse were paddocks, the second element of the stud. Studs needed a central building where mares could foal and horses could be watched, but most of the horses were kept in paddocks where they could run loose. In 1776, William Ward recommended that the field in which the foal was put out should be as large as possible, but most commentators recommended dividing such fields into easily managed paddocks of between one and three acres.[45] (In 1841 Tattersall considered the four-acre paddocks at the royal stud at Hampton Court unnecessarily extravagant.)[46]

182 George Dance, detail design for the stalls at Camden Place, Kent, 1807 (by courtesy of the Trustees of Sir John Soane's Museum).

Horses left in paddocks for long periods needed some form of shelter. At its simplest, this might be a lean-to with a roof and a rack and manger where horses could be fed. The Duke of Newcastle illustrated a simple shelter in his treatise of 1658, the only unusual feature being a pair of doors allowing it to be closed (fig. 183).[47] But the hovels or boxes that Ward referred to were more substantial structures, fully enclosed and properly roofed. Some of the earliest representations can be found in

183 Field shelter, from William Cavendish, *Methode et invention nouvelle* (Antwerp, 1658), pl. 11 (detail) (© The British Library).

paintings by Stubbs of horses at stud. *The Duke of Ancaster's Bay Stallion Spectator*, painted at Grimsthorpe, a renowned eighteenth-century stud, in about 1762–65, and *Dungannon, The Property of Colonel O'Kelly*, dated 1793, have good examples of hovels (figs 184 and 185). Interestingly, though the hovel in the Grimsthorpe picture is a rather crude thatched building, that in the painting of Dungannon is a more regular, tiled construction. As thirty years lie between them, it is possible that this should be seen as part of the rising standards of stable design.

By the turn of the century, considerable thought was being put into the design of studs, as can be seen at Hackwood Park, Hampshire, where Lewis Wyatt designed an integrated stud for the second Lord Bolton in 1817. Here, the stable quadrangle opens into a large paddock, flanked at each end by further paddocks, with brick and slate hovels placed between them so that they could serve either paddock. The rear range of the stable quadrangle was designed as a long covered ride with octagonal turning areas at each end. This could be used for exercising racehorses and hunters.

Lambert Marshall's *Sultan at the Marquess of Exeter's Stud, Burghley* of 1826, shows the regular arrangement of hovels and paddocks typical of the well-organised stud (fig. 186). Similar stables on a much grander scale were designed about the same time at the Royal Stud at Hampton Court, the largest of the nineteenth century. Hampton Court had been a royal stud since the reign of William III, but the Prince Regent revived the breeding of bloodstock there in 1812. By 1815, expenditure of £23,834 had been authorised under John Nash's direction. New paddocks were formed – twenty-six are mentioned – with brick walls instead of wooden pales. New hovels were built and old ones repaired. George IV handed the stud to his brother the Duke of York on his accession in 1820, but resumed possession on the

184 George Stubbs, *The Duke of Ancaster's Bay Stallion Spectator*, 1762–66 (private collection).

185 George Stubbs, *Dungannon, The Property of Colonel O'Kelly*, 1793 (private collection; photo: Bridgeman Art Library).

186 Lambert Marshall, *Sultan at the Marquess of Exeter's Stud, Burghley*, 1826 (Yale Center for British Art, Paul Mellon Collection; photo: Bridgeman Art Library).

latter's death in 1827. A design by Lewis Wyatt dated 1831 for two hovels 15 ft square with 25 ft-square yards, authorised at a cost of £450, survives among the papers of the Board of Works.[48]

The Hampton Court stud is described, generally favourably, by George Tattersall, whose description of Willesden Paddocks, his own stud on the outskirts of London, is the most complete account of an early nineteenth-century stud (fig. 187).[49] Tattersall designed this in about 1840 and the arrangements were considered by Knight in 1844 to be 'among the most perfect of their kind'.[50] Tattersall does not give a general plan of the stud, but it is possible to trace the buildings he shows on the earliest Ordnance Survey maps of the area, the 6 in of 1864 and the 25 in of 1865.

Tattersall argued that the best size for the boxes was 14 ft square. This would allow a mare and foal or two young horses. Each box had its own yard, 30 ft by 25 ft, where the horse could move about when not allowed out into the paddocks, and the yards opened into the paddocks, each paddock serving several boxes. To achieve an efficient use of space and keep down the cost of construction, boxes were most economically arranged in blocks, with several boxes under one roof. Tattersall preferred blocks of four boxes and condemned an eight-box unit at Hampton Court as wasteful of space.[51]

The site of the Willesden stud has long since been built over. Paddock Road in Dollis Hill is its only relict, but the Hampton Court paddocks and boxes survive, hidden behind a high brick wall, though the royal racing stud was moved to Sandringham in 1894.[52] The network of paddocks at Sledmere, Yorkshire, one of the largest, if not the most successful, nineteenth-century studs, also survives. When the entire stud was sold in 1863 there were 313 horses, including 111 brood mares and 55 foals.[53] A patchwork of paddocks with boxes often abutting two or more paddocks can be made out to the west and south-west of the house in the first 6 in Ordnance Survey map of 1850–51, with more having been added by the second edition of 1904. Those that survive are plain, symmetrical brick buildings, very similar to Tattersall's functional designs.

Boxes were also needed to house stallions. These were often built in pairs, 14 ft square with yards 20 ft by 25 ft. Tattersall stressed that these should be made of the best materials, strong and well-seasoned, and that they should be built as solidly as elephant cages at the zoo, or else they would be constantly in need of repair. He suggested that they be placed as far from the mares as possible.[54]

There was also a demand for stables around the centres of the racing industry, such as Newmarket and Epsom. Newmarket had been a focus for racing since the reign of James I and indeed Inigo Jones's earliest recorded buildings are a series of three stables built there for James I, Sir Thomas Compton and Mr Dupper.[55] None survives. Nor do the large stables built for Charles II after the Restoration on Palace Street.

Newmarket's fortunes declined under George I and II, but revived after the foundation of the Jockey Club in 1752, which formalised and enforced the rules of horseracing and gambling. With the headquarters established in the town, a sequence of aristocratic owners built houses and stables, also in the town. John Chapman's plan of Newmarket of 1768 lists five dukes, one marquis, four earls and a viscount. But despite this concentration of wealth, most eighteenth-century Newmarket stables were simple affairs.[56] In 1769 the second Marquis of Rockingham, a passionate racehorse owner who had just began work on the great stables at Wentworth Woodhouse, rebuilt the stables on the house he had bought at

187 Elevation of the paddocks at Willesden, Middlesex, from George Tattersall, *Sporting Architecture* (London, 1841) (© The British Library).

Newmarket.[57] But these bore no comparison to the great quadrangle at Wentworth Woodhouse. Instead, like most of the early stables at Newmarket, they were laid out in the narrow burgage plot behind his house. An idea of what they were probably like can still be seen in the surviving, though altered, eighteenth-century stables at Cadland House. Even the stables built at Queensbury Lodge for the Earl of March, later the notorious fourth Duke of Queensberry ('Old Q') – the spelling of the lodge was changed in the twentieth century – lack any architectural ambition, despite the involvement of the architect John Johnson in 1771. Lord March built the grandest eighteenth-century house in Newmarket, the only one with its own landscape garden, and his surviving stableyard across the High Street was equally extensive, though with neither the symmetry nor the elaboration of contemporary country-house stables.[58] Clearly, display was not considered a significant function of the racing stable, as it was in country-house and hunting stables, perhaps because betting made security, and therefore access to racing stables, a much more sensitive issue in places such as Newmarket.

The first important break with the Newmarket tradition of plain buildings set behind the house on the narrow burgage plot may have been the stables built by the Prince of Wales. The prince was passionate about racing, winning 185 races, including the Derby, between 1788 and 1792.[59] His stables can be seen in William Faden's 1787 revision of John Chapman's map on the north edge of the town on what is now Exeter Road. Their isolated position on the edge of the heath and their apparently regular three-sided quadrangular shape anticipated the nineteenth-century fashion for substantial racing stables set apart from the town, conveniently placed near the gallops.

Stable writers tended to be ambivalent about stables at Newmarket, while acknowledging that many innovations emerged there. Darvill, writing in 1828, was distinctly lukewarm: 'There are tolerably good stables, both public and private, at Newmarket, and at many other places where racehorses are trained . . . yet there are some of these stables so very badly arranged with regard to convenience as to render them very inferior to others.'[60] George Tattersall was even more damning in 1841: 'A stranger in the racing world, would naturally turn his eyes towards Newmarket, as the metropolis of the turf, expecting there to find everything done in superior style. But far from this, Newmarket, like the guide post, shows the way it goes not, and though much, nay, everything may be learned in the way of good advice at Newmarket, no model can be shown, which it would be wise to follow.'[61]

Quite why Darvill and Tattersall were so scathing can be understood by examining Bloomsbury Cottage in Fitzroy Street, a good example of the sort of small-scale early nineteenth-century training stables that were once common at the rear of many Newmarket houses. The appearance of the stable has more in common with farm stables than with the grander racing establishments that were emerging.

One debate that continued for much of the nineteenth century was about the relative merits of stalls and loose boxes for racehorses. Eighteenth-century racing stables had stalls with robust wooden partitions, such as those that can be seen in James Seymour's painting *The Stables and Two Famous Running-Horses Belonging to His Grace the Duke of Bolton* of 1747 (fig. 188). A number of these survive in early stables, including those at Cadland House, though all have subsequently been converted to loose boxes. In 1798, Lawrence declared that loose stables were generally

adopted in sporting stables, but as late as 1828, Darvill had to argue defensively in favour of as many as eight loose boxes in a public training stable with twenty-six stalls. For private training stables he recommended loose boxes for most horses.[62] Though individual noblemen and gentlemen owned the more famous stables, public stables run by a professional groom were well accepted by the early nineteenth century.[63]

Darvill makes it clear that he has no doubt about the superiority of the loose box for racehorses, a good early example of which can be seen in Jacques-Laurent Agasse's *No 3, Wellesley Arabian* (fig. 189). Here the racehorse is shown in a very

188 James Seymour, *The Stables and Two Famous Running Horses Belonging to His Grace the Duke of Bolton*, 1747 (Yale Center for British Art, Paul Mellon Collection; photo: Bridgeman Art Library).

189 Jacques-Laurent Agasse, *No 3, Wellesley Arabian*, *c.*1810 (Yale Center for British Art, Paul Mellon Collection; photo: Richard Caspole).

plain loose box, with solid wooden walls and a high-level corner hayrack. Darvill goes on to explain the contemporary argument that three or four delicate horses standing together would feed better than if kept alone in a loose box. He also notes concerns that horses that travelled around the country to races that were used to loose boxes might find it hard to sleep in the stalls, which were all they could expect at an inn. Despite his preference, even in his design for an ideal training stable he provides only eight loose boxes to twenty-six stalls.

The loose boxes Darvill shows differ from what was becoming the convention for loose boxes in country-house and hunt stables. These, like traditional stalls, were placed within the stable, opening off a corridor. Darvill's loose boxes, like the early loose stables shown in Soane's stable designs, each have a single door opening onto the courtyard. Early surviving examples can be found at Exeter Lodge, which is probably contemporary with Darvill's design. It was to be a long time before loose boxes were universally accepted, but individual boxes were to become increasingly common at Newmarket.

Exeter House also had the first, and for many years only, covered ride in Newmarket, which Tattersall illustrated in *Sporting Architecture* (fig. 190).[64] Covered rides were built to exercise horses in poor weather, particularly when the ground was hard with frost, thus maintaining their high state of fitness. These rides are further evidence of the thoroughbred-driven late Georgian revolution in stable design.

The earliest covered rides, like that at Hackwood, were long, thin structures placed generally at the rear of the quadrangle in which the horse could be trotted up to one end and then back again. John Soane was particularly keen on rides,

190 The covered ride at Exeter House, Newmarket from George Tattersall, *Sporting Architecture* (London, 1841) (© The British Library).

incorporating them in his designs for stables for Mulgrave Castle, Yorkshire, of 1786, Skelton Castle, Yorkshire, of 1787 (figs 192 and 193), Lees Court, Kent, of 1789–90 and Tyringham, Buckinghamshire, of 1793.[65] There was a particularly fine example, now divided up, at the stables designed by Henry Hakewill for Charles Savill Onley in 1825 at Stisted Hall, Essex.[66] Here the ride ran along an open veranda at the front of the stables, with the horses turning at each end in circular protruding bays. Though economical in its use of space, the ride must have made access to the stalls impossible when horses were being exercised.

It must have been difficult to exercise more than one horse at a time in a long thin ride and they were soon superseded by covered circuits. Where the long thin rides had been completely enclosed, these were normally open in the centre. John Carr included what may be the prototype of such rides around the inside of his great circular stables at Buxton, Derbyshire, in 1780, where those taking the waters at the spa could exercise on horseback in poor weather. Lord Belmore considered a similar ring in an early design for his stables at Castle Coole, probably in the 1790s, but this remained unexecuted.[67]

Lawrence was the first writer to recommend the covered ride in his ideal stable of 1798,[68] while the circular Gothic stable that Loudon devised in 1809 for Richard Mytton of Garth, Powys (demolished in 1947) had a 'veranda' round the outside for exercising horses (fig. 191).[69] Both were circular. Detached circular structures can still be found at Belvoir Castle and Exton Hall in Leicestershire, both important hunting stables. Another, now gone, was built at Brocklesby Park, Lincolnshire, the home of the Brocklesby Hunt, in 1866.[70] The Belvoir Castle ring, which was 12 ft wide, was designed by James Eaton with the aid of the estate carpenter. He set out the principles behind it in 1819 when he wrote advising on the possible construction of such a ring at Stoneleigh Abbey, Warwickshire. Twenty horses could be exercised at a time, he explained, and the stable lads said they felt no dizziness as they rode round.[71]

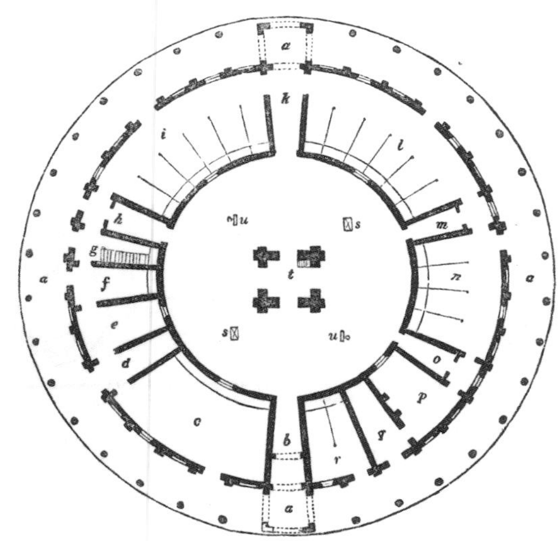

191 Plan of Richard Mytton's circular Gothic stable and exercising ring at Garth, Powys, 1809. From J. C. Loudon, *An Encyclopaedia of Cottage, Farm and Villa Architecture* (London, 1846) (© The British Library).

192 Sir John Soane, the stables of 1787 at Skelton Castle, Yorkshire (William C. Rolf, 2000).

The alternative was a rectangle with rounded corners, sometimes built around the inside of a quadrangle, sometimes in a yard behind it, sometimes completely detached. Joseph Bonomi was one of the first architects to include covered rides in his stable designs. His design for a stable for George Finch Hatton at Eastwell Park, Kent, of 1800, has an 18 ft-wide ride around the centre of the quadrangle. The stable for the Earl of Dartmouth at The Orchard, Blackheath, Kent, of 1801,

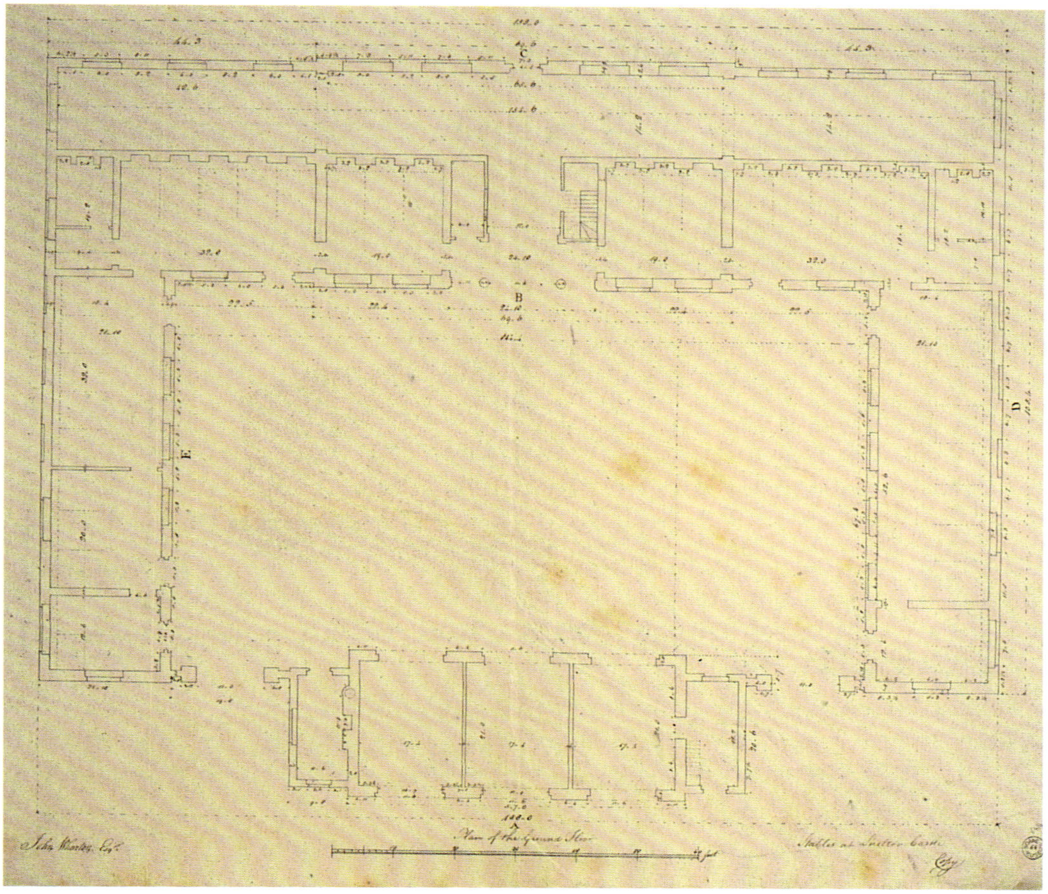

193 Sir John Soane, design for the plan of the stables at Skelton Castle, Yorkshire, 1787 (by courtesy of the Trustees of Sir John Soane's Museum, 64.4.64).

has one attached at the back.[72] The stable at Toddington Park, Gloucestershire, designed and built from 1820 to 1835 by Charles Hanbury Tracy for himself, has a good surviving covered ride and that built for the Duke of Northumberland by Anthony Salvin in the 1850s at Alnwick Castle, Northumberland, is still in use. The Alnwick ride was built to exercise hunters, but covered rides were a more common feature of studs. In 1841, Tattersall included one in his ideal racing stable, although of the stables he recorded at Newmarket, only the Marquess of Exeter's stables had one (fig. 190).[73]

A more flexible, though more expensive, alternative was a riding house. In the eighteenth century, riding houses had principally been the preserve of enthusiasts for *haute école*, the army and occasional public riding school. Around the turn of that century they became symbols of aristocratic extravagance, included principally for show, perhaps in emulation of the Prince Regent who had large riding houses at Carlton House, London (1788–91) and the Royal Pavilion, Brighton (1804–08; see figs 194 and 195). It was probably the scale of the riding house that appealed to the Prince Regent, who made it clear that the grandest stable, if it was to be considered complete, needed a riding house not tucked away, but at the centre of the stable where its bulk could be appreciated. This may explain why so many of the great stable projects of his day were designed with riding houses, although in the end prudence often meant that they were not built. Both Benjamin Wyatt's and

194 Exterior of John Nash's stables of 1804–8 at the Royal Pavilion, Brighton. From John Nash, *The Royal Pavilion at Brighton* (London, 1826).

C. R. Cockerell's designs of 1815–18 for the grandest architectural scheme of the early nineteenth century, the palace for the Duke of Wellington, included a riding house. Similarly, Wyatville's grandiose early design for the stables at Longleat, Wiltshire, of 1800, and Cundy's first designs for stables at Middleton Park, Oxfordshire, of 1805, and Tottenham Park, Wiltshire, of 1816, all included riding houses that were excluded from the final design.[74]

In practice, most riding houses were built only where they were needed. Sir Henry Harpur in 1768 at Calke Abbey, Derbyshire, was the first private individual to build a riding house that was not intended for *haute école*.[75] Harpur was a keen breeder of racehorses and as he did not build his riding house for *haute école*, he must have built it to help train his racehorses. The same is true of the second Marquis of Rockingham, whose extensive stable complex at Wentworth Woodhouse, Yorkshire, included a riding house built in 1780–86, and the Duke of Richmond who built a riding house in 1783 at Itchenor House, Sussex, his marine villa and racing stables on the coast near Goodwood.[76] At Woburn Abbey, Bedfordshire, Henry Holland's work for the fifth Duke of Bedford (of whom it was said that 'the turf seems to have engaged his early and earnest attention'[77]) not only included a large riding house in 1790 but paddocks, hovels and a new house for the stud groom.

Elsewhere, riding houses were built by keen hunting men. The Earl of Lonsdale, who kept fifty hunters, included a new riding house when Robert Smirke rebuilt Lowther Castle in 1806–10. Stapleford Park, Leicestershire, is in the middle of the country's best hunting land and the riding house there was probably built to exercise hunters by the fifth Earl of Harborough. The riding houses at Hornby Castle, Yorkshire, built before 1838, and the early nineteenth-century Firle Place, Sussex, were also probably built by keen hunting men early in the century.[78] For these men, riding houses and covered rides were alternatives, as can be seen in the way James Leigh toyed with the idea of a covered exercising area at Stoneleigh Abbey, Warwickshire, in 1819, though he ended up building a riding house because it was more adaptable. But a riding house was more expensive than a ride and at Newby Hall, Yorkshire, Earl de Grey considered building a riding house in 1823 before settling on a covered ride instead, probably for reasons of expense.[79]

The early nineteenth century also saw the development of specialised hunting boxes and hunt stables. There was nothing new about the idea of a hunting lodge, a small, convenient house with attached stabling built by a landowner at a distance from his main house or on a secondary estate used principally for hunting. An important early example was the early seventeenth-century lodge at Wothorpe, Cambridgeshire, built by the Earl of Exeter. Another, on a rather grander scale, was John Webb's Amesbury Abbey, Wiltshire, in the middle of Salisbury Plain. Built in about 1659, this developed the idea of a single double-height central room (for post-hunt feasting), flanked by apartments. William Kent published this plan in the *Designs of Inigo Jones* in 1727[80] and subsequently used it for the grandest of eighteenth-century hunting boxes, Wakefield Lodge, Northamptonshire, in 1748.[81]

195 The riding house in John Nash's stables of 1804–8 at the Royal Pavilion, Brighton. From John Nash, *The Royal Pavilion at Brighton* (London, 1826).

Wakefield Lodge was designed for the Duke of Grafton, who was particularly fond of hunting, and indeed died in 1757 after a fall while hunting. This was a house devoted to the joys of the chase, as the chimneypiece in the hall with the head of a fox and a badger, the hunters' prey, makes clear. Kent also built extensive

stables, which, with fifteen bays fronting a large quadrangle, comes close to dwarfing the house (fig. 196). Though the central three bays were for coach houses most of the building was given over to stabling. The plan was unusual, with a series of individual stables opening not into the yard but to the front of the building with the stalls placed along the side walls, not the rear wall as was conventional. Above were numerous rooms off a central corridor. Their number suggests that they were intended not so much for hunt staff but as overnight accommodation for the duke's hunting guests, presumably the reason for the unusual stabling below, also only intended for brief occupation.

Wakefield Lodge lay in the long tradition of the aristocratic hunting lodge set alone on an estate owned by its builder. Less conventional was the lodge built for the Duke of Richmond at Charlton, Sussex, in 1732 (fig. 197). In the early eighteenth century, Charlton was home to the most fashionable pack of hounds in the country, its popularity with visitors from beyond the locality anticipating the fashionable Shire packs of the nineteenth century. These visitors were forced to hire rooms in the village, but the duke built himself a small Classical lodge so that he could sleep in comfort.[82] It was a model that would become common a century later when keen hunting men took to passing the winter away from home in hunting country. Sir Thomas Mostyn's estates, for example, were in Wales, but during the hunting season he lived at Swift's House near Bicester in Oxfordshire and served as Master of the Bicester Hunt for the first three decades of the nineteenth century. Nimrod described his stables, which formed a quadrangle with a huntsman's house, kennel and blacksmith's house outside, as uncommonly good.[83] The hunting lodge that Nimrod considered 'the most complete hunting box in England, perhaps in any other country', was Newton House, Yorkshire, built by Ignatius Bonomi for the third Earl of Darlington in 1822. Darlington, created Duke of Cleveland in 1833, was one of the most fanatical huntsmen of the century. 'To the title of sportsman who can have a better claim?' wrote Nimrod. 'It is as legitimate as his Earldom, and he must pay reverence to his name.' He was a master of foxhounds for fifty-three years and hunted a vast country from his Durham seat of Raby down into Yorkshire. Newton House was his southern base and had stabling for about thirty hunters besides coach houses and kennels.[84]

Not only masters of fox hounds built hunting lodges. W. H. Lambton is said to have taken the first house for hunting at Melton Mowbray, the 'Queen of the Shires', in about 1787. This town, whose favoured position meant it commanded the Quorn, Belvoir and Cottesmore hunts, probably had the greatest number of hunting lodges. Jack Brownlow's history of Melton Mowbray lists more than twenty-five.[85]

In such prime hunting land, a man might hunt six days a week in the season, changing horses each day, which meant that he needed a stud of perhaps a dozen horses, as well as covert hacks to take him to the meet. So it is not surprising that dedicated men who spent the whole winter at Melton should build special lodges for themselves and their horses. Some of these were on a scale to rival the stables of the hunt itself. Park House in Melton Mowbray, rebuilt in 1831 by the Earl of Wilton as Wilton House, had stabling for forty hunters. Most hunting lodges were smaller.[86] More typical was Sysonby House, which was advertised for letting in 1789 with stabling for sixteen horses.

196 William Kent, the stables of 1748 at Wakefield Lodge, Northamptonshire (William C. Rolf, 1998).

197 The Duke of Richmond's Lodge of 1732 at Charlton, Sussex (The Landmark Trust).

The development of formal hunts brought with it the need for dedicated hunt stables. Some hunts continued to be run from the country houses of dominant local landowners, such as the Brocklesby, run from the Earl of Yarborough's Lincolnshire seat, Brocklesby Park (figs 241 and 242). More generally, and increasingly, hunts were funded by subscription. Even when a hunt was paid for or run by an individual, that person was no longer necessarily a local landowner with his own stables from which to run the hunt, but often a wealthy outsider with no estate in the area. The hunt establishment had become separated from that of the squire and needed its own stabling.

The need for specialised hunt stables was soon realised by the pattern-book architects. Robert Lugar's *The Country Gentleman's Architect* of 1807 included a design with kennels, hunting stables for twelve horses and a cow yard, though this lacked the degree of specialisation that later came to be expected.[87] In particular, it had neither loose boxes nor sick boxes. Thereafter, designs for hunting stables were a regular feature of books on stables.

198 J. F. Herring, *The Hunting Stud*, 1845 (Tate, London).

In outward form and architectural style, hunting stables had much in common with country-house stables. They were markedly more elaborate than racing stables. The most important distinction was the lack of coach houses and stalls for coach horses, although some provision was occasionally made for the master's carriage. In their place were more stables, which were increasingly fitted out with loose boxes. It was claimed that the great huntsman Assheton Smith had a loose box for every hunter in his stables at Tedworth, Hampshire, in 1829, though this appears to be untrue. Opposite the page in which J. E. Eardley-Wilmot claims this in his reminiscences of Assheton Smith is a plan of the stables with fourteen stalls specifically labelled as for hunters and only five loose boxes (numbers confirmed in the plan Tattersall published in 1841).[88] Nevertheless, hunting stables had an increasing preponderance of loose boxes. Tattersall published two further plans that he recommended, that of the Quorn Hunt's Billesden stables, built by Thomas Smith for Lord Suffield in 1838 with thirteen boxes and twenty-four stalls, and the Warwickshire Hunt stables of 1839, which had ten loose boxes, nine stalls for hunters and four stalls for hacks (figs 199 and 200).[89]

There has always been a distinction between the grand Shire hunts, most of them clustered around Leicestershire, which traditionally attracted 'foreign' hunting men, and local hunts such as the Meynell that were largely patronised by those who lived within the hunt boundaries. Hunting has always been expensive, but the Shire hunts were run on a far grander scale than the local hunts and were therefore a fine way to show off great wealth. Many of the masters who ran such hunts built stables on a scale intended to impress. Sometimes they appear no better than seventeenth-century noblemen anxious to present all their horses in one *coup d'oeil*, contrary to all the practices of good stable design. Tattersall condemned the stables at Quorndon Hall, built by Lord Sefton, master of the Quorn in 1800–1 and 1804–5, which housed twenty-eight horses in an undivided stable.[90] However, the great hunting

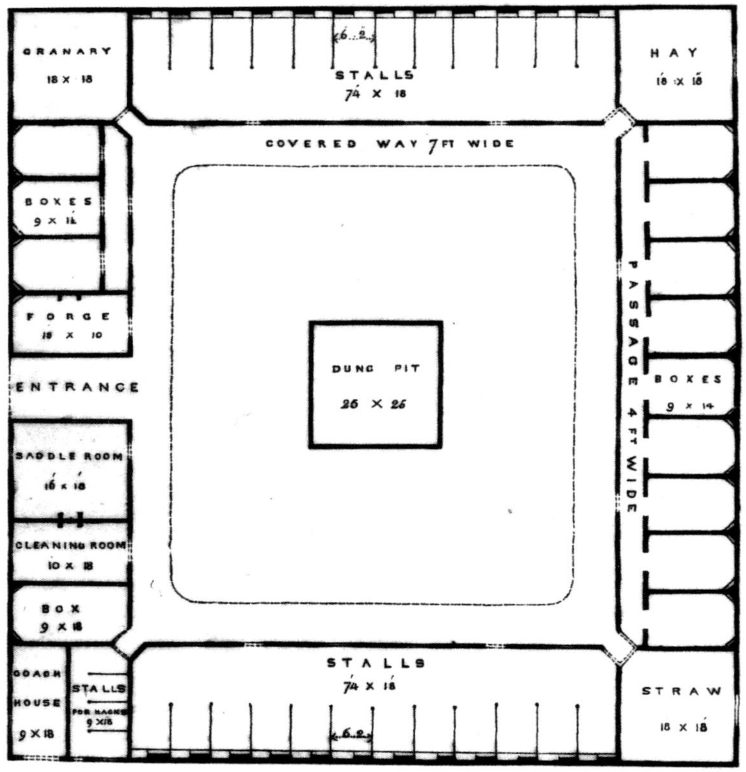

199 Plan of the Quorn Hunt stables of 1838 at Billesdon, Leicestershire, from George Tattersall, *Sporting Architecture* (London, 1841) (© The British Library).

journalist Nimrod talked of these stables with awe: 'When I cast my eye upon it, I could not help fancying myself looking at King Solomon's stud, having never seen so many horses in one stable before.' Sefton's stables were only one element of his extravagance. According to Nimrod, Leicestershire never witnessed greater splendour than during his mastership: 'The splendour of his establishment gave to the spectators more the idea of an imperial hunting party in a foreign country than that of an English pack of fox-hounds. It brought to our recollection Dido's hunting party to the godlike Aeneas.'[91] Tattersall was relieved to say that the Quorndon stables had been divided up by 1841.

But not all hunts could find a master wealthy enough to build stables. The Warwickshire Hunt stables at Kineton, which were designed by Hugh Williams, one of the hunt members without the assistance of an architect, just as Assheton Smith had designed his Tedworth stables, were built in 1839 in a spirit of local cooperation. George Lucy of Charlecote provided the site and costs were kept down to £2,510 for the stables and kennels by 180 farmers providing 553 wagon loads of haulage.[92]

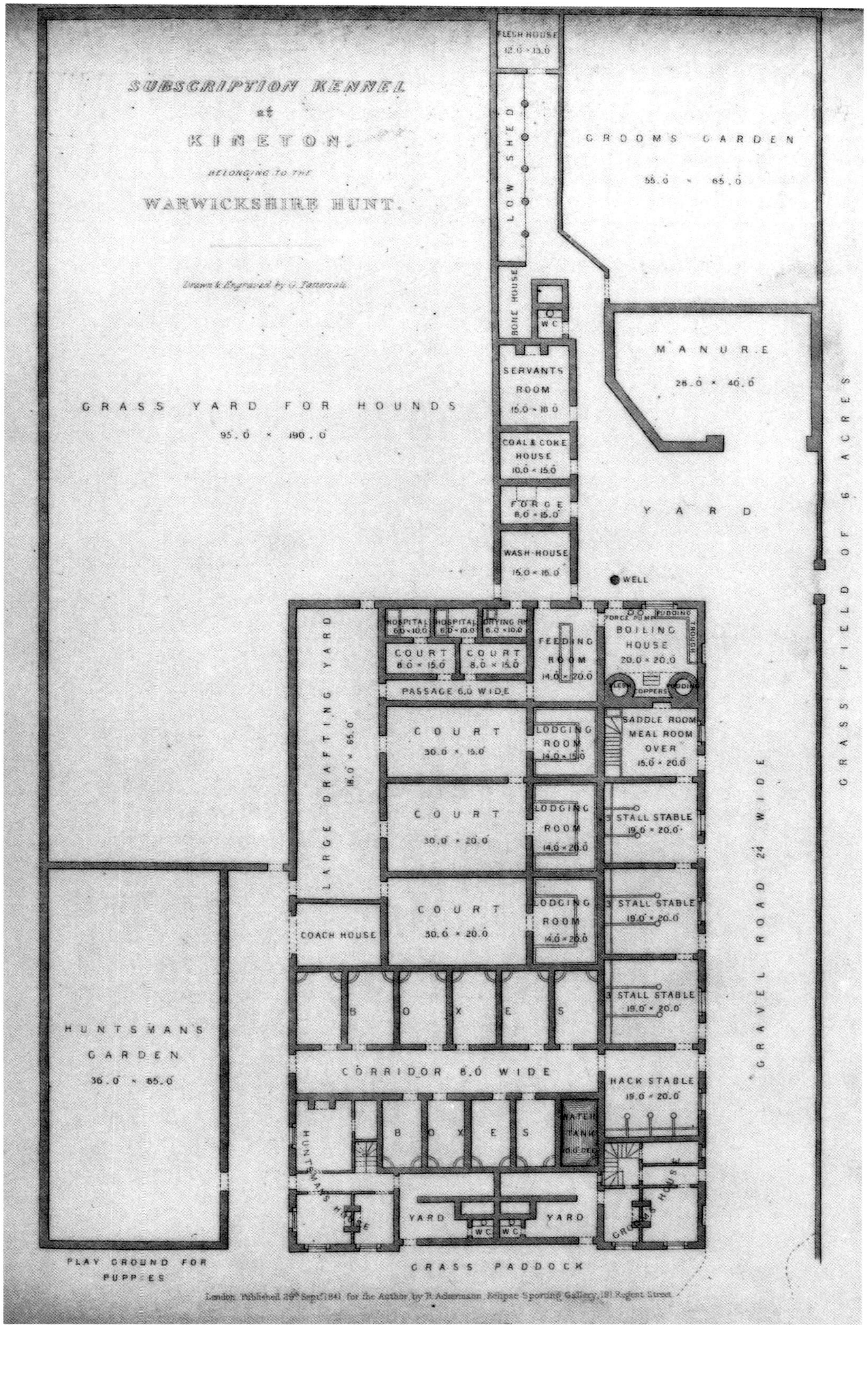

200 Plan of the Warwickshire Hunt stables of 1839, from George Tattersall, *Sporting Architecture* (London, 1841) (© The British Library).

As Britain's wealth and population increased during the nineteenth century, the horse attained an unparalleled level of importance. Though steam engines robbed the horse of its role in long-distance transport and as a power source for machinery, the resulting surge in productivity and prosperity brought ever-greater demand for horses, both directly from industry and transport, and indirectly thanks to the wealth that accrued. Until the invention of the internal combustion engine at the end of the century, the horse was the primary means of short-haul transport, whose volume increased exponentially with the success of the railway. Farming, enjoying a prosperity seldom if ever equalled in the middle years of the century, was dominated by the horse. The fast-expanding cities required increasing numbers of horses to pull the omnibuses and cabs that made life in them practical.

9

COMMERCIAL, INDUSTRIAL AND MILITARY STABLES

The army, maintained at a size unparalleled in peacetime (if small by comparison with Continental armies), numbered its horses in tens of thousands. And growing numbers of the upper and middle classes could afford more carriages and hacks, hunted more often, and with more horses, and owned more racehorses. All these horses needed to be housed, sometimes in substantial numbers. Before 1800, the history of the stable is dominated by the country house, but though in the nineteenth century country-house stables achieved an opulence that exceeded anything found before, in many ways the most interesting developments came elsewhere, in industrial, commercial and military stables.

The statistics are imprecise, but in 1811 there were probably 1,287,000 horses in England: 800,000 used in agriculture, 251,000 used commercially and 236,000 in private ownership. By 1871 this had risen to 2,112,000, with 1,254,000 on farms, 444,000 in commercial use and 414,000 privately owned. Numbers peaked around 1901 with 3,276,000 horses, 1,511,000 on farms, 1,166,000 in commercial use and 600,000 privately owned. In late Victorian Britain, one horse was needed for every ten people. As Professor Thompson notes: 'The railway age was in fact the greatest age of the horse, albeit in terms of its total contribution to the economy rather than in terms of any heroic qualities in its unaided achievement.'[1] Although the number of horses on farms nearly doubled and the number of privately owned horses more than doubled, the dramatic increase was in the number used commercially.[2] These more than quadrupled.

Of all these uses, farming was the most ubiquitous, though given the nature of farming it was also the least impressive in terms of stabling. In the eighteenth century, horses had largely replaced oxen, which did not need stabling, as the main draught animals. Horses were needed on all farms, to plough, and in due course to pull mechanical reapers and for transport. It has been calculated that one horse was needed for every twenty arable acres, though the lighter the soil and the larger

201 Charles Cooper Henderson, *Interior of a Post-Horse Stable with Horses Feeding* (detail of fig. 215).

C
62
A
A

the farm, the fewer the horses needed per acre. Smaller farms still needed horses, so even a fifty-acre farm would usually have had four horses.[3] On large farms they could easily be in double figures. Arthur Young, writing in 1804, cites one 1,000-acre farm on the Holkham estate in Norfolk that had one horse for every thirty acres, which meant more than thirty horses.[4] Breeding horses for sale was also a significant part of an arable farm's economy, as Collins showed in his analysis of the way farms supplied cities with their haulage horses in the nineteenth century.[5] All needed to be housed. As John Hudson, a great advocate of progressive farming, wrote in 1850: 'The keeping of cart horses is a very important item in the economy of the farm: they require at all times to be in good condition and always ready to do a good day's work.'[6]

Despite the example of the stables at Holkham Hall, Norfolk, built for the first Earl of Leicester and illustrated in Matthew Brettingham's *The Plans, Elevations and Sections of Holkham in Norfolk* in 1773, which included stabling for carthorses, country-house stables were seldom built as one unit with the home farm in England (fig. 202).[7] John Martin Robinson's gazetteer of Georgian model farms includes only a single example of a combined home farm and domestic stable in England, at Moccas Court, Herefordshire, built in 1783–84 by Anthony Keck for Sir George Cornewall.[8] To this could be added Timothy Lightoler's stables at Burton Con-stable, Yorkshire, of about 1760, which were designed with a double quadrangle, the first for the domestic stables, the second for farm buildings.[9] In Scotland and Ireland, the home farm and stables were often built together, reflecting a major difference in agricultural practice, with the home farm exercising an importance in Scotland and Ireland that it did not have in England.

Scottish examples of combined country-house stables and home farms are numerous, with Linlathen, Angus, built for Robert Graham in 1770, Playfair's Ardkinglas House, Argyll, built in 1790 for Sir Alexander Campbell, and Keithick, Perthshire, built in about 1818 for James Wood by David Whyte, among the better documented.[10] Examples of double quadrangles combining stables and farm buildings, like those at Burton Constable, can also be found in both Scotland and Ireland. Among them are the stables at Kennet House, Clackmannanshire, built for Robert Bruce in about 1793, James Gandon's extremely fine stables at Carriglas Manor, County Longford, of about 1790–95 and the magnificent complex at Castle Coole, County Fermanagh.[11]

In Scotland, a single quadrangle often combined the stables and steading. In these cases it was normal for the front range to be devoted to the domestic stable, and the quadrangle behind to be given over to farm buildings, which were often dominated by a doocot. James Ramsay's 1797 design for the steading at Rednock House, Perthshire, shows this well. On each side of the central arch lay a four-horse stable, beyond which were coach houses opening forwards rather than into the quadrangle. Behind, around the quadrangle, lay a common stable, a Dutch byre, a car house, hen houses, pigsties and a brewhouse. In the middle was the dung pit.[12] This allowed the combined steading and stable to attain an architectural distinction that, particularly on smaller estates, the stables or the farm building on their own would not have achieved.

On eighteenth-century model farms, much attention was paid to the architectural composition of the farm buildings, in which the stables were one important,

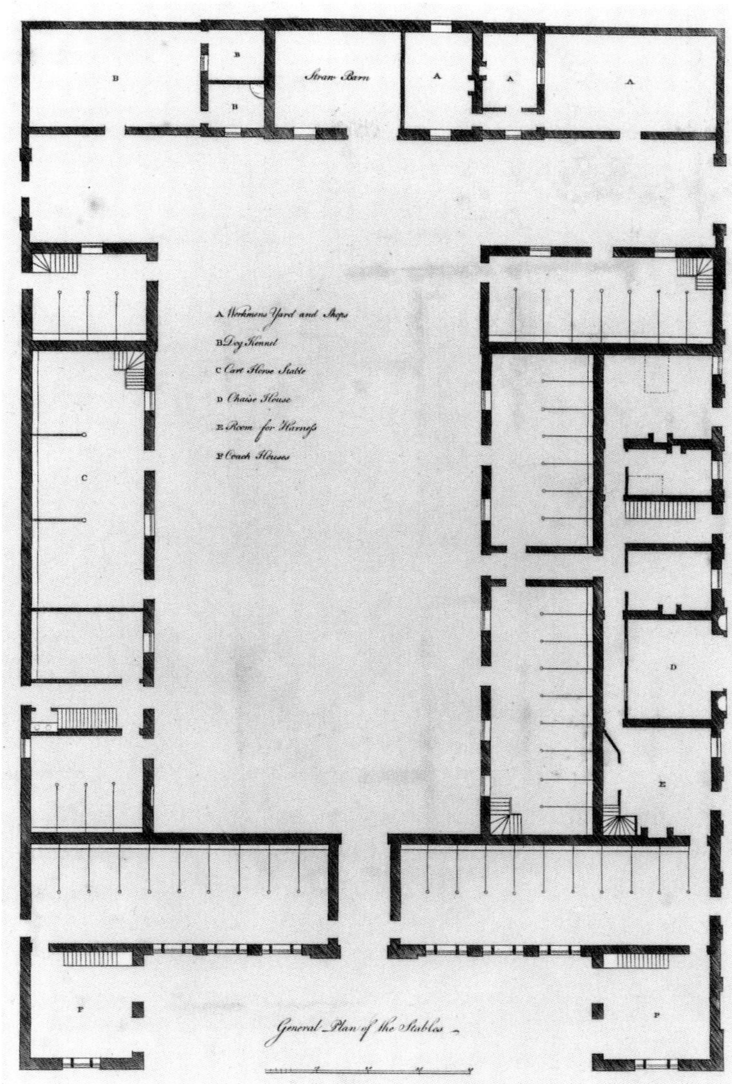

202 Plan of the combined house and farm stables at Holkham Hall, Norfolk, from Matthew Brettingham, *Plans, Elevations and Sections of Holkham in Norfolk* (London, 1773), pl. 65 (© The British Library).

though not predominant, unit. At Culzean Castle, Ayrshire, Robert Adam's ambitious stable block for the Earl of Cassilis's home farm, built in 1777–79, housed thirteen horses. Cadland Farm, Hampshire, designed by John Soane while he was still working for Henry Holland in 1777, had two carthorse stables for sixteen horses as well as separate riding-horse stables (fig. 203). In the nineteenth century, attitudes changed. Architectural display in farm buildings was seen as extravagant and was discouraged in favour of more utilitarian buildings, though Susanna Wade Martins notes one particular architectural trait in her survey of Norfolk farm stables, the use of semi-circular lunette windows.[13] This was probably a vernacular transfer of the Diocletian windows common on neo-Palladian stables, most famously at Houghton Hall, Norfolk.

Large numbers of horses were held on the bigger nineteenth-century farms. Outchester Farm, Northumberland, on the Greenwich Hospital Estate, had stalls for fourteen carthorses and five riding horses in 1837; Northbrook Farm, Kirklington,

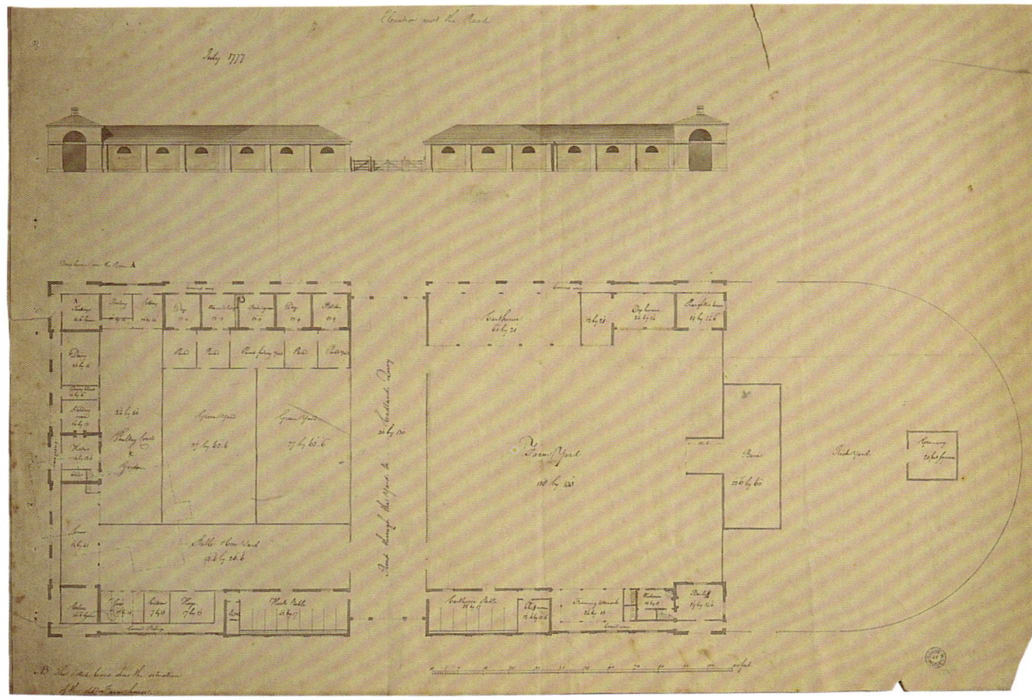

203 Sir John Soane and Henry Holland, plan of Cadland Farm, Hampshire, 1777 (by courtesy of the Trustees of Sir John Soane's Museum, 64.6.41).

Oxfordshire, of 1858 had stalls for fifteen carthorses; Flemish Farm in Windsor Great Park of 1863 had thirteen stalls and a loose box.[14] Perhaps surprisingly, despite the increase in the amount of machinery available to farmers after 1850, which might be expected to lead to a demand for more stabling, there is little sign of extra stables being added. On many farms in the nineteenth century, the existing eighteenth-century stables remained adequate, though where new stables were built after 1800 they tended to be single storey and not to have a hayloft above, the result of a growing concern for ventilation already seen in country-house stables.[15]

Most of the stables on model farms were designed as single units with a long row of stalls, much like a coach-horse stable at a country house. The alternative was to keep horses in smaller units, with the mangers along the side wall or walls rather than the back wall. This had the advantage of minimising the space needed for stabling, but carried with it the danger that the horse furthest from the door and window would get little air. This system was commonly found on smaller farms, but occasionally even on the most ambitious farms, such as Egmere Farm on the Holkham Estate in Norfolk, built between 1850 and 1856 at a cost of £5,500, the most expensive farm built on that keen agricultural estate. Here, all the stabling was divided up into separate units, with one stable for three riding horses, four loose boxes (one of them double, presumably for breeding) and three separate cart-horse stables. These must have housed about eighteen horses, although it is difficult to be certain as, unlike the riding-horse stable, these did not have partitions between the horses.[16]

As Egmere Farm shows, it was not unusual in cart horse stables for two horses to share one, slightly wider, stall, or for several horses to be tethered together. This was particularly common on smaller farms and continued throughout the nineteenth century. William Flint, writing in 1815, considered it normal for two horses to share a stall and not uncommon for a farm stable to have no partitions.[17] Shared stalls were considered acceptable because farm horses always worked in pairs and therefore would not create a disturbance if stalled with their team mate. On Welsh farms, single stalls were only marginally more common than double stalls.[18] But single stalls were always preferred by theorists, as Dickson makes clear, to prevent the risk of horses kicking each other.[19]

Economy, not display, was the driving force behind farm stables as can be seen in George Morland's painting *Old Horse in a Stable* (fig. 204). Fittings were generally robust and inexpensive, though those for riding horses, needed when farmers were supervising work in the fields, were often more elaborate. Instead of the glazed windows standard in country-house stables, sliding wooden shutters were often used, which were both cheap and allowed ventilation to be controlled. As carthorses would usually be out working during the hours of daylight and as their appearance did not need to be so carefully prepared, good lighting was not as essential as it was in a country-house stable. Surprisingly, James Paine had recommended sliding wooden shutters for the windows to the courts and passages of the stables at Chatsworth in 1758 as a cheaper and more durable alternative to sashes and glass. Given the grandeur of the Chatsworth stables, probably the most architecturally ambitious private stables ever built in Britain, it seems a strangely penny-pinching suggestion.[20]

The early nineteenth century was also the high point of the canal and coaching trades, both of which were dependent on stabling vast numbers of horses. Until

204 George Morland, *Old Horses with a Dog in a Stable*, *c.*1791 (Yale Center for British Art, Paul Mellon Collection).

the widespread use of powered boats early in the twentieth century, horses pulling boats were vital for canals. Many stables were built along the canals to accommodate horses overnight or at canal depots, particularly at wharfs, docks and long flights of locks.

Eighteenth-century canal stables had often been simple affairs, open-sided thatched shelters of a type that survived on the River Thames until the end of the nineteenth century. On nineteenth-century canals the standard was higher, with solid brick structures, similar in many ways to farm stables. Many were small, but at key sites they could be much larger. On the Shropshire Union Canal, stables varied in size from four to twenty-two stalls, but on the busy Birmingham Canal Navigation, substantial stables were built with stalls on two levels. Canal stables were often built by the companies running the canal, or by companies that used them, such as Pickfords, which always had well-designed stables. On the Birmingham Canal Navigation, the Incorporated Seamen and Boatmen's Friendly Society even provided a series of free stables at busy lock flights.[21]

Before the invention of the railway, the horse was the only means of long-distance travel for individuals and inn stables were as necessary to service horses as the petrol station is for the car today. These stables had two functions: to provide temporary accommodation for travellers' horses and permanent stabling for the horses used by public coaches or hired by private individuals for a single stage. The number of private travellers and their horses using inns is hard to estimate, but the amount of stabling required must have been substantial.

The impact of the public carriage horse is more easily assessed. In the second half of the eighteenth century, improved roads and coach design revolutionised the speed and comfort of coach travel. In 1761, a coach to London from York travelling in summer took 72 hours. By 1772, a flying coach took 36 hours. That had been cut to 20 hours by 1836. The inevitable result was a rapid increase in coach

205 Charles Cooper Henderson, *Interior of a Post-House Stable, with Horses Feeding*, n.d. (Yale Center for British Art, Paul Mellon Collection; photo: Richard Caspole).

206 Thomas Rowlandson, *The Stable of an Inn; Four Horses Being Fed, Watched by a Man, a dog and Two Stable Boys*, *c.*1790 (Yale Center for British Art, Paul Mellon Collection).

travel. In 1796, there were five mail and nine other coach services operating from York, most of them daily. By 1823, these had increased to nineteen and seventeen respectively.[22] The number of horses required to service these coaches was astonishing. Each stage, the distance that one team of horses could work daily was ten miles. Therefore a fresh set of four horses (or rather five because each horse rested one day in five) was needed every ten miles. On routes where a coach set off from each end of the journey every day, this meant that approximately one horse was needed for every mile of the route. For the 158-mile journey of the Shrewsbury 'Wonder' to London, 150 horses were needed. Anthony Bird estimated in *Roads and Vehicles* that 150,000 horses were being directly employed by the coaching companies in 1835.[23] No wonder there were doubts as to whether the country could afford to feed so many horses, nor that inn stabling was a major industry. A good contemporary view of an inn stable was made of the Spread Eagle coachyard in Wandsworth on 16 July 1839 by George Scharf (fig. 207).

Some idea of the volume of stabling required can be seen by examining Oxford, an important position on the coach route north-west out of London. During the 1820s and 1830s, ten coaches left the Angel inn at eight o'clock every morning. Most of these were horsed by the proprietor of the Angel, Richard Costar, who must have had upwards of three hundred horses at work.[24] He was not alone. Many of the other inns in Oxford also had substantial stabling. The Bear in the High Street was sold in 1801 with thirty bedrooms and stabling for thirty horses.[25] Similarly, a valuation of the Star in 1861 recorded stabling for thirty horses as well as coach houses and lofts.[26] Thus an average-sized inn might provide more stabling than most country houses.

207 George Scharf, *The Spread Eagle Coachyard, Wandsworth*, 16 July 1839 (British Library, Add. MSS 36489A fol.103; © The British Library).

Few inns survive that have not been substantially altered, and this is particularly true of their stables, which have been redundant for the better part of 160 years. But surveys of three Oxford inns made at the height of their prosperity give a good idea of the layout of a typical inn. These are of 1829 for the Angel in the High Street (now the site of the Examination Schools) and of 1832 for the Star in Cornmarket and its neighbour the King's Head. (These were later united to form the Clarendon Hotel, which was demolished in 1954–55.)[27]

The pattern they followed could be found all over the country and had changed little since the Middle Ages. All three inns had narrow street frontages but deep sites. Immediately behind the street were the main domestic quarters of the inn, perhaps focused on a small courtyard. A passageway ran through this part of the inn from the street to the backyard, where the stables were clustered. These would appear to have been built without any particular order, presumably at different times and with varying sizes of stalls. As they were hidden from public view, they would have been architecturally undistinguished. In all three cases there was a narrow yard immediately behind the main inn buildings. This was probably the original stable-yard, and at the Star it was still being partly used as such. Beyond lay a larger yard, an extension of the original stabling that may be the result of increased travel during the eighteenth century.

The introduction of the coach had a major effect on the layout of the inns. The original entrances of both the Star and the Angel were designed to take only horsemen and those on foot, not vehicles. To accommodate the latter, it was necessary to design a new entrance to the stableyards. At the Angel, this required leasing more land; at the Star, access across the King's Head yard. In both cases the original entrance was then closed in to become the entrance hall.

The quality of inn stables was probably as varied as that of the accommodation provided for the traveller. As travellers' reports are regularly interspersed with complaints about the poor quality of their accommodation, there can be little doubt that their horses suffered equally. Sir Matthew Decker's experiences in 1728 were no doubt typical: 'We sett out for Norwich, and met on the road with such bad Inns that our Coachman, rather chuse to leave his horses before the coach, and give them some hay and straw there than to put them in what they called a stable.'[28]

Little had changed by the nineteenth century when inn stables of dubious quality inevitably feature in Surtees's works. In *Hawbuck Grange* Tom Scott was not impressed by the Goldtrap Arms: ' "You've got room I suppose?" "*Plenty* of room!" replied the man; and truly when Tom lighted on the veteran dung heap at one stable door and saw the broken panes and gaping deals of the other, he didn't wonder at it.' He was a little more sanguine at the Fender and Fire-iron Arms, where Sleekpow the groom 'was "sure he didn't know what they should do. The stable wasn't fit to put a dog horse in, let alone such horses as *ours*"; and after divers lamentations led the way into a sort of cross between a stable and a cowhouse – not very good to be sure, but a place that might have been a good deal worse. The ceiling was very low, and formed of loose rafters for the support of hay; but that was a deal better than no ceiling, or a roof with holes in it; and though the stalls were merely formed of swing bars [or bales], that is of little consequence with horses that know each other. The stable was warm and dry, and there was plenty of clean straw.'[29]

It would be a mistake to condemn all inn stables, and from the groundplans of the Oxford stables it is clear that the quality of stable varied not only between inns but within individual inns. There is a wide range in the width and depth of stalls, the ease of access, and the ventilation. What the surveys do not reveal is the height of the stall, a matter of equal importance, but it has frequently been suggested that the short life of the coaching horse had as much to do with the poor quality of its stable as the fierceness of its exertion.

An exception could be made for newly built inns designed to take advantage of the increasing volume of traffic. One, for which the original drawings by Robert Adam survive, was for the Veillistor Hotel at Quarndon in Derbyshire, built in 1760 (fig. 208). The Cleveland Tontine in Yorkshire was built as a speculation following the completion of the new turnpike road from Thirsk to Yarm, lying at the junction of the road to Guisborough. On its erection in 1804, the Post Office was successfully petitioned to start a daily mail post from Thirsk to Guisborough. The inn, which cost £2,500, 'was designed on an extensive and elegant plan', the chief mover being Mr Scarth of Castle Eden, 'under whose more immediate direction the plan was designed and erected'.[30] The 'elegant plan' included a regular stable quadrangle with handsome Gothick detailing (perhaps inspired by nearby Mount Grace Priory). Unlike traditional inns, the Cleveland Tontine stables were clearly visible and, as competition between isolated inns must have been fierce, the erection of

208 Robert Adam, plan of the Kedleston hotel, Derbyshire, 1760 (by courtesy of the Trustees of Sir John Soane's Museum, 40.58).

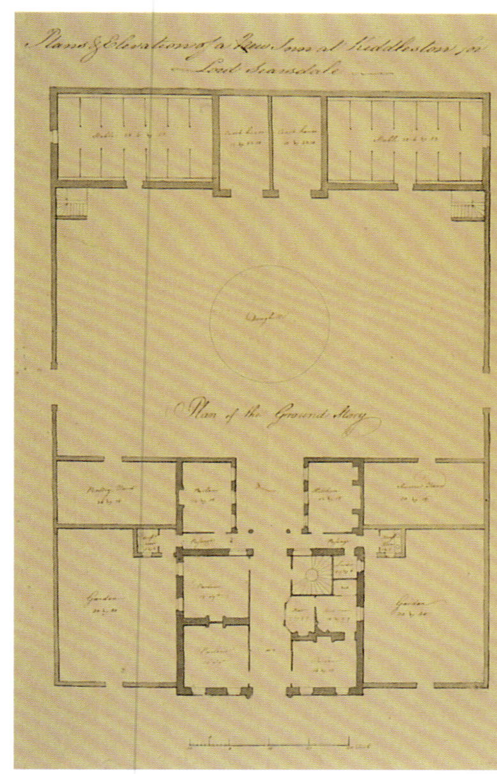

stables fit for a gentleman's house, with distinctive detailing, advertised the quality of the inn and served to set it apart from its competitors.

The greatest inn stables of the eighteenth century eclipsed even the largest country-house stable. 'No one would ever imagine, from the appearance of the building', declared the anonymous correspondent of *Ackermann's Repository of the Arts* of John Carr's stables at Buxton in Derbyshire, 'that it was intended for such a purpose on account of its grandeur.'[31] As part of the Duke of Devonshire's scheme to turn this remote town into a northern rival to Bath, Carr built the Crescent, which combined two hotels, assembly rooms, lodging houses and a house for the duke. To service this and the rest of the spa he also built a large octagonal stable, completed in 1789 at the enormous cost of £16,470 3s 10½d. This had room for about 120 horses and about 20 coaches as well as a circular courtyard 180ft across, around which ran a columned ride.[32]

William Macritchie considered the Buxton stables the most magnificent in the country, and they can hardly have failed to impress the visitor with the grandeur of the resort.[33] But not everyone reacted favourably. 'Up early to find my Cavalry;' declared John Byng in June 1790, 'who are lodged in a most ill-contrived, magnificent mews, where all things are in common, and where they and the furniture must be hourly watched, nothing like a quiet stable to be call'd your own – snug lodging houses, with adjoining small stables were more necessary and comfortable, than useless, ill-contrived grandeurs.'[34] Carr's stables were converted into a hospital in the nineteenth century and the circular courtyard was roofed over, but retaining the circular ride.

The coaching boom did not last long. The opening of the Liverpool and Manchester Railway in 1830 announced the doom of the coaching trade. By 1854, 6,000 miles of railway were complete and another 3,200 authorised. By then, the prosperity of the inns had been destroyed, leaving their stables empty. The useful life of the Cleveland Tontine must have been short, and it cannot have been long before it was a 'forlorn edifice, only partially inhabited'.[35] The short-sighted and the inflexible were driven to bankruptcy or penury, but the clear-headed realised the success of the railway and embraced it whole-heartedly.

William Chaplin was one of the greatest coach-horse contractors, with nearly two thousand horses stabled in five different yards in London. Instead of fighting the railway, he welcomed it. As it made its way north he operated a shuttle service around incomplete sections, sold his coaches on the Continent and invested heavily in the London and South Western Railway, of which he eventually became chairman. He joined with Benjamin Horne, another big London coach proprietor, to found Chaplin and Horne, cartage agents for the London and Birmingham Railway. Pickford's was another firm that adapted to the railways. Unlike Chaplin, its interest had lain in the canals, which had expanded in the late eighteenth and early nineteenth centuries with a rapidity equal to that of the coaching networks, and had generated a similar demand for horses to tow the barges. In the 1820s Pickford's had maintained four thousand horses to run canal fly-boats between London and Birmingham. These soon proved insufficient to cope with the new local town traffic generated by the London and Birmingham Railway, which opened in 1838.[36]

The tales of Chaplin and Horne and Pickford's point one way forward for the dramatically different horse economy of the later nineteenth century. The other trailblazer was the military stable. Before the nineteenth century, military stables were one of the few places where horses were housed in larger numbers than in country-house stables, but parliamentary distrust of a standing army meant that large bodies of troops were not maintained in peacetime. The regular troops that did exist were generally billetted out, not housed in barracks, as much because the Government was not prepared to find the money needed for barracks as for reasons of constitutional propriety.[37] The major exception was in London, where the monarch's Life Guard needed accommodation and stabling at every royal palace.

The original Horseguards in Whitehall was built in 1663–64 with stabling for 108 horses, though by 1713 this had been reduced to 62.[38] When Kensington Palace was bought in 1689, barracks and stables were built for £5,002, including the thirteen-bay range that survives at the south end of Palace Green.[39] Light Horse Court records the Horse Guards Barracks at Hampton Court. These are shown in a plan of 1710–13 in Sir John Soane's Museum, London, with the ground floor given over to stables. Twenty-four stalls for troopers' stables were placed back to back in the centre of the building, with a pair of four-stall stables, presumably for the officers' mounts, at each end.[40] None of these guard stables was very large, and in scale and design they differed little from other royal stables. The double-pile plan of the old Horseguards is very similar to that of the stables built later in the same decade for Queen Catherine of Braganza at Somerset House.

The one exception was the groin-vaulted stable of the new Horseguards (figs 209, 210 and 211). This replaced the old building, which by 1745 was in such a decayed state that the army requested its replacement, and was built to William Kent's designs in 1750–59 with housing for sixty-two horses.[41] Groin vaulting was usually reserved for the finest stables, and its presence at the Horseguards was not the result of any special military requirements but rather of the architectural ambition of Lord Burlington and William Kent, whose stables at Houghton Hall, Norfolk, had been similarly vaulted. The symbolic importance of the Horseguards, both as the formal entrance to the Palace of Whitehall and the headquarters of the army, justified the lavish architectural treatment of the stables and the rest of the complex. There was also a practical justification for the vaulting as it would have reduced the smell of the horses in the extensive offices and other accommodation placed above the stables. The combination sounds impractical, but the Household Cavalry is still housed in the original stables at Horseguards.

The stables in the immediate vicinity of the palaces were only large enough to accommodate the troop on duty. To house the rest of the Horse Guards, there were a series of other 'Guard Stables'. Rocque's map of London shows four of these in mid-eighteenth-century Mayfair, two on either side of Chapel Street, built in around 1730, one rather larger one in Woods Mews, built around a quadrangle in 1738 and intended for the second troop of Horse Guards, and one just south of Chesterfield House in 1751–52. Riding House Street in Marylebone commemorates the now-demolished stables and riding house built for the second troop of Life Guards in 1752.[42] In 1792, these were replaced by a large new barracks on the edge of Hyde Park, designed by James Johnson for the first and second troops of Life Guards. This was built around a quadrangle, which in 1805 held 384 soldiers and their horses, with the officers and their horses in separate buildings.[43]

The French Revolution, which brought with it fears of civil unrest, and the subsequent Napoleonic Wars stimulated the Government to embark on a massive

209 (*above left*) William Kent, Horseguards, Whitehall, London, 1750–59 (William C. Rolf, 1999).

210 (*above right*) William Kent, Horseguards, Whitehall, London, 1750–59 (William C. Rolf, 1999).

programme of barrack building from 1792. Many of the cavalry barracks were deliberately sited in industrial towns because mounted troops were particularly well suited to putting down urban unrest. During July and August 1792 alone, the Deputy Adjutant General Colonel Oliver de Lancey made arrangements for cavalry barracks to be built in Sheffield, Nottingham and Birmingham. Others were built in Coventry, Norwich and Manchester, the barracks in Manchester being the largest, housing six troops of cavalry, with stabling for 264 horses, at a cost of £7,900 (fig. 212). The outbreak of war with France in 1793 led to a string of cavalry barracks concentrated along the south and east coasts. Thirteen small barracks for a single troop of fifty-eight men and their officers were linked by larger regimental headquarters, at Edinburgh, York, Winchester, Leicester, Ipswich, Canterbury, Brighton, Dorchester, Exeter, Perth, Hamilton and Northampton.[44] By 1805, there were seventeen thousand cavalrymen with their horses in permanent or hired barracks.[45]

Provision also had to be made for the artillery's horses, with a sequence of new artillery barracks, one of the grandest being the Brompton Artillery Barracks at Chatham, designed by James Wyatt in 1804 and completed in 1806 with stabling for two hundred horses and sheds for thirty gun carriages. The largest of these was the Royal Artillery Barracks at Woolwich. Built between 1786 and 1808, it was the largest barracks in the country, holding 1,718 horses and 3,492 men.[46]

Most of the Napoleonic barracks were demolished after the war, but a typical example of the larger cavalry barracks, much of which survives, was that at York. This was built in 1795–96 at a cost of £25,000 for a regiment of 232 men by the War Office Barracks Department architects James Johnson and John Sanders.[47] As at Manchester, the barracks were laid out around three sides of a quadrangle, with the officers' quarters on one side flanked by two ranges of stables, above which were the men's barracks served by galleries on cast-iron columns. An impression of what life must have been like can be found in Thomas Rowlandson's *Life in Cavalry Barracks* (fig. 213). Military authorities showed little concern for barrack-room conditions in the eighteenth century and so the disagreeable effects of so many men sleeping above a stable were ignored. By contrast, artillery horses were housed separately.[48] Barracks were built in brick in the conventional restrained Georgian style typical of late eighteenth-century military establishments, that at York being dominated by a large Coade stone achievement of the royal arms in the pediment of the officers' wing.

The decades that followed the end of the Napoleonic Wars were years of low military expenditure, though new barracks were built for the Horse and Life Guards in London at Regent's Park and St John's Wood in the 1820s. Elsewhere, fears of civil unrest meant that the inland cavalry barracks were retained at Birmingham, Manchester, Nottingham, Coventry, Hounslow, Ipswich, Brighton, Dorchester and York, with new ones built at Burnley and Leeds.[49]

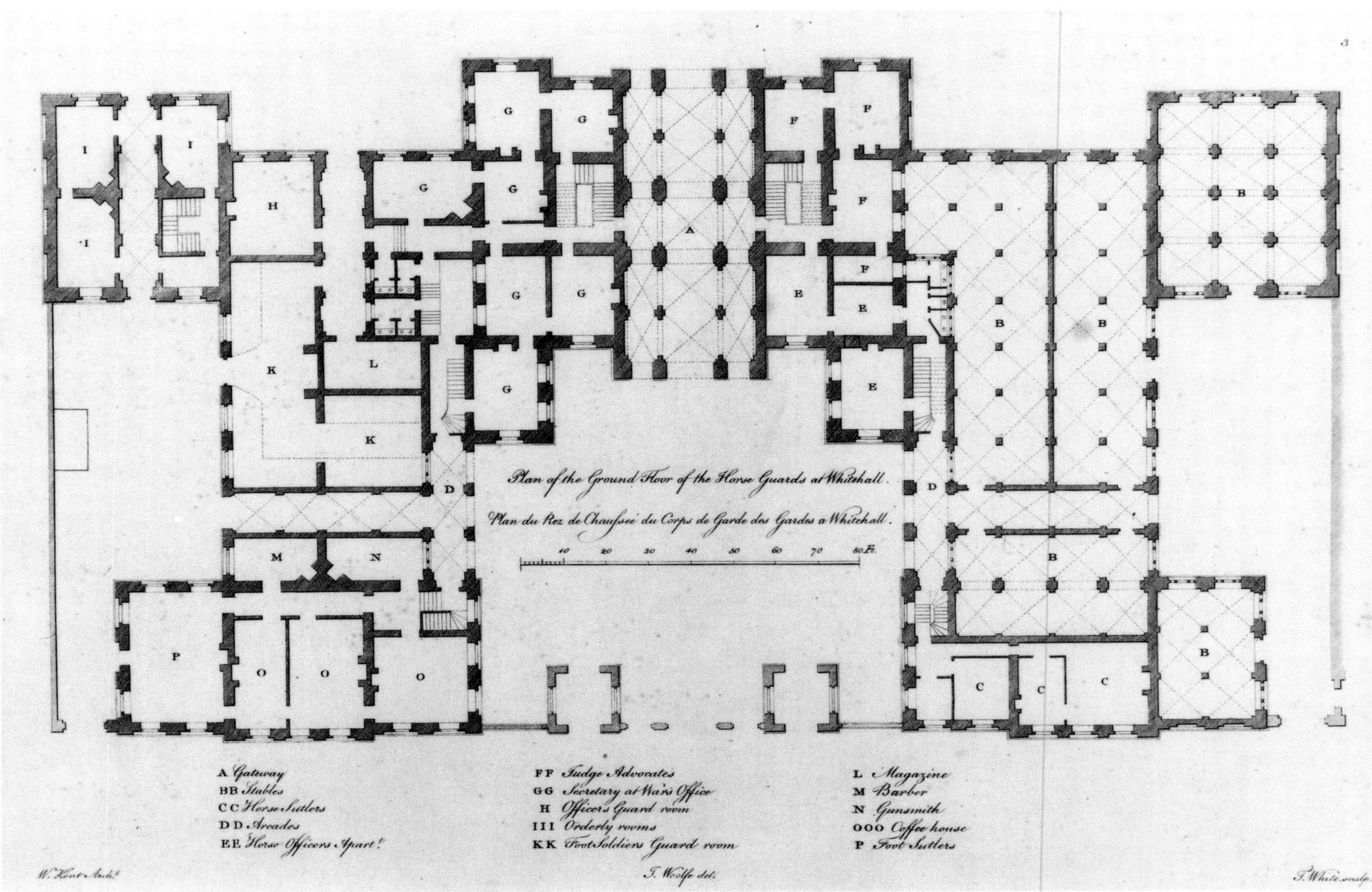

211 Plan of William Kent's Horseguards, Whitehall, London, of 1750–59, from John Woolfe and James Gandon, *Vitruvius Britannicus* II (London, 1771), pl. I (© The British Library).

212 Plan of the cavalry barracks at Hulme, Manchester. Redrawn from an original drawing in the National Archives by Sue Goodman (National Archives HO4220, National Monuments Record).

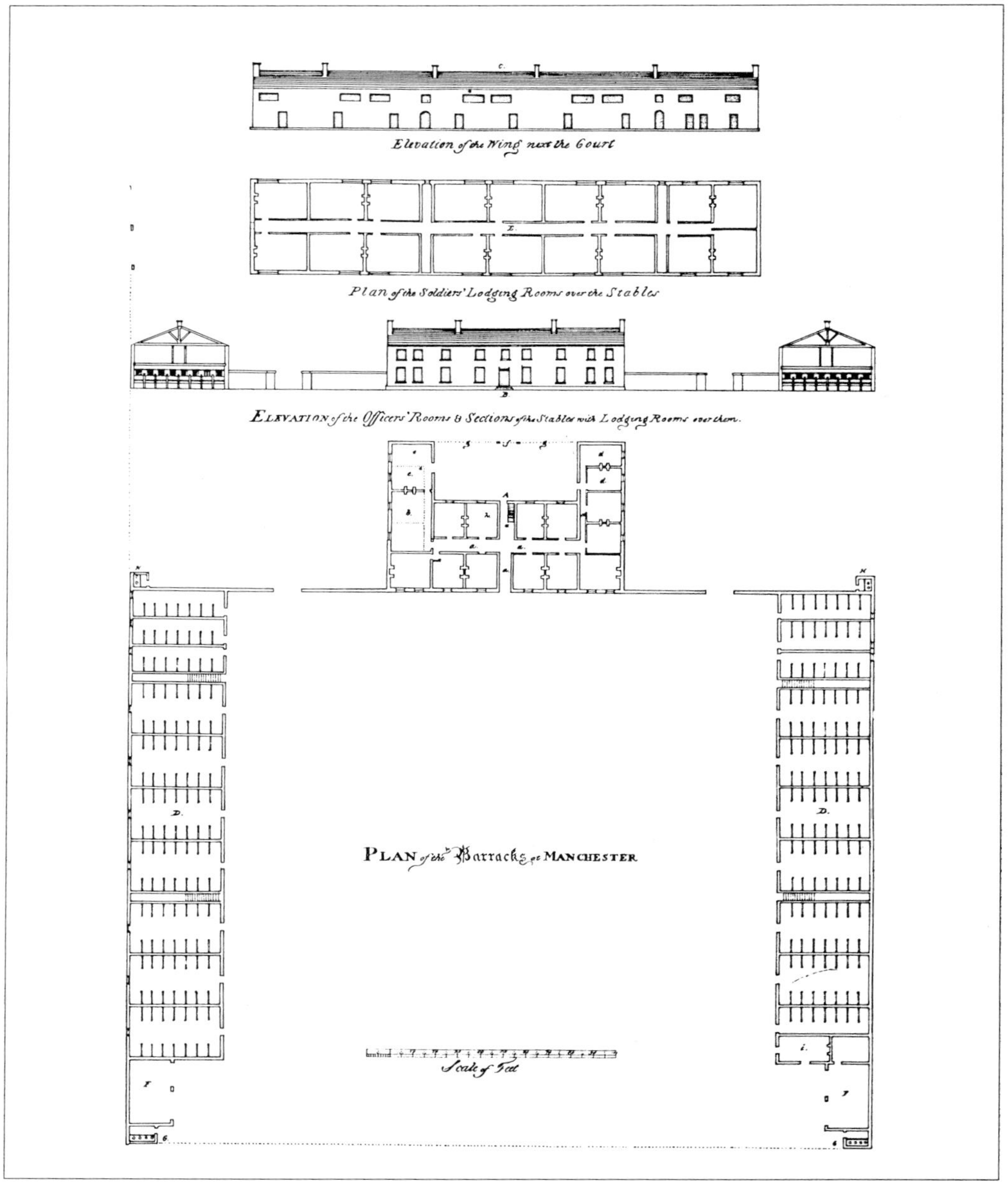

It was not until the death of the Duke of Wellington in 1852, and the military incompetence revealed in the Crimean War, that serious thought was again applied to barrack building. The most important result was the creation of a large permanent barracks at Aldershot, aimed not at internal security but at improving military training. These included three cavalry barracks, originally East, West and South Cavalry Barracks but later named Beaumont, Willems and Warburg Barracks. Built between 1854 and 1859, each was designed to hold a regiment of cavalry of eight troops, with its own riding house. In all, there was accommodation for 1,872 cavalrymen and 1,584 horses. These were not only the largest cavalry barracks built at the time, they were also the most elegant, with long Italianate elevations and a particularly handsome rusticated riding house for the Beaumont lines (fig. 214).

213 Thomas Rowlandson, *Life in Cavalry Barracks*, n.d. (Yale Center for British Art, Paul Mellon Collection; photo: Richard Caspole).

Unlike the earlier barracks, these were not built around a central parade ground but in long terraces, though they did follow the traditional pattern of ground-floor stables with troopers' accommodation above.

Growing concerns about stable hygiene can be seen in the design of the large Le Cateau barracks built in Colchester in the 1860s. Though troopers were still housed over their horses, the floor dividing men from horses was built of much

214 The riding house of 1854–59 at Beaumont barracks, Aldershot, Hampshire (National Monuments Record, 1963).

more substantial concrete, supported by iron columns and beams, and the stables were well ventilated. The wider ground-floor stables formed a flat-roofed terrace along one side of the barrack rooms.[50] The issue of ventilating cavalry stables while keeping costs to a minimum was one that had troubled successive generations of cavalrymen. John Stewart commented in 1838 that the introduction of Professor Coleman's system of ventilation tubes must have saved the Government thousands of pounds a year in horseflesh.[51]

Insalubrious stables were only one issue in a much wider condition about barracks as a whole and in 1857 a Royal Commission was set up on the subject. Chaired by Sir John Sutherland, an army physician and sanitary expert, this subsequently became the Standing Army Sanitary Commission. Its comments on military stables in 1861 were crushing. Sutherland found much to criticise, but his main objection was that most cavalry barracks were built with the men's quarters above the stables. 'Every cavalry barracks we have seen constructed upon this plan is saturated throughout with ammonia and organic matter; and in the cases where the barrack rooms have been shut up and unoccupied for some time, the putrescent odour experienced on entering them is indescribably offensive.'[52] Stale air inevitably percolated into the barracks above, while the stable below could not be adequately ventilated. 'It is impossible to ventilate satisfactorily a stable accommodating a large number of horses, if anything beside the roof is interposed between the stable and the outer air, and . . . it is equally impossible to keep the air in the men's rooms over the stables pure and free from stable odour.'[53] The combined stables and barracks of the 1790s, such as those at York, represented everything of which he disapproved.

As a palliative measure it was recommended that stables like those at York should have ventilating shafts added, but the main recommendation for stables was the abolition of the stable/barrack and its replacement with single-storey stables, preferably with louvres along the crown of the building.[54] Finally, in 1864, the Army Sanitary Commission decided that quartering men over horses was unacceptable, requiring the construction of separate stables.[55] The effect that this could have is shown in Fitzwygram's calculations in 1869 of the different cubic footages per horse found in the cavalry barracks at York. The original 1790s stables had 740 cubic feet, though the preferred minimum was 1,200 cubic feet. Of more recent stables, those with rooms overhead only gave 900 cubic feet, but those with no rooms above and louvres provided 1,500 cubic feet.

The result can be seen in a view by T. P. Chapman, dated 1889, of the stables of the tenth Hussars being inspected by officers and ladies (perhaps after church parade) (fig. 215). The view, showing horses arranged back-to-back either side of a long central aisle, illustrates the utilitarian nature of military stables. With its exposed brick walls, functional iron columns holding up an exposed roof structure, supporting hooks and bars for the saddle and bridle, and with metal bales dividing the horses, the stable has none of the functional elegance of the gentleman's stable. Nevertheless, it is light and airy. It was this type of stable that the great industrial and commercial stables would try to emulate.

Horses played a major part in industry. Before the invention of the steam engine, the horse had provided the primary motive power of much machinery. By the late 1600s, horses were being used to raise coal from working levels to the pit head by

215 T. P. Chapman, *The Stables of the 10th Hussars Being Inspected*, 1889.

means of cog and rung gins. *The Compleat Collier* of 1718 calculated that a pit sixty fathoms deep would mine about ninety tons on a good day and would need at least eight horses to perform that work.[56] As Jennifer Tann has shown, the horse was even able to compete in the short term with the steam engine and according to Butterworth's *Historial Sketches of Oldham*: 'The greater part of the earliest cotton mills were moved by horse power.' Indeed, Richard Arkwright himself began factory production in Nottingham with one thousand spindles powered by nine horse wheels. Arkwright's horse wheels, used on three-hour shifts, would have called for considerable numbers of horses – if there were four shifts, totalling twelve hours, and horses only worked one of them, then thirty-six horses would have been needed. One of the major attractions of the steam engine was that the cost of stabling and maintaining horses was avoided. Even so, in some areas of industry, such as small under-capitalised mines using horse gins, horses continued to be a motive power into the second quarter of the nineteenth century.[57]

Little evidence remains of these industrial stables, and it is unlikely that they were ever particularly distinguished, but this was not always the case. Perhaps the greatest monument to the part horses played in the Industrial Revolution was the stable built for John Josiah Guest's Dowlais ironworks at Merthyr Tydfil in 1820 (fig. 216). This was designed on an enormous scale, being 250 ft wide and housing up to two hundred horses. It was sited to dominate the skyline of Guest's growing industrial town, its size and handsome Classical style – cupola, central pediment, centrepiece and pavilions with ashlar quoins, dressings and string courses, the whole clearly inspired by country-house stables – are proof of Guest's appreciation of the horse's key role in the industrial process.[58]

Guest was a pioneer in the growing nineteenth-century trend for large industrial stables, perhaps housing hundreds of horses under a single roof. Though the

216 The 1820 stables at Dowlais Ironworks, Mid Glamorgan, 1976. (National Monuments Record for Wales, 711457).

most ink was spilled over the design of fashionable country-house stables, the real significance of advances in nineteenth-century stable technology was in the development of sophisticated systems of ventilation and drainage that allowed the great urban stables of the haulage companies to be built. As Coleman pointed out, there were considerable differences between building stables for a handful of expensive hunters and for large numbers of more robust military and commercial horses that spent little time in the stable during the day and could tolerate simpler systems of stabling.[59]

The increasing industrialisation of the horse brought with it problems of scale. Horses were brought together in larger numbers than ever before and stable design had to adapt to cope. The army, which owned more horses than any other organisation, had faced this problem more acutely than anyone else, but its problems were much the same as those that faced large-scale commercial stables, particularly the stables of omnibus and major haulage companies, which formed the backbone of the Victorian transport system. It is not surprising, perhaps, that the greatest Victorian authority on stables was a soldier, Frederick Fitzwygram.

William Gordon's *Horse World of London* of 1893 is a mine of information about horse use in the capital as the horse-drawn economy reached its climax. According to Gordon there were at the time about 22,000 horses for licensed omnibuses, just under 10,000 tram horses, about 15,000 cab horses and about 25,000 carriers's horses.[60]

The importance accorded to commercial stables is shown by the extensive report on the new stables of the brewers Truman, Hanbury, Buxton and Co. published in 1837, in the first issue of one of the first architectural magazines, the *Civil Engi-*

neer's and Architect's Journal (fig. 217). The stables, built at a cost of £10,000 by R. Davison in Brick Lane, Spitalfields, in the East End of London, and described as 'the most complete of any establishment in London', survive, and their sixteen-bay blind-arcaded façade is an excellent example of early nineteenth-century industrial architecture. However, the report was more interested in the stable fittings, which do not survive. The building was formed of four main stables lying parallel to each other, each 69 ft by 30 ft with twenty-four stalls back to back. What most impressed the magazine was the extensive use of cast iron. The heel posts that sup-

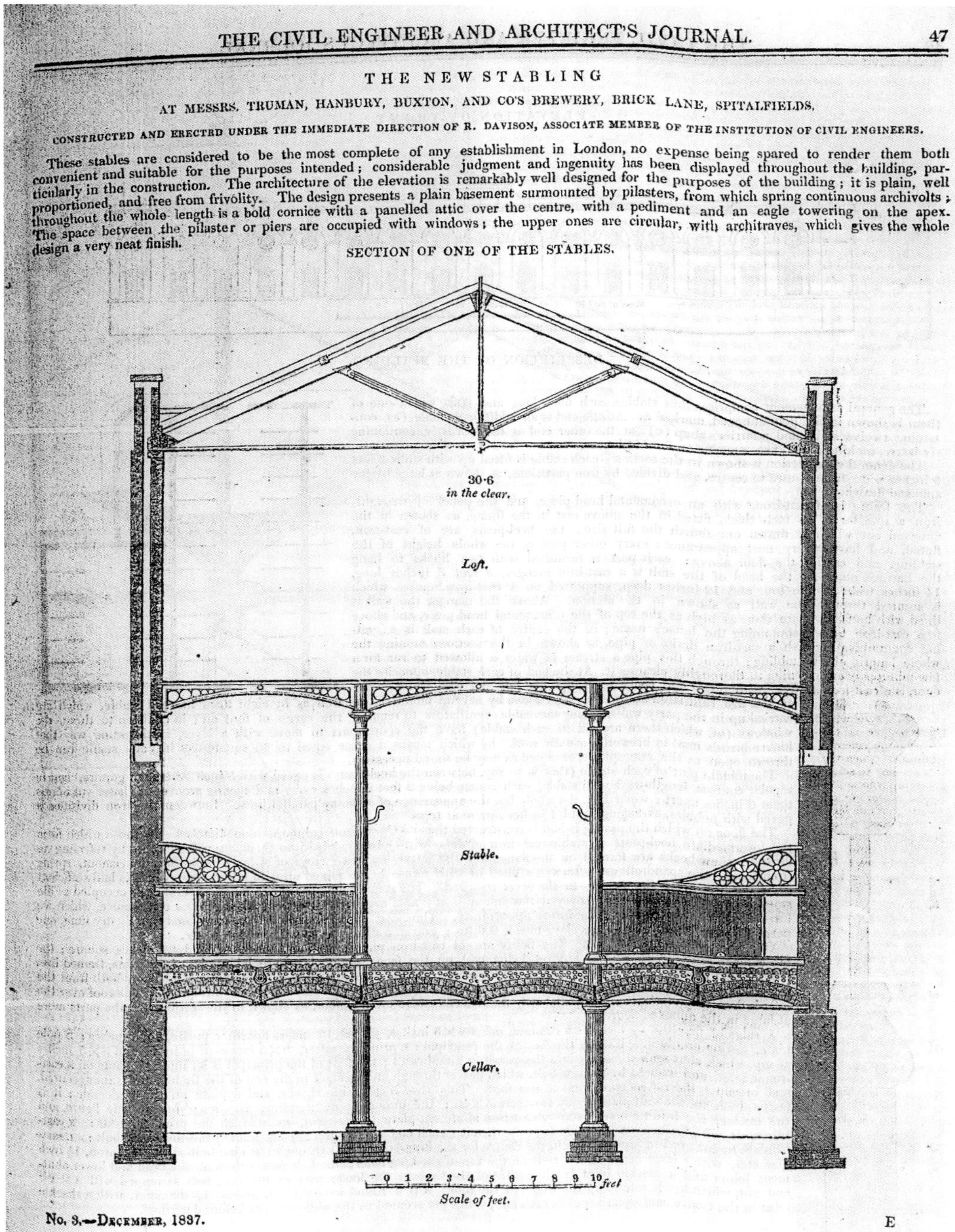
THE CIVIL ENGINEER AND ARCHITECT'S JOURNAL. 47

THE NEW STABLING

AT MESSRS. TRUMAN, HANBURY, BUXTON, AND CO'S BREWERY, BRICK LANE, SPITALFIELDS,

CONSTRUCTED AND ERECTED UNDER THE IMMEDIATE DIRECTION OF R. DAVISON, ASSOCIATE MEMBER OF THE INSTITUTION OF CIVIL ENGINEERS.

These stables are considered to be the most complete of any establishment in London, no expense being spared to render them both convenient and suitable for the purposes intended; considerable judgment and ingenuity has been displayed throughout the building, particularly in the construction. The architecture of the elevation is remarkably well designed for the purposes of the building; it is plain, well proportioned, and free from frivolity. The design presents a plain basement surmounted by pilasters, from which spring continuous archivolts; throughout the whole length is a bold cornice with a panelled attic over the centre, with a pediment and an eagle towering on the apex. The space between the pilaster or piers are occupied with windows; the upper ones are circular, with architraves, which gives the whole design a very neat finish.

SECTION OF ONE OF THE STABLES.

No. 3.—December, 1837. E

217 The stables of Truman Hanbury and Buxton, Spitalfields, London, engraving. From *Civil Engineer's and Architect's Journal*, October 1837.

218 'Interior of No 2 stable' at the Albion Brewery, Mile End Road, London. From Alfred Barnard, *The Noted Breweries of Great Britain and Ireland* (London, 1889) (© British Library).

ported the ceiling, the partitions, and the racks and mangers were all made of cast iron.[61]

Alfred Barnard's tour of breweries published in 1889 repeatedly stressed the scale and the quality of the stables of the metropolitan breweries. At Messrs Barclay, Perkins and Co.'s Anchor Brewery in Southwark, the principal stable was 300 ft long with individual stalls, gas lighting, ornamental iron and timber stalls, and ventilation from circular louvres in the ceiling. No wonder the veterinary surgeon and manager of the stables Mr Gibbing, 'with great pride and satisfaction, showed us the magnificent and valuable stud under his care, and the arrangements made for their health and comfort'. Combe and Woodyard's Brewery had two-storey stables. Charrington's Anchor Brewery at Mile End Road had a special twelve-stall horse hospital 'fitted with every modern appliance'. But it was Mann, Crossman and Paulin's Albion Brewery in Mile End Road that really impressed Barnard (fig. 218). These stables, completed in 1885 and considered 'the best arranged stables in any

London brewery', were so renowned in the East End that 'any person on presenting his address card is allowed full liberty to inspect the premises'. For these great breweries, the horses that pulled their wagons were symbols of corporate pride – 'the firm pride themselves on the quality and condition of their horses, and own some of the finest quadrupeds in the district' – in a way that any Stuart nobleman would have recognised, and their stables were designed accordingly.[62]

The first omnibus was introduced to London in 1829 and the first horse-drawn tram, in 1861.[63] Provision of buses outstripped population growth, rising from one for every thousand head of population in 1850 to one for every 350 by the end of the century.[64] As it required an average of eleven horses to keep one bus in service, omnibus and tram companies became one of the largest employers of horses in the country.[65] The London Omnibus Company, with ten thousand horses, was the largest horse user in London in 1893.[66] Such companies had the advantage that their depots were usually at the end of the line, where the cost of land was lower. Thus Byng Giraud's depots for the North London Tramway Company at Ponder's End, for the London General Omnibus Company at Poplar, and for the Cambridge Tramway Company, which he published in 1891, were generously laid out with single-storeyed, louvred stables (fig. 219).[67] These were placed to either side of the car shed so that the minimum time was wasted starting in the morning or putting the horses to rest in the evening.

A London horse-drawn-bus stable of about 1885 illustrated in Roger Whitehouse's *A London Album* is very similar to Chapman's view of a cavalry stable, with two long rows of horses separated by bales, and an open roof with a central louvre (fig. 220).[68] Charles Dudley illustrated a comparable photograph in about 1901 of the stables of Southwark Borough Council in Walworth – 'Well lighted, adequately

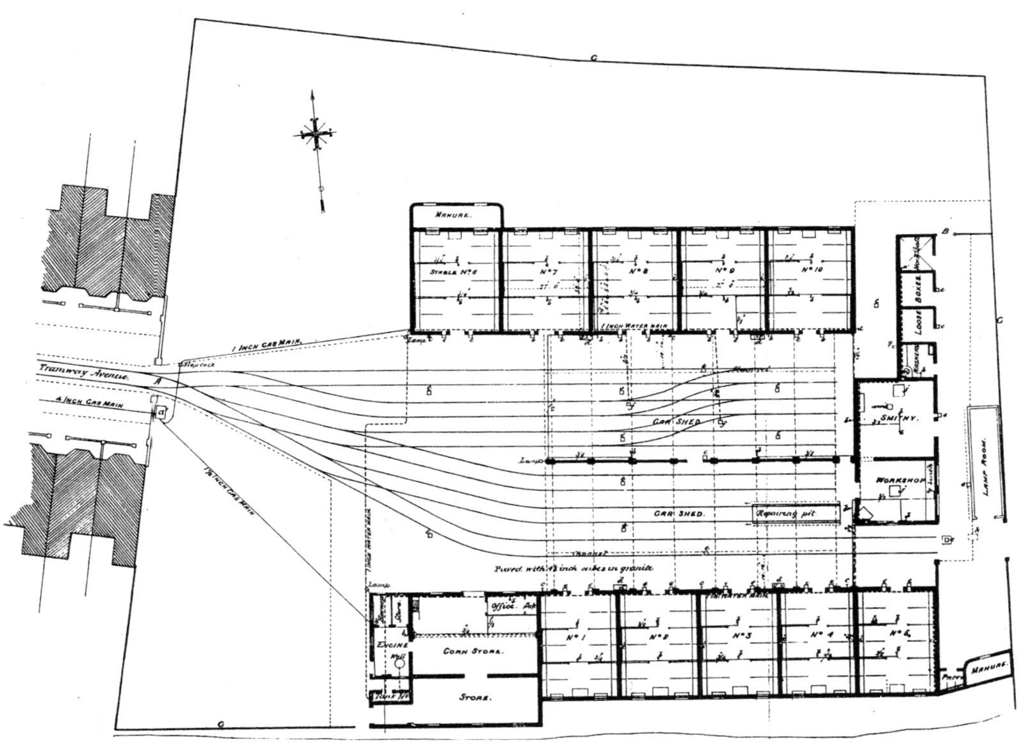

219 Plan of the stables of the North London Tramways Company at Ponders End, London. From Byng Giraud, *Stable Building and Stable Fitting* (London, 1891).

220 A London horse-drawn bus stable, *c.*1911 (London Transport Museum).

ventilated, provided with every necessary convenience' – again with back-to-back stalls, a louvred roof and fixed partitions.[69] Bales continued to be used where space was at a premium. They also had the advantage of allowing freer movement of air. But the best stables, such as those of Tilling's of Peckham, one of the city's largest jobmasters, had fixed partitions.[70] Gordon's 1893 account of equine London carefully distinguishes between those stables with bales and those without. Most had bales.

Generally, there were a few loose boxes, which were used solely to rest sick horses. Giraud recommended one for every twenty stalls, but this varied greatly: the Crosse and Blackwell stables of 1876 had one for 35 stalls, the North London Tramways Company had three for 120 stalls and Pfeil, Stedall and Sons' warehouse of 1884 had four loose boxes for 12 stalls.[71]

But even the omnibus and tram companies could not always afford to keep all their horses at ground level. Gordon considered the Farm Lane depot of the Road Car Company, the second largest omnibus company in the capital, to be the finest omnibus yard in the country. But its 700 horses had to be stabled on two levels around a central yard that held 60 cars.[72]

Horses and stabling were thus major expenses for omnibus and tramway companies and they were always on the lookout for alternatives. When the Newcastle upon Tyne Tramway Company built a new depot in Percy Street in 1880 it included two ranges of stables, one of which was designed so that it could easily be converted into an engine shed should steam come into general use for tramways.[73] Steam did not replace the horse, but the internal combustion engine did, and the changeover was rapid. In 1903 there were 13 motor buses and 3,623 horse buses in London. In 1913 there were 3,522 motor buses and 142 horse buses.[74] The last horse bus was withdrawn the following August with the outbreak of war and horse requistioning.[75]

Closer to the centre of cities, the pressure on space was much greater. By 1900, the number of cabs had grown virtually tenfold from the 1,265 in 1830.[76] Each cab needed two horses, and these were usually kept in mews. During this period there were large numbers of mews that were not wanted by the houses to which they were attached, and other mews were built independently of specific houses. Mews also housed many of the other small-scale horse-owning businesses. But some of the larger cab firms had specially built stables, one or two on more than one floor.[77] The trend towards verticality in stable design was the result of increasing land values. In April 1876, *The Builder* illustrated R. L. Roumieu's new stables for Crosse and Blackwell in Crown Street, Soho (fig. 217). The façade of the building had a striking neo-Norman entrance arch curiously set under a slight pediment, and flanking neo-Norman windows, but what the magazine noted was the trend towards verticality that it illustrated. The building was designed on two levels, with recesses for eighteen vans at the ground level around a central court, with thirty-five stalls on the first floor, approached by a gentle ramp. These were arranged around the central court and were ventilated by a louvred roof.[78]

Because there were no further floors, the Crosse and Blackwell stables were as well ventilated as they would have been in single-storey stables. This was not always possible. In 1884, *The Builder* illustrated a warehouse built by James Peacock in Charles Street off Drury Lane for Messrs Pfeil, Stedall and Co.. Here, the ground floor was given over to pits for stacking iron, with stables on the first floor and two floors of warehouses above.[79] The Charles Street stables, with room for only sixteen horses, were small. By contrast, the Mint stables at the Great Western Railway depot at Paddington, which were new in 1893, housed about 600 horses on four floors (fig. 222). There were 50 in the basement, 260 on the ground floor, 228 on the first floor and 66 on the second floor.[80]

Railway stations generated vast amounts of work for horses – a certain amount of shunting, but principally delivery and collection of railway goods – and big railway companies came to require about as much space for stables as for engine sheds.[81] In 1893, there were 27,826 horses owned by railway companies in Britain.[82] As they were built as close to the city centres as possible, space was always restricted, especially for horses. Some horses even had to be housed under the station, like the three hundred horses of the North Western Railway under Broad Street Station, or under railway arches, as at Bristol Temple Meads and Paddington Station, where Gordon mentions stabling for 140 horses and three dozen stalls for invalids, 'sensibly ventilated and lighted electrically'.[83] But multi-storey stables were common at large railway termini such as Paddington, or College Station, Glasgow, where the Bell Street stables of the Glasgow and South Western Railway, built about 1900, were three storeys high and seventeen bays wide. Multi-level stables could also be found at the headquarters of large haulage companies, such as Carter Paterson, whose Goswell Road stables were on three floors.[84]

The question always remained whether such large stables were healthy. Many commercial stables could not possibly be so described: underground stables in coal mines, the stables of shunting horses set inside dark and smokey tunnels.[85] For these

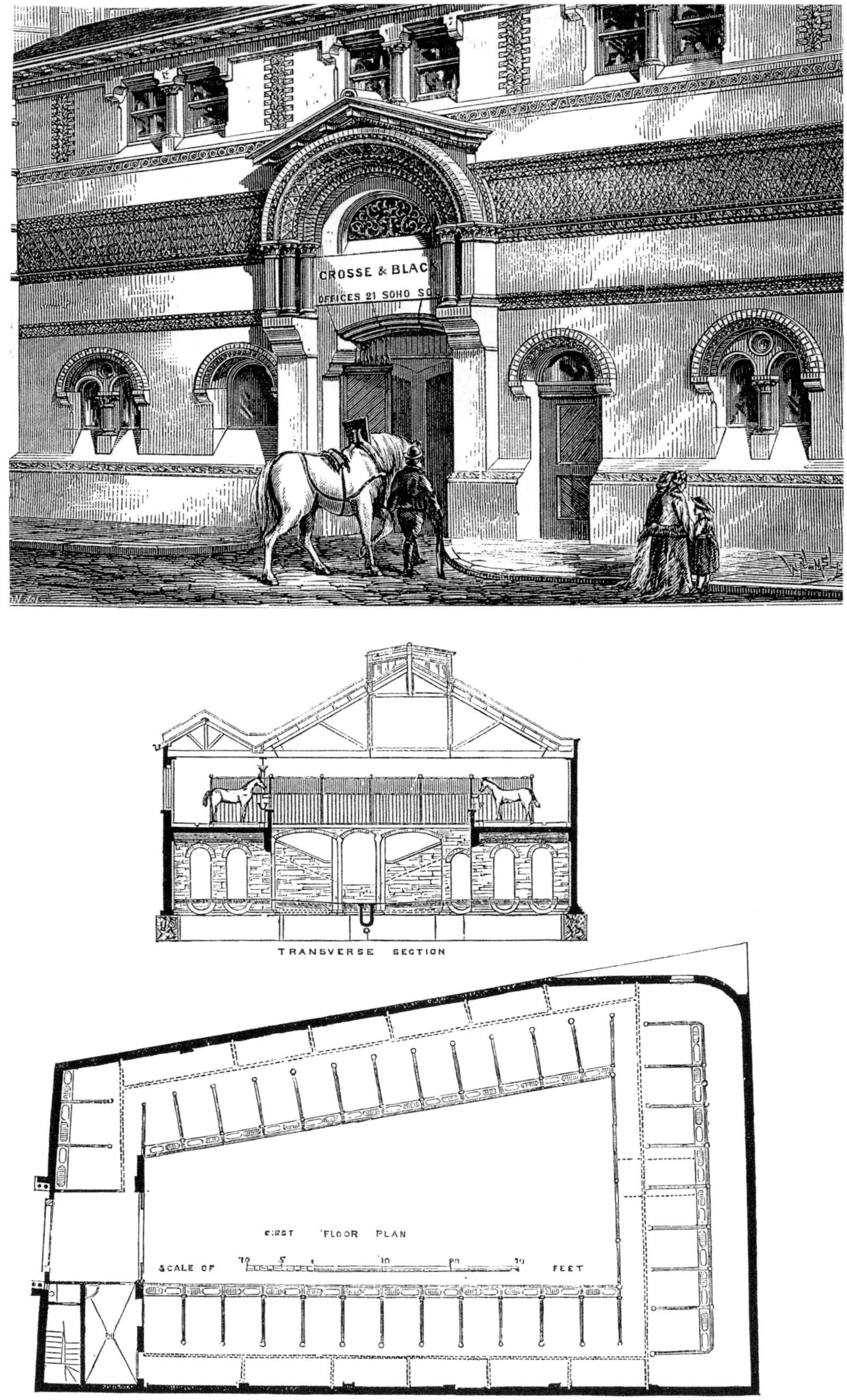

NEW STABLES, CROWN STREET, SOHO.——Mr. R. L. Roumieu, Architect.

221 R. L. Roumieu's Crosse and Blackwell stables, London. From *The Builder*, 15 April 1876.

222 The Mint Stables of 1893 at Paddington Station, 1922 (National Monuments Record).

there were always violent critics such as Birch in 1892. But Gordon, whose knowledge of the London horse was unrivalled, was more sanguine. He describes the horses of the jobmasters Tilling of Peckham as being 'as well cared for as the plate at a silversmith'. Admittedly, such horses were the cream of the carriage trade, but the stables for haulage company Carter Paterson were described as 'clean and roomy' and coal merchants' stables as 'capital examples, in fact, of a cart-horse stable in which a few horses are kept under the best of conditions for health and efficiency'. He goes on to write that 'Here and there, among the second-rate men, the old dark, stuffy den is the horses' home; but . . . the average is treated much better than he used to be; he is better fed, better housed, and more intelligently looked after, and he lives longer, works more and is better worth looking at than in the merry days of the past.'[86] Housing large numbers of horses in a crowded city could never be easy. But in London in the 1890s, horses were probably better looked after on a larger scale than at any time in history.

Horace Smith's father was a horse dealer before the First World War:

> Everything in the well-run old-fashioned stable of those days was very decorative and spotlessly clean, and highly polished brass was to be seen wherever it would show up to good advantage. The more energetic were the stablemen, the more rows of dirt there were to be seen while grooming was in progress. My father used to tell me that if the mangers were clean and the brass rings used for the white head-ropes and the brass on the head-collars were shining there was no need to look any further; the man was a good stableman. There were plaited straw mats, made up with red, white and blue ribbons, lying around the stall posts and inside the door as one entered the stable; and wonderful window-boxes of geraniums, lobelias and daisies in the same colours as the ribbons in the straw mats.[87]

Nothing gives a better picture of the commercial stable at its best, at the peak moment of the horse's importance, and just before its rapid eclipse.

223 The stalls at the Horseguards, Whitehall, London (William C. Rolf, 1999).

1ST GATE SENTRY
JNCO
GUARD COMMAND
COH

10

THE VICTORIAN COUNTRY-HOUSE STABLE

GIVEN THE CENTRALITY OF THE HORSE TO VICTORIAN BRITAIN, it is not surprising that in the nineteenth century a wave of books on stables appeared. At least one relevant work, usually more, was published every decade from 1788 until 1909. The high point was the 1860s when ten appeared, marking the climax of interest in stable building, and in particular remodelling, that left such a marked impact on the British country house. More than fifty separate works wholly or with significant sections devoted to stable design can be identified between James Clark's pioneering *A Treatise on the Prevention of Diseases Incidental to Horses* in 1788 and V. K. Shaw's *The Encyclopedia of the Stable* in 1909.[1] The information available in such books, and in magazines, drawings and the vivid descriptions of Victorian novelists, particularly Surtees, combined with the vast numbers of Victorian stables that survive, often with their fittings intact, means that knowledge of nineteenth-century stable design is very complete. Perhaps inevitably, the greatest attention was paid in publication to the prestigious and expensive gentleman's stable, but commercial stables were big business, and were much discussed in books and journals.

For much of the first half of the nineteenth century, information on stables is usually found in veterinary works. William Haycock's *Gentleman's Stable Manual; or a Treatise on the Construction of the Stable*, of 1859 – a book on stable design with a sub-section on the health of the horse – reversed this trend. The following year, William Miles was the first to devote an entire book to the theory of stable design: *General Remarks on Stables and Examples of Stable Fittings*. More general works on horsemanship and stable economy, with sections devoted to stable design, remained common. The most popular was Fitzwygram's classic *Horses and Stables*, first published in 1869 and reaching its fifth edition by 1901. During that time the author rose from being Colonel F. W. G. Fitzwygram to Lieutenant-General Sir Frederick Fitzwygram Bt, Inspector of Cavalry.

Stables were regularly featured in the increasingly popular architectural journals, beginning in 1843 with *The Builder* but including in particular *The Architect* and *Building News*. As well as descriptions of newly-built stables, with comments on any particularly novel aspects, they included frequent practical notes on stable design. In three successive volumes of *Building News*, for example, volume seventy had two articles on noiseless paving stones, volume seventy-one, a long article on the sanitary construction and fitting of stables and volume seventy-two, a series on stable construction.

New books were sometimes paraphrased by the journals – William Haycock's *The Gentleman's Stable Manual* appeared in *The Builder* of 5 November 1859, William Miles's *General Remarks on Stables* in the edition of 23 March 1861. Books on stable design were reviewed, sometimes critically. Byng Giraud's *Stable Building* was

224 John Kinross, the stables of 1895 at Manderston House, Berwickshire (William C. Rolf, 1998).

described in *The Builder* of 11 April 1891 as 'somewhat commonplace and much open to improvement'. Magazines also sought to influence stable design through occasional competitions, such as those run for young and trainee architects by the *Building News* Design Club, which offered prizes for the best design of a set subject and published comments on the entries. Stables were popular subjects because of their relative simplicity.

Despite the advances in stable design, not all were complimentary about contemporary stables. Edward Mayhew, writing in 1864, was vehement in his criticism:

> Stables, as they now exist, are tainted with all the evils of antiquity. Improvement has changed the homes of the people and has even amended the cage of the imprisoned songster; but it has completely skipped over the gaol of the horse. The place, and the people about it, smack of a time when corruption was the rule and filthiness was a fashion. The question, therefore, to be considered is, not what stables are, but what they should be.[2]

One must take such books with a certain pinch of salt. Self-advancement, rather than a balanced view of affairs, motivated many of the books most critical of contemporary stables. After all, if there was nothing wrong with contemporary stables why would anyone buy the author's book or commission him to design them a stable? When Thomas Knightley claimed in 1862 that there was 'no work existing that has professed to deal with the subject sufficiently either with respect to the design or construction' of stables, he ignored three such works published in the three preceding years. But then Knightley's real aim was partly to sell 'Knightley's Patent Improved Stable Floor'.[3]

Concern about stable hygiene, already strong in the early nineteenth century, became an obsession in the middle years of the century. At times, ideas grew dangerously complicated, for simplicity remained the key to good ventilation and drainage. However well designed the ventilation, it was of little value if unused, or deliberately blocked. The theorists had little doubt that the grooms were unsympathetic to ventilation, continuing to denounce them throughout the nineteenth century. When Haycock advocated a ventilation system based on twin-tubes in 1859, he added ominously that 'stablemen require watching, or many of them, unless prevented, will keep the tubes permanently closed. They want a close stable, and cannot be prevailed upon to adopt any other method.'[4] The grooms' objection to ventilation came partly from a fear of draughts but also from laziness. Grooms saw a close stable as a short cut to a fine-looking horse as it produced a superficially glossy coat that at first sight made the horse seem healthy.

Surtees, as ever the sharpest observer of equine life, highlighted this attitude in *Mr Sponge's Sporting Tour* (published in 1853). 'Soapy' Sponge, arriving at the inn at Laverick Wells with a couple of half-bought horses of dubious character that he intends to sell for profit, is naturally anxious that they should look good if nothing else. He visits the stable where they are housed:

> Mr Leather [the groom], of course, had the valuable shed under lock and key, with every crevice and air-hole well stuffed with straw as if they had been the most valuable horses in the world. Having produced the key ring from his pocket, Mr Leather opened the door, and having got his master in, speedily closed it,

lest a breath of fresh air might intrude. Having lighted a lucifer, he turned on the gas, and exhibited the blooming coated horses . . . Mr Sponge stood eying them for some seconds with obvious approbation.[5]

Whatever the theorists might write, and whatever the architects might build, while there were still grooms like Mr Leather, horses could expect little fresh air even in the best designed stables unless strict control was maintained by the owner.

The simplest, and therefore often most effective, form of ventilation remained the window, as many writers such as Fitzwygram, Kerr and Coleman acknowledged. Sidney recognised that it was better to have a crude stable with a window than a luxurious stable with impracticably ambitious ventilation when he wrote of the sale of the Earl of Stamford's stud in 1881 following his resignation as the Master of the Quorn Hunt. It was generally admitted, he commented, that never had so many horses been brought to the hammer in such fine condition, although their lodgings had been no better than unimproved cart horse stables. Many a healthy stud, he added, used to old-fashioned buildings that let in the wind by roof, doors and windows, had pined and sickened when removed to buildings erected regardless of cost.[6] It seems the Earl of Stamford had put more effort into architectural appearance than stable technology at his vast complex of stables and kennels at Bradgate House, Leicestershire, designed in 1856 by M. J. Dain. The stables, which housed forty-seven horses as well as hounds and staff, were entered through an immense, elaborate gatetower worthy of a great Jacobean house.

If ventilation was one Victorian stable obsession then drainage ran it a close second. Arguments between closed drains and open drains raged throughout the century. Giraud strongly recommended open gutters because they could be kept clean by constant brushing, and because it would not be necessary to disturb the ground to get to the drains if they became blocked. Richardson criticised open gutters for the labour cost in keeping them clean, the waste of straw and the general unhealthiness of urine-soaked bedding. He would have preferred the mid-nineteenth-century stalls at Althorp, Northamptonshire, where narrow drains were closed by short runs of metal plates punctuated by holes through which the urine could run.

This debate had changed little since the eighteenth century, but it was made more complicated by the invention of patent drains, whose complexity was not always matched by their efficacy. Thus Miles claimed that good drainage was impossible until the abolition of 'that abomination in a stable "the bell trap" or as it is more appropriately designated "the stink trap"', which kept out smells from the drains but ensured the permanent smell of stale urine. However, Richerson argued that every stall and loose box should have a regularly scrubbed bell trap.[7]

As in ventilation, the key to good drainage seemed to lie less in the particular system used than in the standard of stable management, the diligence with which the stable was cleaned. This led Birch to argue that the simpler the drainage the better, and to suggest that the best drainage was no drainage. He believed this to be possible in well-laid loose boxes, but that ordinary stalls and loose boxes would need some form of drainage. The main disadvantage of this scheme was the large amount of straw needed to absorb the urine, which would have to be changed regularly, and the need for good ventilation to keep the atmosphere sweet.[8] The

225 Nineteenth-century loose box in the stables at Althorp House, Northamptonshire (William C. Rolf, 1999).

value of simplicity and solidity was certainly the message visitors would have picked up from Fitzwygram's own stables at Leigh Park, Hampshire, which he bought in 1875 (fig. 226).[9] Though not designed by Fitzwygram, the great stable expert would undoubtedly have chosen his stables with care. Their basic drainage and simple, robust partitions probably reflect his experience of military stables.

Much depended on the quality of the flooring. Clinker bricks, particularly hard-baked bricks, which had the advantage that they could be moulded to allow drainage channels and to prevent slipping, were almost universally recommended for flooring. For smoothness of surface, durability and grip, a properly laid clinker brick floor was preferred throughout the century, the only main objection being one of cost. Only Fitzwygram, writing in the third edition of his book in 1886, criticised their practicality, arguing that it was impossible to bake bricks equally hard throughout and that as a result when the outer surface was chipped or rubbed away the surface wore into holes.[10]

Cheaper bricks such as Blue Staffordshire or Broseley were sometimes suggested. On farms ordinary bricks were common, although these had little to recommend them except their low cost, as they were porous and liable to become uneven under

226 Interior of the stables at Leigh Park, Hampshire (William C. Rolf, 2000).

constant wear. Even cheaper was pebble pitching, presumably the successor of the seventeenth-century cobble. This was criticised by most writers, including Miles who said that in 1860 it was the most common form of flooring.[11] Other alternatives included concrete (increasingly popular towards the end of the century), wooden blocks, granite paving pitchers, cork and even India rubber – although the last seems a temporary fad in the early 1860s and never caught on.

Ninteenth-century racks in good stables were made of cast iron which Coleman praised because it was non-absorbent, clean in use, strong, could not be chewed, had a good appearance and was easy to clean. He believed that cast iron had largely superseded wood, which he considered unhygienic, liable to harbour disease and to damage from gnawing.[12] Iron was also the common material for mangers, although, as with racks, wood never became entirely redundant.

Other materials were sometimes used for mangers, including stone, which Knightley criticised, and slate, which was vigorously marketed by the Penrhyn slate-works. Neither was popular, although Tattersall describes the 'super-excellent' stables at Tedworth, Hampshire, of Assheton Smith, owner of the Vaynol slate quarries, as being largely fitted with slate.[13]

227 John Nash, the quadrangle of the Royal Mews of 1822–25 at Buckingham Palace, London (William C. Rolf, 1999).

As the catalogues of the St Pancras Ironworks Company show, partitions between the stalls could be of iron, but the conventional Victorian partition had an iron frame infilled with wood. Nineteenth-century writers did not share earlier beliefs that horses should be unable to see each other. Indeed, it was felt that because of their sociability it would be better if they could. The favoured nineteenth-century combination was a wooden partition with open grille above, which Mayhew claimed to have invented and can still be seen in the Royal Mews, a rare example of a great Victorian stable still used to its fullest extent (figs 227, 228 and 229).[14] At the Royal Mews, the partitions were raised a few inches off the ground and set back a little from the wall to ease the flow of air in the stable.

Both Miles and Fitzwygram found that elm, though tough, tended to warp, and preferred good red deal in partitions.[15] Fitzwygram believed that oak was best, but expensive, while Axe preferred teak rubbed smooth and oiled.[16] The use of teak at Sir Richard Wallace's stables at Hertford House in London was justified for its great strength – Ambler, the architect, said that its use as a backing to the iron of warships had recommended it – but the great connoisseur was clearly as interested in its rich colour. *The Builder* recorded that the rich walnut colour of the teak,

228 Interior of John Nash's Royal Mews of 1822–25 at Buckingham Palace, London (William C. Rolf, 1999).

FLAGSHIP
TOLERANCE
CYPRUS

229 Interior of John Nash's Royal Mews of 1822–25 at Buckingham Palace, London (William C. Rolf, 1999).

230 William Theed, *Hercules Taming the Horses of Diomedes*, 1860, in the pediment of John Nash's Royal Mews of 1822–25 at Buckingham Palace, London (William C. Rolf, 1999).

together with the chocolate-coloured ironwork and the china slabs of glazed turquoise on the walls, gave the stables at Hertford House a feeling of warmth and good taste.[17]

This understated opulence was characteristic of the best stables. The combination of dark wooden partitions, brightly polished brass fittings and above all lustrous tiles shows that Miles's dictum that 'A stable is no fitting place for the display of great taste or unmeaning ornament'[18] did not mean that stables needed to be utilitarian in appearance.

Tiling – known but uncommon in the seventeenth century – was a particularly attractive feature of high-class stables in the nineteenth century in the gap between the rack and the manger. It appealed to the Victorians because it was easily kept clean and looked attractive. Subdued colours such as greys or bluey-greens, such as those used in the Royal Mews, were preferred (fig. 229). White was avoided because it was too bright. The decorative quality of tiles was emphasised by the use of varied shapes, including squares, hexagons and octagons, and by the addition of an ornamental border, often Greek key or rope pattern. As alternatives, Coleman mentions marble slabs, plain or white slate slabs and enamelled iron plates.[19] Rusconi marble of a bluish-grey tinge was used in the Marlborough House stables, fitted up for the Prince of Wales in 1863.[20] This was taken to an extreme at some stables, where all the walls were covered in ceramic tiles, as in the stables at Witley Park, Surrey, built by H. Paxton-Watson for the crooked company promoter Whitaker Wright, in 1896 in a free Tudor style. These were the acme of self-indulgence, with carefully chosen ceramic-tiled walls supporting a barrel vault, and Tudor-style decorative strapwork on the end walls. But then Wright believed in extravagance and also built an underwater glass-ceilinged billiard room at the bottom of his lake.[21]

The improved quality of stable fittings, and more scrupulous attitudes towards hygiene were important reasons why many Victorian landowners chose to upgrade their Georgian stable interiors, but the principal reason was to incorporate the loose box, which was lacking in the Georgian stable. By 1861, Walsh and Lupton could write that every private horse owner who could afford the extra space had adopted loose boxes.[22]

Certainly, by the 1870s it was standard for every hunter to have a loose box. The Meynell Hunt stables at Sudbury in Derbyshire of 1873, a very good example of a fully developed Victorian stable of one of the less grand hunts, had twenty-three loose boxes to eight stalls (fig. 231).[23] The classic arrangement of grand establish-

231 George Devey, plan of the Meynell Hunt stables, Sudbury, Derbyshire, 1872 (RIBA Library Drawings Collection).

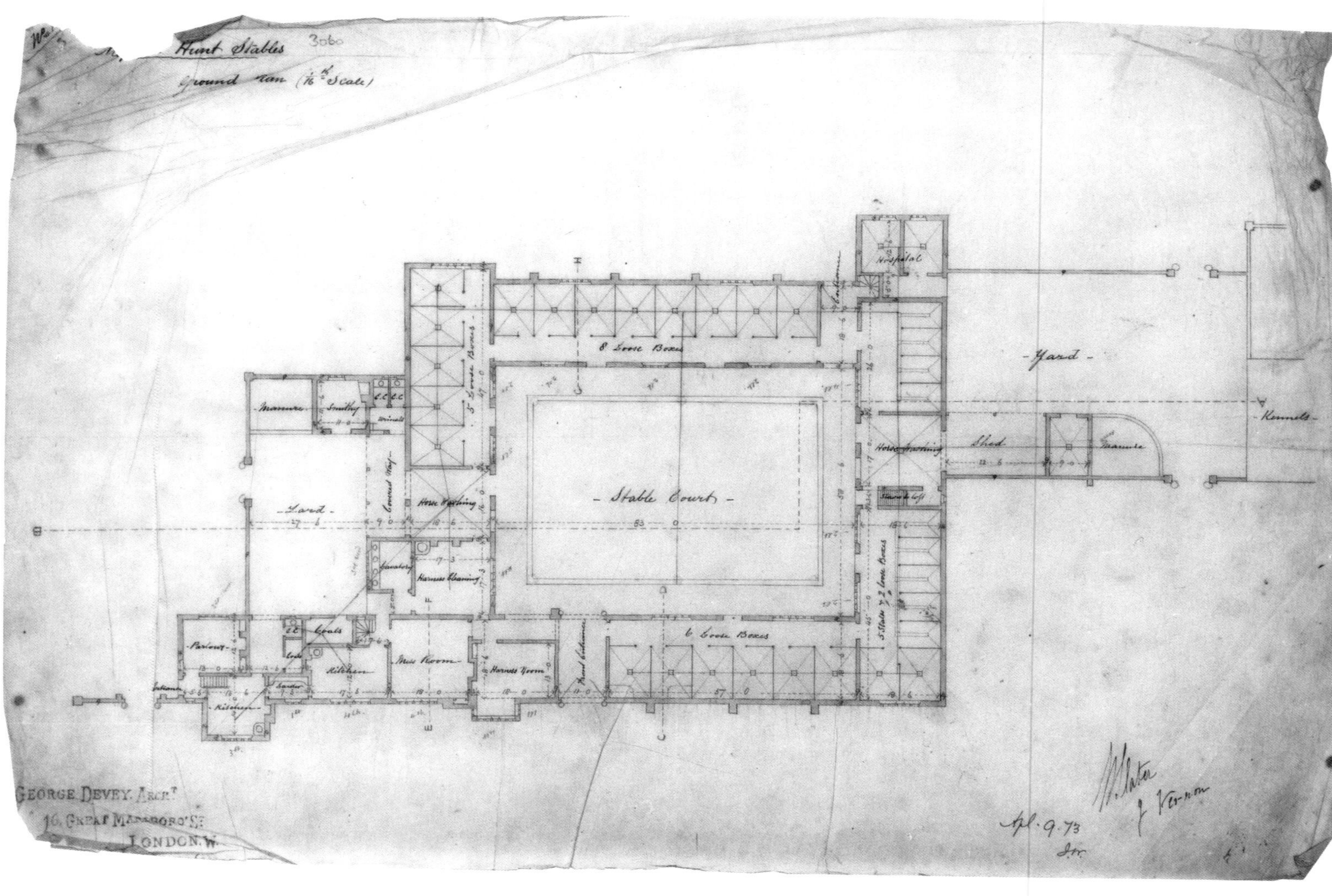

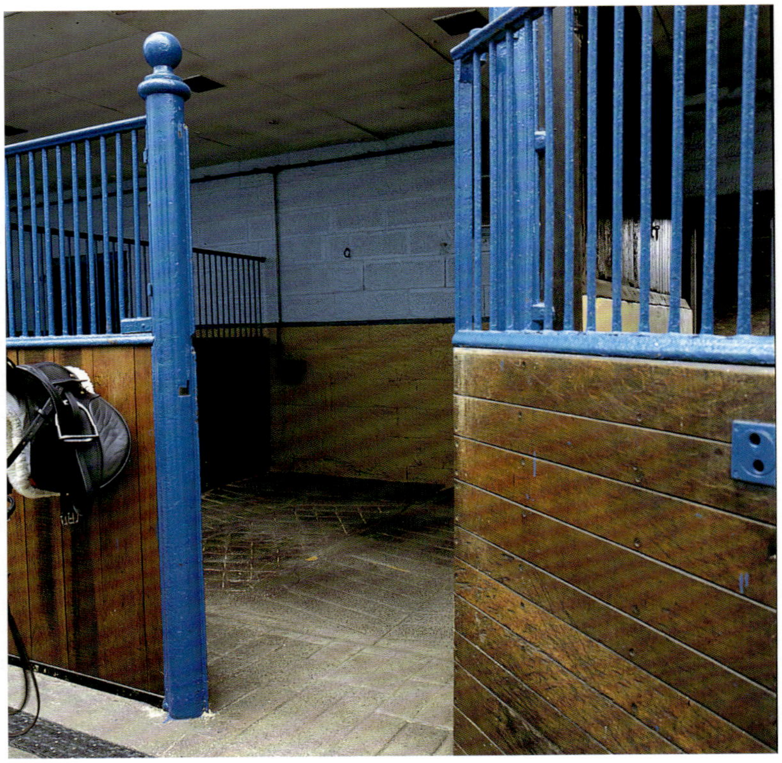

232 (*above left*) Anthony Salvin, the aisle and loose boxes in the stables of *c.*1852–61 at Alnwick Castle, Northumberland (William C. Rolf, 1998).

233 (*above right*) Anthony Salvin, loose box in the stables of *c.*1852–61 at Alnwick Castle, Northumberland (William C. Rolf, 1998).

ments built in the second half of the nineteenth century was a long corridor with boxes opening off it. An early example of this was the Duke of Northumberland's stables at Alnwick Castle, Northumberland, designed by Anthony Salvin as part of his massive rebuilding of the castle between 1852 and 1861. Set within a suitably medieval-looking structure, this has a glazed central corridor that is still painted mustard yellow and blue, as it probably appeared when first completed (figs 232 and 233).

But though Fitzwygram claimed that the difference in building costs between stalls and loose boxes were small, most owners felt that the extra space required, and the cost of larger buildings and more expensive fittings, was only justified on hunters and other more valuable horses.[24] Carriage horses were normally housed in stalls as it was thought that working horses, taking regular exercise, did not need loose boxes. At the Horseguards, the officers' horses were given loose boxes, but not the less expensive troopers' horses. The stables at Lyme Park, Cheshire, where all bar four of the twenty-two horses were stabled in boxes in 1864, remained the exception.[25]

The introduction of loose boxes was not the only difference between the Georgian and the Victorian stable, for the elements included in a Victorian stable increased as rapidly as they did in the Victorian country house, and new technology was swiftly applied.

Little concern had been shown in Georgian stables for the grooms' quarters, but in large Victorian stables these became extensive. There was usually married accommodation for the head groom, chief coachman or huntsman – at the Meynell Hunt stables at Sudbury, the huntsman had a parlour, kitchen and two bedrooms.[26] In more important establishments, particularly studs, a separate cottage was often provided, either on its own or overlooking the stableyard. The grooms and other stable hands usually slept in a barracks or dormitory, generally placed over the coach houses not the stables so as not to disturb the horses. Some stables, such as the excellent hunting stables at Stapleford Hall, Leicestershire, built by Peter Dollar in 1899, had separate rooms for the grooms.

Some of the new elements in Victorian stables were subdivisions of existing facilities. Thus, just as the grooms' chamber became a number of separate bedrooms together with mess room, kitchen and WC, so the harness room might be complemented by a saddle room, cleaning room and tool room. Horses could be cleaned in their stalls, but superior establishments had a separate cleaning area. In 1859, Haycock thought that 'a good stable cannot be considered as complete without a commodious porch' for cleaning. Kerr was more expansive, believing that a superior establishment should have a grooming shed and a shed for rubbing down horses, as well as a horse-bath and shower.[27] The Meynell Hunt stables had two horse-washing sheds, rather than the usual one, to cater for the return of the hunters all together.

Lighting, always expensive before the invention of gas and electricity, was one critical issue, particularly because of the risk of fire with naked flames and large quantities of inflammable straw. In 1534, Hall's *Chronicle* recorded the destruction by fire of the king's stable at Charing Cross, 'wherein was brent many great horses and great store of hay', and in 1683 a fire breaking out in a stable in Newmarket 'by the carelessness of a Groom smoaking Tobacco' caused the destruction of half the town.

> It first broke out in a Stable, having seized upon the Litter, Straw, Hay and Roof of the Stable before it was discern'd, and then too violent to be overcome, till it had seized upon the next House, and so from one to another till it had run through all the Houses on that side of the Town . . . That which made the worse for the Town was the many Thatch'd Houses, Stables and Hayricks, which took Fire from one another like a train of Gunpowder . . . Many brave Horse, Coaches and Chariots, with all their Rich Furniture were lost.[28]

Despite the risk, lighting was essential given that so much activity, particularly in hunting stables, took place in the early morning or after dark. Surtees's novels again highlight divergent contemporary practices. In *Hawbuck Grange*, Tom Scott comments on the numerous lanterns in the stables at Dawdle Court, the home of the rich and extravagant Lord Lazytongs. In contrast, the stables of the Goldtrap Arms not only had broken window panes and piles of dung but were only lit by candles.[29] Effective lighting took on a particular significance to the Victorians as it encouraged a clean stable, or rather allowed the lazy groom no corners lost in the gloom.

Gas was the Victorian solution to the problem of (relatively) safe, effective light. Loudon was still recommending candles in lanterns on cords in 1836, but the fact that the inn stable that Surtees's Mr Sponge visited in Laverick Wells had gas suggests that this must have been fairly commonplace by 1852.[30] Miles recommended gas lighting in 1860, although advising a separate provision of air for fear of using up all the oxygen in the stable. He also gave alternative designs for oil lamps.[31] By 1869, Armatage saw gas lights as perfectly common, only commenting that they

234 and 235 Anthony Salvin, details of the door furniture in the stables of *c.*1852–61 at Alnwick Castle, Northumberland (William C. Rolf, 1998).

should not be placed in loose boxes as horses had been known to turn them on by mistake.[32] Electricity, which eliminated the need for a naked flame, was obviously the ideal light source for stables. It was installed in the stables at Ingestre Hall, Staffordshire, built before 1892, but was still not common in 1897 when Coleman recommended electricity as the safest and best artificial light.[33]

Running water laid on to each stall was another improvement noted in the 1850s and 1860s, though grooms usually preferred to water their horses by bucket so that they knew how much each horse had consumed.[34]

All writers agreed that stables needed to be kept warm, particularly for high-bred horses such as hunters, steeple chasers and racehorses. These required greater cosseting than more robust horses and because of fashion were not allowed to grow shaggy coats. The traditional way to keep horses warm was to cover them with a blanket, but in 1862 Knightley suggested a less conventional system: 'in cold seasons stables may be colder than desirable, this may be remedied by hot-water pipes under the manger'.[35] To modern ears, the idea of centrally heated stables sounds absurdly extravagant, but both Mayhew in 1864 and Coleman in 1897 recommended it. An article on stable design in the *Building News* of 27 March 1908, said that it was not uncommon, where the horses were of special value, to provide artifical heating. The Duke of Westminster's grand stables at Eaton Hall, Cheshire, of 1870–73, still have evidence of central heating behind the stalls as do those built as part of George Skipper's lavish remodelling of Sennowe Park, Norfolk, in 1905–7.

Central heating in the stalls remained exceptional. Muthesius noted that while stables were sometimes artificially heated, either by a hot-air system or in small stables, by warm water, this was rare.[36] It was useful, however, in the tack room, for drying wet harnesses or saddles. The tack room at Badminton is still ringed with large radiators, with further radiators down the middle of the room and special racks over them for drying saddles (fig. 236).

Central heating, or a cosy coal fire, made tack rooms pleasant masculine retreats, as Anthony Trollope shows in *The Little House at Allington*, published in 1864. Adolphus Crosbie, staying at de Courcy Castle, is visited by the Hon. John de Courcy: 'Bring your [cigar] case with you', de Courcy declares, 'and come down to the harness-room, if you won't smoke here. I've had a regular little snuggery fitted up there; and we can go in and see the fellows making up the horses.'[37] The self-conscious grandeur of the early seventeenth-century stable interior may have gone, but, as Trollope reminds us, keen horsemen would visit their stables regularly and display was still an important element in stable design. At the great hunting centre of Melton Mowbray, rival hunters were keenly watched. The recreation of Sunday mornings, when there was no hunting, was to walk around the different stables discussing horses. This meant any innovation was quickly known.[38]

Such traditions survived into the twentieth century. Christopher Hussey, writing in *Country Life*, recalled a visit to Robert Tristram Combe at Earnshill, Somerset, an early eighteenth-century house, in 1938.

> We were sitting in a cosy room in the east end of the first floor when, our wives having retired to bed, he expressed anxiety about the condition of a horse on which he had been hunting that afternoon and proposed that I should inspect it with him. We descended a back staircase and, to my surprise, immediately

236 The tack room in the stables of 1878 at Badminton House, Gloucestershire (William C. Rolf, 1998).

found ourselves in the stables, where I was deputed to walk the sweating hunter up and down the drive for half an hour by moonlight.[39]

Hussey also recalled, when writing about Brocklesby Park, Lincolnshire, in 1934, how 'Till within recent times it was the custom after dinner, when the port had circulated, for the butler to announce 'The horses are bedded, my Lord', whereupon the gentlemen trooped out along the pazey [covered passage linking house to stable] to assure themselves of their mount's comfort.'[40]

One unexpected luxury found in some Victorian stables from the late 1850s was the Turkish bath. John Lawrence's ideal stable of 1798 had included a warm bath, but it was over half a century before they were seriously discussed. Walsh and Lupton recorded a new fashion in 1861 to sweat horses in a Turkish bath, perhaps inspired by the erection of a Turkish bath by the Hungarian Count Batthyay at his Melton lodge in 1861.[41] Two years earlier, the *Dublin Builder* had reported that Turkish baths had opened in Dublin and noted that 'a large apartment, with separate entrance, is provided for the application of the system to horses and other animals'.[42] Knightley referred to Turkish baths in 1862 and in 1866 Gamgee described them as being 'of late years introduced'.[43]

Gamgee, and Armatage who wrote three years later, show that the initial reaction to the Turkish baths had been over-enthusiastic. Gamgee criticised those who exaggerated their virtues for delaying their adoption, and Armatage wrote of their great popularity and of the bad results that had arisen from their indiscriminate overuse. It was not only horses that benefited from them. The Duke of Portland recalled how his trainer, Mat Dawson, had had one installed. Dawson did not always approve of horses using it, but commented to the duke: 'Fred Archer finds it a'fully useful, and spends hours in it; and when I think any of the stable-boys are o'er stoot, I send them in, too.'[44]

As ever, the most expensive extra was a riding house. Though covered rides were a cheaper way to exercise hunters and racehorses under cover, the riding house was always more flexible. As M. H. Hayes put it in his *Stable Management and Exercise* of 1900: 'A riding house is one of the greatest conveniences a hunting man, trainer or owner of polo ponies can possess; for it is not only very useful for the breaking in of saddle horses and ponies but it also supplies its possessor with the invaluable luxury of a dry and comfortable place in which to exercise his animals in wet, snowy and frosty weather.'[45]

Riding houses were particularly valued where the stud was close to the house or shared the stable. This was the case at Burton Constable, Yorkshire, in the 1840s, where Sir Clifford Constable was such an enthusiastic racing man that he built a racecourse in his park, and at Ladykirk, Berwickshire, built for David Robertson in 1850–53, whose Little Wonder won the Derby in 1840 at 50–1 (fig. 237).[46] But in 1904 it was polo that lay behind the new riding house at Moreton Hall, Warwickshire, bought by the wealthy polo-playing American Charles Garland, who founded the Moreton Morrell Polo Club.[47]

For the grandest establishments, riding houses were simply the largest of the extras that stables gained with their increasing specialisation, as was the case at Windsor Castle, Berkshire (1839) (fig. 238), Eaton Hall, Cheshire (1850), Mentmore, Buckinghamshire (1869) and Thoresby Park, Nottinghamshire (1872).[48] There were

237 H. S. Ridley, riding house of 1850–53 at Ladykirk House, Berwickshire (Mark Fiennes).

238 (*below*) Sir Jeffry Wyatville, plan and sections of the riding house at the Royal Mews, Windsor Castle, 1839 (National Archives).

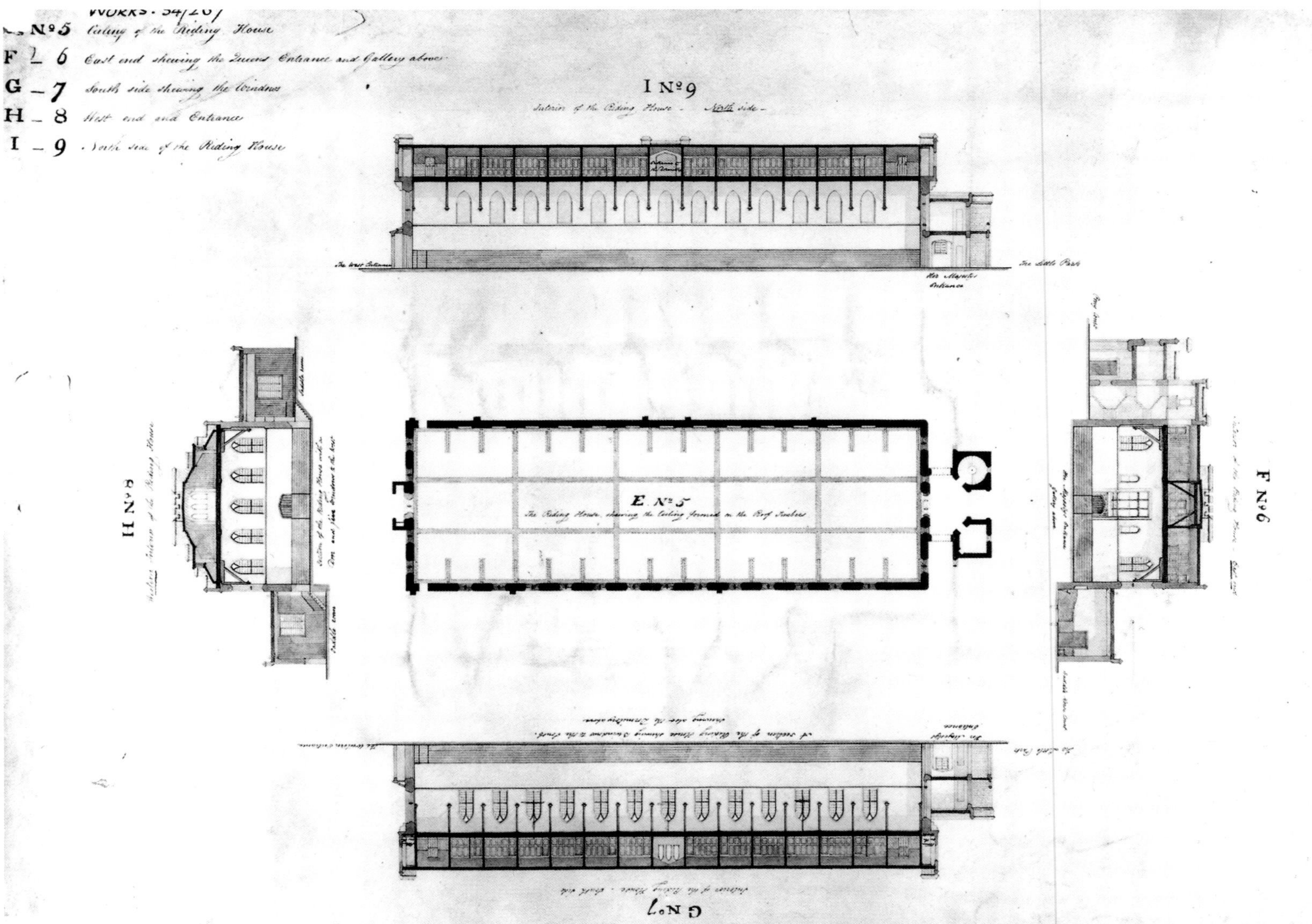

239 George Soudon Bridgman, stabling and exercising pavilion at Paignton, Torquay. from *The Architect*, 27 June 1874.

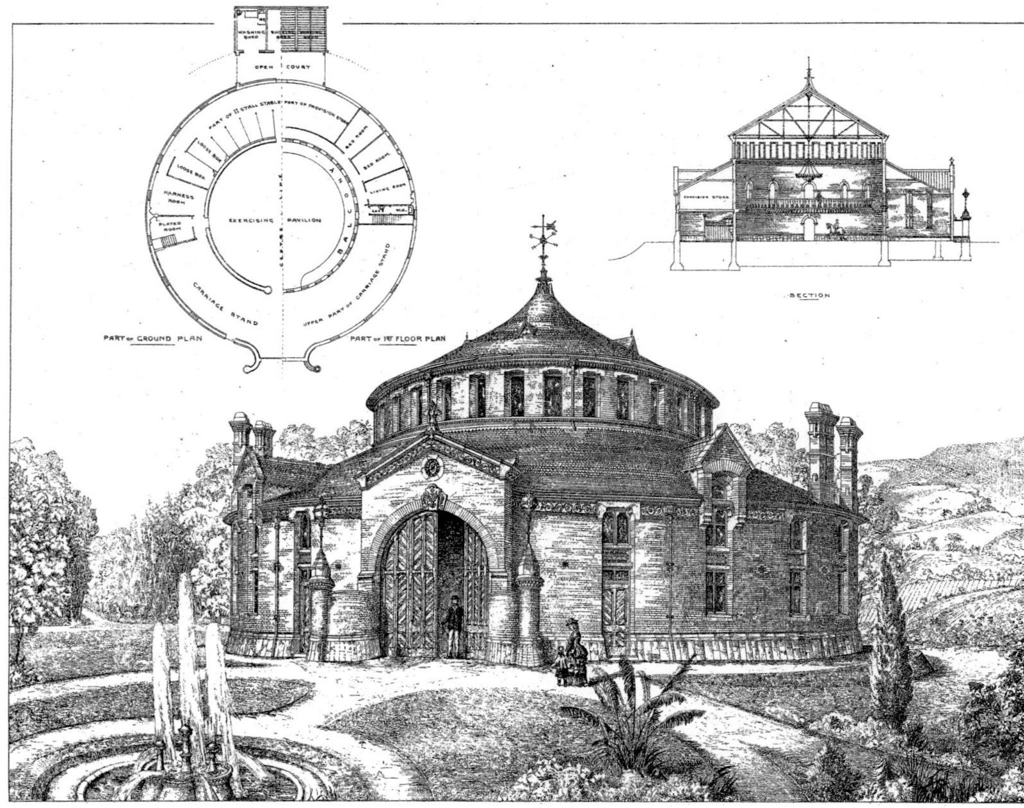

studs at both Eaton Hall and Mentmore, but the riding house was attached to the house stables. Nor were they built for display, often being hidden behind the stables, as at Eaton Hall and Thoresby Park. Instead they were a useful extra that their wealthy owners could easily afford, but unlike in the eighteenth century, they were no guarantee of an obsession with horses. Sometimes they were soon found surplus to requirements, as with that at Raby Castle, which Nimrod had described in 1835 as 'a most spacious riding-school attached [to the stables] for exercise in bad weather'. This was dismissed by the Duchess of Cleveland in 1870 as 'large and dark, with small unglazed windows on each side; no horses are ever exercised there now, and it is used only as a receptacle for the numberless waifs and strays that accumulate so rapidly in a great house'.[49]

Nineteenth-century riding houses were generally larger than those in the previous century. Hayes argues that their minimum width should be the same as that of a circus, about 43 ft. Few eighteenth-century riding houses had been as wide as 40 ft but in the nineteenth century this was fairly common and widths of 50–60 ft were not exceptional. The riding house at Burton Constable was 47 ft wide; that at Durdans built by George Devey for the fifth Earl of Rosebery's racing stables in 1881, 50 ft; the Duke of Wellington's in Knightsbridge 61 ft.[50] Most riding houses continued to be built with timber roofs using traditional collar beams and king and queen post roofs, although Salvin at Thoresby and Devey at Durdans used more complicated vaulted timber roofs. Others had metal roof structures. The Duke of Wellington's riding house, whose iron roof was singled out for description by *Building News* in 1857, was the widest unsupported private riding house ever built in Britain. An alternative was to follow the glass and metal technology being evolved for railway stations and exhibition halls, and build a barrel-vaulted top-lit riding house. The first example was that designed by C. J. Richardson for Sir Edward Filmer at East Sutton Park in 1846. Most impressive of all was the riding house at Welbeck Abbey, Nottinghamshire, built for the eccentric fifth Duke of Portland in 1862 and linked by a tunnel to the stables. At nearly 100 ft wide and 400 ft long this was the largest riding house ever built in Britain and reputedly the largest in Europe after one in Moscow (fig. 240). Such a width could not have been built by conventional means and had to be spanned by a great iron vault with glass panels, supported on either side by columns.[51]

Given this revolution in stable technology, it might be expected that large numbers of new stables would have been built during the nineteenth century, as had happened during the previous revolution in stable technology in the late sixteenth and early seventeenth centuries. New stables were built in considerable numbers, but principally for new houses or where an existing house was rebuilt. Eighteenth-century country-house stables had been solidly built on a generous scale, both in internal dimensions and in the number of horses and carriages they could hold. They could be brought up to nineteenth-century standards with a minimum of structural alteration so it would have been unnecessary and extravagant to rebuild them. As a result, the dominant impression when visiting British country houses is that the stables are Georgian, but with Victorian fittings, as at Brocklesby Park (figs 241 and 242), Lincolnshire, home of the earls of Yarborough and the Brocklesby Hunt. Here, the stable quadrangle was extensively remodelled in 1866, shortly after the succession of the third earl. From the house, the stables still appear to be a simple early eighteenth-century range, but inside are the long row of looseboxes so characteristic of the mid-nineteenth century.

The remodelling of Flitcroft's twin eighteenth-century quadrangles at Woburn Abbey, Bedfordshire, in 1903, was felt to be important enough to be described in some detail in *The Builder*:

240 The riding house of 1869 at Welbeck Abbey, Nottinghamshire (National Monuments Record).

241 Exterior of the stables at Brocklesby Park, Lincolnshire (William C. Rolf, 1999).

> The fine range of stables at Woburn Abbey, well known to visitors, which were somewhat antiquated in their fittings and arrangements, have been refitted and modernised. Many of the old stalls have been removed and in their places new loose boxes of ample dimensions have been substituted with massive iron and brass fittings. The walls have been lined with boarding and green tiles, and the old paving has been removed and adamantine clinkers laid instead. The three stables have by slight structural alterations been thrown into one. The whole of the work has been carried out by the St Pancras Ironworks Co, under the supervision of Mr G. E. Whitcombe.[52]

Despite the prevalence of remodelled stables, enough new stables were built to show a variety and invention that makes the nineteenth-century country-house stable at least as interesting as its eighteenth-century predecessor, and certainly more varied. A changing attitude towards the relationship between a house and its stables was evident in newly built houses. The eighteenth-century fashion had been to design a country house and its grounds as if the house sat alone in its landscape. This meant both offices and stables were planted out or set some distance away. Humphry Repton led the move away from this, writing in 1806, 'One of the greatest errors in Modern Gardening has been that of placing a large house, not only

242 Ninteenth-century loose boxes in the stables at Brocklesby Park, Lincolnshire (William C. Rolf, 1999).

243 'View from the Stable Yard', at Toddington Manor, Gloucestershire. From John Britton, *Toddington, Graphic Illustrations with Historical and Descriptive Accounts* (1841), pl. VIII (© The British Library).

on a naked lawn, but in the centre of it.'[53] Ten years later he expanded on his theme: 'It may perhaps be observed by the trim imitators of Brown's defects, that the stable, barns, gardens and the other appendages, ought to be removed to a distance. I have, in my former volume, endeavoured to show the folly of expecting importance in building without extent of appendages, and the absurdity of banishing to a distance those objects which are necessary to the comfort of a country residence.'[54]

The famous contrasting views of the Classical and the Picturesque country house in Richard Payne Knight's *The Landscape*, published in 1794, showed that what distinguished the Picturesque house was its variety of massing. By adding offices to a house, the symmetry was broken and a diversity of heights and shapes assured. This argument also held true for stables. By placing them as part of the spreading outbuildings, they were both conveniently to hand, rather than several hundred yards away, and added to the picturesque effect of the whole.

Toddington Manor, Gloucestershire, is a classic example of the sort of house that Repton was advocating (figs 243 and 244). One of the great Gothic houses of the early nineteenth century, it was built in 1820–35 with a keen eye for accuracy of detail by Charles Hanbury-Tracy for himself. In seeking to give his house the feel of a medieval abbey, Hanbury-Tracy designed the building as three quadrilaterals of varying heights and sizes, incorporating house, offices and stables, joined at the corners, giving a variety of mass that is truly Picturesque.

Toddington also shows how Repton and nearly all his successors cheated when it came to designing houses. One of the principal arguments they used to support their idea of the Picturesque house was that old houses had always been set among spreading outbuildings and that it was therefore absurd and unhistorical to build a house as if it sat alone. But they had no intention of recreating the forecourts of medieval and Tudor houses. The nineteenth-century owner might have been happy to have the bulk of his house emphasised by the proximity of the stables, but he had no more desire that his guests should arrive through his stableyard than that his servants should dine in the great hall.

Toddington established what became standard Victorian practice. While the exterior of the stable might be visible to the visitor approaching the front door, the stableyard and all its activities were hidden away. Shadwell Park, Norfolk, where the house and stables were designed as one unit by S. S. Teulon in 1855–60, is a typical example of a Victorian house with adjacent stables (fig. 244). The bulk of the stable, and in particular the spire over the gateway, acts as a counterpoint to the greater mass of the house, but in no way imposes upon it.[55] As always, the Victorians sought a sanitised version of the Middle Ages.

Where Victorian architects did build detached quadrangles, they tried to improve on the simple forms of their Georgian predecessors. Fitzwygram's small stables at

244 S. S. Teulon, Shadwell Park, Norfolk, of 1855–60 (*Country Life*, 1964).

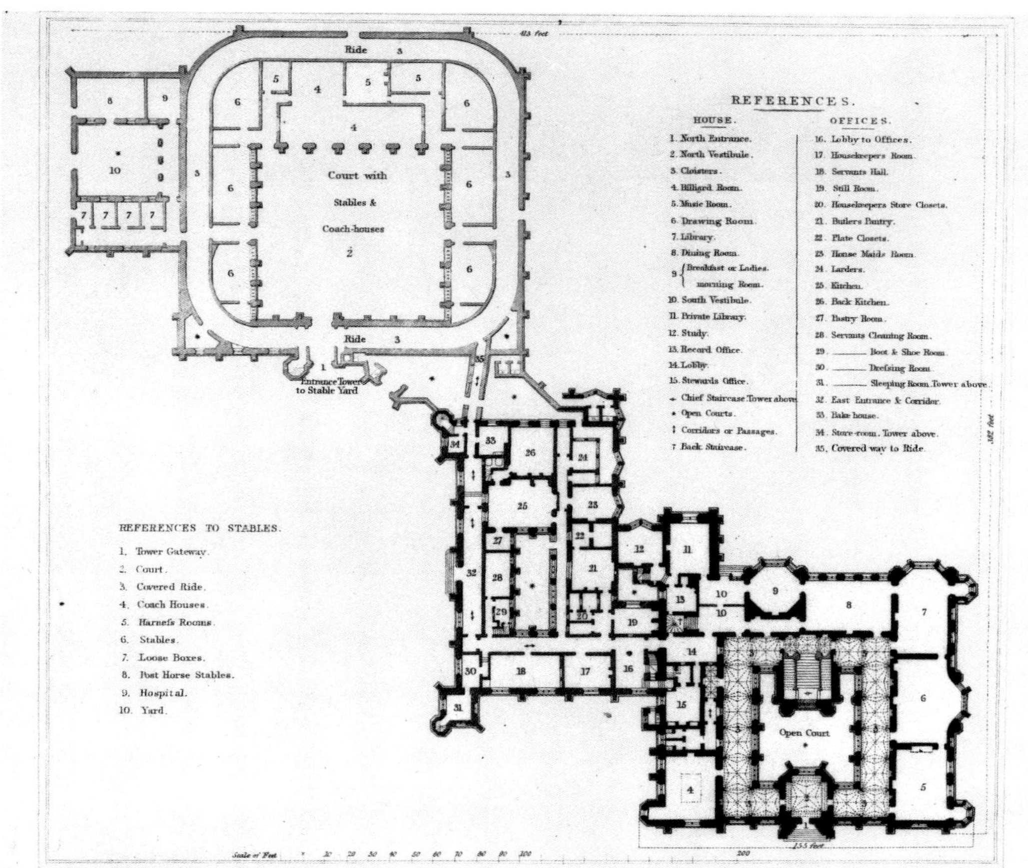

245 Plan of Toddington Manor, Gloucestershire. From John Britton, *Toddington, Graphic Illustrations with Historical and Descriptive Accounts* (1841), pl. 1 (© The British Library).

246 Exterior of the stables at Leigh Park, Hampshire (William C. Rolf, 2000).

Leigh Park combine a romantically vernacular approach to design with an iron-framed, glazed canopy sheltering half the yard that would not look out of place on the platform of a country railway station (fig. 246). At Mottisfont Abbey, Hampshire, where the stables were rebuilt by Sir John Barker-Mill, a noted master of hounds despite being a clergyman, in 1836, the roof oversails dramatically into the yard, giving the building a carefully contrived 'primitive' feel, as well as practical, rain-free access to the stables (fig. 248).[56]

At Wimpole Hall, Cambridgeshire, H. E. Kendall self-consciously set out to rival the great Georgian quadrangles (figs 251 and 252). Here, unusually, the original 1690s stable block in front of the house had survived the eighteenth-century fashion for demolishing forecourt stables and replacing them with a distant quadrangle. This finally happened in 1851 when Kendall built the handsome Baroque Revival stables as part of major improvements to the house for the fourth Earl of Hardwicke. With its dramatic tower recalling John James's St George's Church, Hanover Square, its vibrant red bricks set off with stone, and its lively skyline punctuated by the Hardwicke lion and a stag, segmental pediments and stone balls the Wimpole stables certainly give their eighteenth-century predecessors a run for their money.

247 (*facing page top left*) The stables of 1836 at Mottisfont Abbey, Hampshire (William C. Rolf, 2000).

248 (*facing page top right*) The stables at Mottisfont Abbey, Hampshire (William C. Rolf, 2000).

249 (*facing page bottom left*) Aisle and loose boxes in the stables at Mottisfont Abbey, Hampshire (William C. Rolf, 2000).

250 (*facing page bottom right*) Loose box in the stables at Mottisfont Abbey, Hampshire (William C. Rolf, 2000).

251 H. E. Kendall, the stables of 1851 at Wimpole Hall, Cambridgeshire (William C. Rolf, 2000).

252 H. E. Kendall, the courtyard of the stables of 1851 at Wimpole Hall, Cambridgeshire (William C. Rolf, 2000).

253 Alfred Waterhouse, the stables of *c.*1870–82 at Eaton Hall, Cheshire (William C. Rolf, 1999).

Inside it used the latest technology to create a cast-iron arcade providing convenient covered access to the stables in bad weather.

But impressive though the stables at Wimpole are, nothing can compare with the ducal splendours of the Duke of Westminster's stables at Eaton Hall, Cheshire, part of the magnificent complex created by Alfred Waterhouse at a cost of £600,000 in 1870–82 (figs 253–56). The duke, who as a young man was the Master of the Cheshire Hunt and later in life owned the largest and most successful racing stables in England, winning the Derby five times, built truly magnificent stables that survived the tragic demolition of the house in 1963. Everything in the redbrick and

half-timbered stables is of the highest quality, from the carefully designed brass door furniture to the massive equestrian statue by Sir Joseph Boehm rearing in the centre of the quadrangle. No luxury is missing in a stable that had its own riding house, heated stalls and large glazed roof over the carriage yard.

By 1900, it was thought that about 200,000 horses were kept exclusively for hunting with 300 different hunts.[57] Many of them would have been in hunt stables or hunting lodges. The grandest hunt stables were paid for by an individual. The stables at Billesdon in Leicestershire, which Tattersall praised in 1841, were built in 1838 by Thomas Smith for Lord Suffield, who managed to bankrupt himself after one year as master of the Quorn.[58] Those of the brewer Sir Gilbert Greenall, first Lord Daresbury, Master of the Belvoir from 1896 to 1912, must have been among the most impressive. Taking over the hunt, which had previously been run by the dukes of Rutland from Belvoir Castle, in March 1896, he discovered that there was no suitable house for him. So he built one at Woolsthorpe, Lincolnshire, with stabling for seventy horses, a circular exercising ring, valeting rooms, a forge, cottages for the stud groom and others, an underground telephone line to the kennels and even a special siding off the railway line. The stables were ready for the horses by 1 November.[59] But he had rivals. His neighbour at the Cottesmore from 1907, Lord Lonsdale, had over one hundred hunters in his stables.[60]

The stables of the dukes of Beaufort at Badminton House, Gloucestershire, were no less impressive (figs 257, 258 and 259). The stables of the seventh duke, who succeeded in 1835, held thirty-four regular hunters and ten horses 'able to go a-hunting', including twelve for the duke, eight for the huntsman and fifteen for whips and friends.[61] These were replaced by the present stables in 1880 by the ninth

254 (*above left*) Alfred Waterhouse, the carriage house of *c.*1870–82 at Eaton Hall, Cheshire (William C. Rolf, 1999).

255 (*above right*) Alfred Waterhouse, interior of the stables of *c.*1870–82 at Eaton Hall, Cheshire (William C. Rolf, 1999).

256 Detail of door furniture in the stables of *c.*1870–82 at Eaton Hall, Cheshire (William C. Rolf, 1999).

257 The courtyard of the stables of 1878 at Badminton House, Gloucestershire (William C. Rolf, 1998).

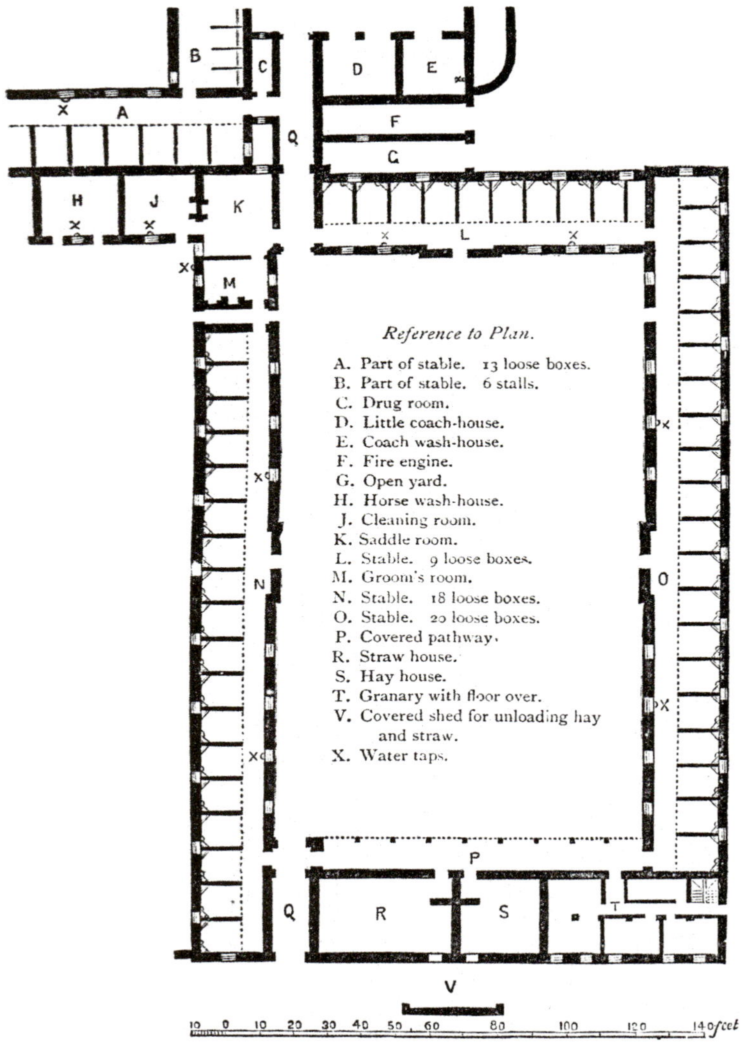

258 (*right*) Plan of the stables at Badminton House, Gloucestershire, from C. F. Somerset and the Duke of Beaufort, *Hunting* (London, 1889).

259 (*facing page*) Loose boxes in the stables of 1878 at Badminton House, Gloucestershire (William C. Rolf, 1998).

duke, master of the Beaufort from 1871 to 1908. He hunted four to six days a week for forty odd seasons and liked to have seventeen or eighteen horses for his own use. The stables had 60 loose boxes, enough for all the hunters, together with stalls for hacks, making about eighty horses in all, looked after by forty grooms. Extravagant as this may seem, no money was wasted on external architectural ornament.[62] Naturally, when the duke came to write his manual on hunting it was the Badminton stables that he featured as his model.[63]

Not all hunts could rely on a rich individual to provide the hunt stables. The death of Hugo Meynell-Ingram, who had borne all the expenses of the Meynell Hunt himself, led to the appointment of joint masters in 1871 and forced the building of new hunt stables. The committee advertised in *The Builder* and from the many replies chose the plans of the architects Giles and Brookhouse as having the lowest estimate. However, when Lord Vernon, who was providing the site, saw the proposed elevation he objected strongly and declared that Giles and Brookhouse's estimates were manifestly untrustworthy. He suggested his own architect George Devey. This was accepted and the stables and kennels were built at a final cost of £10,008 (£5,095 for the stables). Unfortunately, this was £5,324 more money than the hunt had raised, leaving it seriously in debt for several years.[64]

Among others, Knightley and Birch both featured model hunting stables, and accounts of hunting stables, especially those built with no expense spared for the grandest hunts, proliferated.[65] The wealthy brothers Horatio and Julius Behrens hired a special train to bring their forty hunters and grooms from Cheshire to Newport Lodge when they took it in 1869.[66] Birch's design for a hunting lodge in 1892 had stabling for fifteen horses, including two hacks.[67] Another of Birch's hunting-box designs was rather smaller with stabling for only four or five horses.[68] The last of the great hunting boxes built at Melton Mowbray was Hamilton Lodge, built for Lord Hamilton of Dalzeel in 1902.[69]

In the Shires, even large country houses with substantial estates were often little more than glorified hunting boxes. This was certainly the case at Stapleford Park, Leicestershire, built by the wealthy brewer John Gretton in 1894, who was desperate to be accepted into society (figs 260, 261 and 262). The best way to do so was to take up hunting. For that reason he bought Stapleford, which was only five miles from Melton Mowbray, built extensive stabling, and had Micklethwaite remodel his house for entertaining. Unfortunately, he died in 1899, before he had fitted up his new long gallery that would have been fit for a hunt ball.[70]

The second half of the nineteenth century was also the great age of the Newmarket training stable. The stables at Exeter House, built round a small quadrangle on the edge of town, pointed the way forward. Thereafter, quadrangles became commonplace and only the smaller establishments retained their old positions behind houses on the High Street. But though the stables grew in scale, they initially remained plain, utilitarian buildings. Knightley commented, not entirely favourably, on the simplicity of the stables at Palace House, built for Baron Meyer de Rothschild around 1857–60. He also illustrated the eleventh Duke of Bedford's stables, which were well proportioned, built of flint with brick trimming, but not much more elaborate. Both still had a significant number of stalls, eleven stalls to eleven loose boxes in one, and fifteen to twelve in the other.[71] By contrast the racing stables built in 1884 at Chetwynd House, Newmarket, now known as

Machell Place built by W. C. Manning for R. Sherrard, had forty-two loose boxes and only twelve stalls.[72]

Machell Place was discussed and illustrated in the Howard and Craven's *Racing and Steeple-chasing* of 1889, where it is described as 'exceptionally well arranged' (figs 263–66). The stable was built round three sides of a yard, with the trainer's house on the fourth side so he could keep an eye on what was going on. The description picked out key points. Security was obviously important, with the headman's bedroom situated immediately over [the archway] to ensure security of the premises at night. The boxes had four different forms of ventilation: a hit and miss brass ventilator fixed in the outside door just above the bottom rail, a fanlight over the door opening inwards hung on the transom, a ventilator in the back wall over the horse's head with sloping boarded casing to prevent downward draught, and finally a trunk ventilator running along the ridge of the roof to carry off any 'vitiated air'. Each box was lit by a single-jointed gas bracket and the building was constructed of fireproof concrete and iron floors, with three large watertanks in case of fire. In all there were thirty-two loose boxes, all opening onto the yard, though with interconnecting doors for ease of access. The boxes were fitted with slate mangers, the walls lined with boarding six feet high protected with hoop iron. The floor was a perfectly horizontal plane so that the horses could stand in their natural position and avoid any unnatural strain on their frame. A backyard contained 'manure pit, pigstyes and other necessary adjuncts to an establishment of this class, which would detract from the ornamental character of the yard'.[73]

This last point is revealing, for Machell Place is decidedly decorative, with strong elements of the fashionable Queen Anne style and a large Dutch gable with a sculpted horse's head and datestone at the entrance. Newmarket's Edwardian heyday saw a series of grand, architecturally impressive stables in marked contrast to earlier years. Abington Place, built in 1889, is dominated by an elaborately gabled clock tower. But nothing matches the splendour of the stables at Egerton Lodge of 1891, where the Prince of Wales's horses were kept. Here, the central range with its brick basement and half-timbered, gabled upper floor takes on an almost Wagnerian air. The Newmarket racing stable had finally caught up with its country-house rival in grandeur and eclipsed all but the most ducal.

By the 1890s, stables had reached the limits of their development and a consensus had emerged among writers that little could be done to improve them. As Byng Giraud wrote, the existence of horses in European countries 'has become one of greater comfort than in the past, and such stables as were formerly common are now only met with where the owner is heavily handicapped by poverty'.[74] Professor Axe was positively ecstatic in 1907:

> In no other country so much as in Britain is the horse at once the friend and the companion of man, and in no country is he so well housed. The arrangement and construction of a gentleman's stables are of an importance second only to that of the dwelling house itself; indeed it is to be feared that in some cases the accommodation provided for his equine servants claims more thought and care than that provided for his human ones.[75]

The German author Hermann Muthesius, who came to make a serious study of English country houses, at the beginning of the twentieth century, concurred: 'With

260 Peter Dollar, interior of the stables of 1899 at Stapleford Park, Leicestershire (William C. Rolf, 1998).

261 (*right*) Bird's-eye perspective of Peter Dollar's stables of 1899 at Stapleford Park, Leicestershire.

262 (*facing page*) Peter Dollar, the stables of 1899 at Stapleford Park, Leicestershire (William C. Rolf, 1998).

263 Plan of W. C. Manning's stables of 1884 at Machell Place, Newmarket. From H. C. Howard and W. G. Craven, *Racing and Steeple-chasing* (1889).

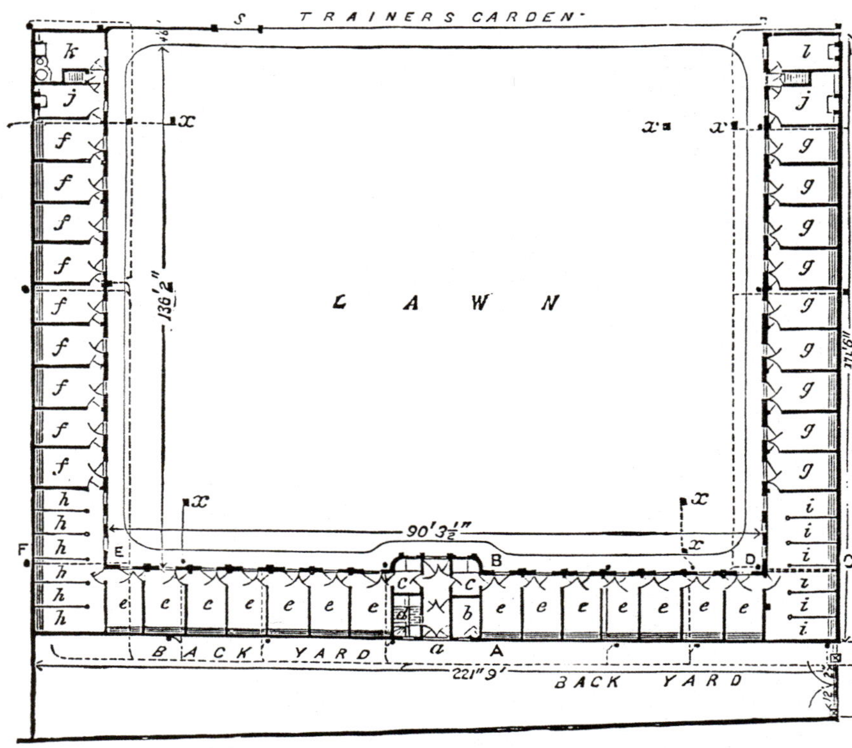

a, entrance passage; *b*, office; *c c*, lobbies; *d*, stairs; *e e*, 14 boxes; *f f*, 9 boxes; *g g*, 9 boxes; *h h*, 6 stalls; *i i*, 6 stalls; *j j*, saddle-rooms; *k*, drying-room and mash-house; *l*, boys' sitting-room; *x x*, gulleys drains, dotted lines.

264 (*right*) W. C. Manning the Clock Tower Gate to the stables of 1884 at Machell Place, Newmarket (William C. Rolf, 2000).

265 W. C. Manning, detail of the Clock tower Gate to the stables at Machell Place, Newmarket (William C. Rolf, 2000).

266 W. C. Manning, doors to the boxes of the stables at Machell Place, Newmarket (William C. Rolf, 2000).

267 John Kinross, the yard of the stables of 1895 at Manderston House, Berwickshire (William C. Rolf, 1998).

all the care that for centuries . . . has been lavished on horse-breeding in England, it is hardly surprising that the best stabling is to be found in England.'[76]

Very large stables continued to be built in the first years of the twentieth century, as can be seen in the pages of the building press. In one single issue, the *Building News* of 9 June 1905, two very substantial quandrangles were featured, both built by Ernest George and Yeates, one for Gerald Hardy at Foston Hall, Derbyshire, and the other for Lionel Dugdale at Crathorne Hall, Yorkshire. On 31 December 1910, the very eve of the decade that ended the primacy of the horse, *The Builder* illustrated the stables built by George Skipper for Thomas Cook at Sennowe Park, Norfolk. With room for nine coaches, twelve loose boxes, nine stalls, a harness room entirely fitted out in teak, and central heating in the stables, these could be described as the swansong of the country-house stable, were that title not more justifiably given to the stables at Manderston, Berwickshire.[77]

When Sir William Miller, heir to a great fortune made from Baltic and Russian trade, began to remodel Manderston he started with the stables designed by John Kinross (figs 267–74). Built around a quadrangle in an elegant Adamesque manner, these were a conscious homage to the great age of the Georgian stable, but with all the luxury associated with the Edwardians. The entrance was through a beautifully cut stone pilastered archway into a single-storey yard that was not economically finished in brick, as its eighteenth-century model would have been, but in beautifully cut ashlar. The tie-up rings are held by fierce lions' heads on swagged

268 John Kinross, the interior of the stables of 1895 at Manderston House, Berwickshire (William C. Rolf, 1998).

269 John Kinross, a stall in the stables at Manderston House, Berwickshire (William C. Rolf, 1998).

270 John Kinross, the tackroom of the stables of 1895 at Manderston House, Berwickshire (William C. Rolf, 1998).

271 (*below left*) A tie-up in the yard of the stables at Manderston House, Berwickshire (William C. Rolf, 1998).

272 (*below right*) A heel post in the stables at Manderston House, Berwickshire (William C. Rolf, 1998).

273 and 274 Details of the friezes on the right and left arches in the yard of the stables at Manderston House, Berwickshire (William C. Rolf, 1998).

pedestals and a scene of horses, hounds and fox enlivens the interior of the arch. As in racing stables, the loose boxes open directly onto the yard, but the high point is the coach-horse stables. This is a noble barrel-vaulted space with a complex system of ventilation above the stalls and in the roof, brass heel-posts and mangers, and the horses' names picked out in gold on marbled plaques above the stalls. This was a stable that even Sir Balthazar Gerbier would have admired.

The stables were begun in 1895. In 1901 Miller began to rebuild the main house. Built to the most modern specification, this included a series of garages, for which less than a decade before there had been no need. Manderston encapsulates the sudden obsolescence of the country-house stable.[78]

For a short time, the horse held off the challenge of the motor car. A motor car was an exciting and adventurous way to make calls, but for matters where punctuality was concerned, the horse and carriage prevailed. In the *Building News* for 27 March 1908, an article discussing the requirements of the modern stable included two garages as well as four coach houses, noting that 'the engine house and garage are modern additions, not always required but now generally found'. This was still essentially a stable, but in 1913 the *Building News* Design Club ran a competition for a 'Motor Garage and Chauffeur's Lodge' (fig. 275). The results were inevitably based on traditional stable design, but whatever the disguise, there were no horses.

Even before the First World War, the horse had been superseded by the motor car as the primary means of personal transport. The effect on the stable was dramatic, for as the coach houses were converted into garages, the establishment of horses was effectively halved. Riding for pleasure was now predominant, and where large stable establishments survived, it was principally for hunting. Even here, in

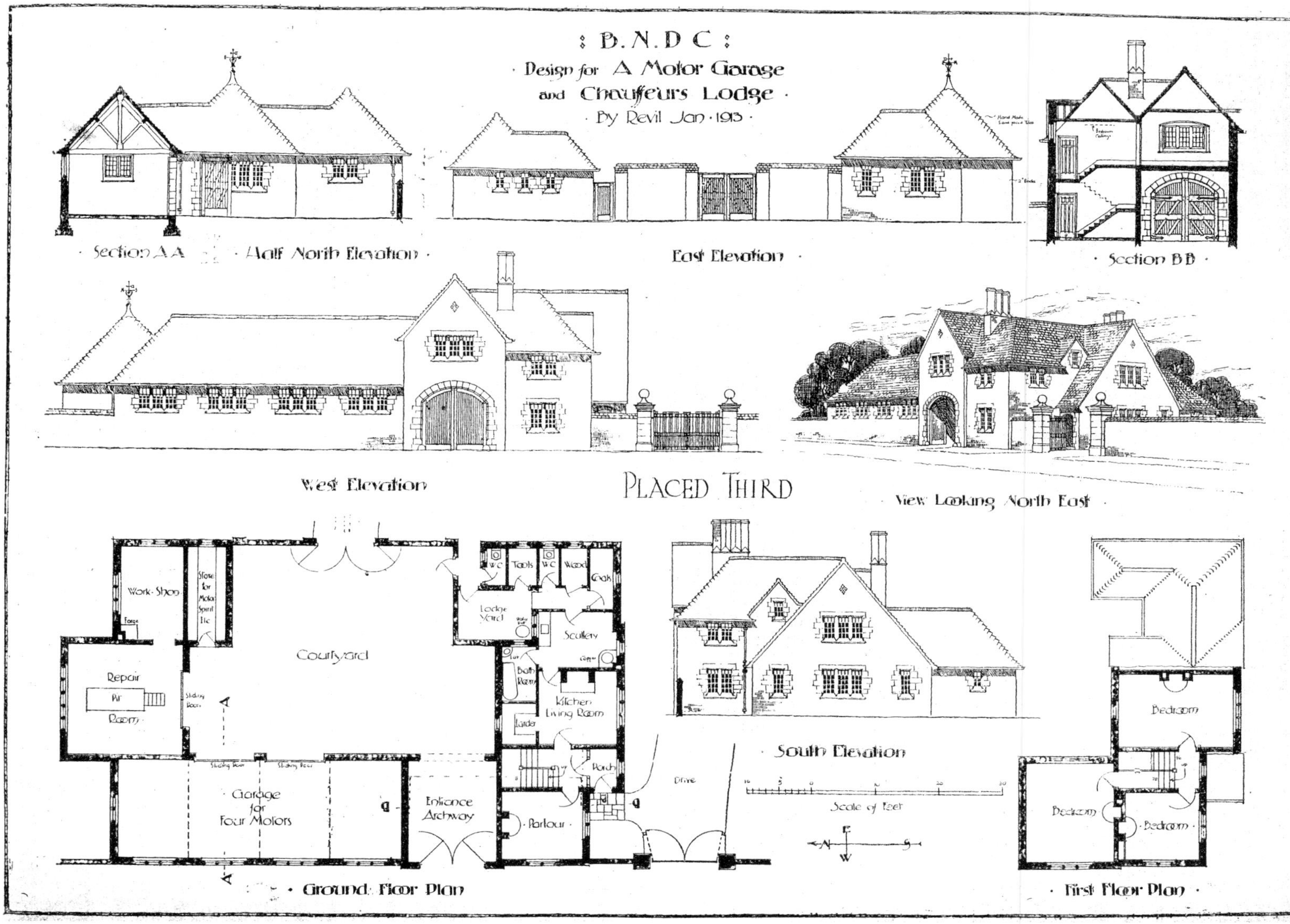

the bleaker post-war economic climate, there seemed little justification for building new stables. The old ones would do. The country-house stable tradition survived strongly into the twentieth century. Then it stopped dead. It has never recovered.

275 'Design for a Motor Garage and Chaffeur's Lodge', from the *Building News* 'Designing Club', 7 February 1913.

NOTES

INTRODUCTION

1 Joan Thirsk, *Horses in Early Modern England: For Service, For Pleasure, for Power* (Reading, 1978), p. 5.
2 François de la Rochefoucauld, *A Frenchman in England*, ed. Jean Marchand, trans. S. C. Roberts (Cambridge, 1933), p. 74.
3 Sir Osbert Sitwell, *The Scarlet Tree* (London, 1948).

CHAPTER I

1 David Barker, 'Excavations at Elstow Abbey, Bedfordshire, 1968–70, 3rd Interim Report', *British Archaeological Journal* VI, 1971, pp. 55–64.
2 P. A. Faulkener, 'Castle Planning in the Fourteenth Century', *Archaeological Journal* CXX, 1963, 225–30, fig. 8.
3 Helen Clarke, *The Archaeology of Medieval England* (London, 1984), pp. 81–83. Even recent survey volumes that do discuss the outer court pay little attention to stables. Both J. Patrick Greene's *Medieval Monasteries* (Leicester, 1992), and Michael Aston's *Monasteries* (London, 1993), discuss stables only in passing.
4 Glyn Coppack, 'Thornholme Priory: The Development of a Monastic Court', in *British Archaeological Reports* CCIII: *The Archaeology of Rural Monasteries*, ed. R. Gilchrist and H. Mytum (Oxford, 1989), pp. 185–222. Despite his comments, however, Coppack's study makes no reference to the stables that must have been an important element in the outer court at Thornholme Priory.
5 Neil S. Rushton, 'Spatial Aspects of the Almonry Site and Changing Priorities of Poor Relief at Westminster Abbey c.1290–1540', *Architectural History* XLV, 2002, pp. 66–67.
6 Anthony Emery, *Greater Medieval Houses of England and Wales, 1300–1500* (Cambridge, 1996–2000). Even where medieval buildings have been traditionally described as stables there is no guarantee that they were originally used for horses. The building described as the Archbishop's Stables at Maidstone, Kent, is likely to have been a great stone barn at the centre of the Archbishop of Canterbury's estates. (E. E. Rigold, 'Some major Kentist Timber Barns', *Archaeologia Cantiana* LXXXI, 1966, p. 2.
7 R. H. C. Davies, 'The Medieval Warhorse', in F. M. L. Thompson (ed.), *Horses in European Economic History* (Reading, 1983), p. 8.
8 C. D. Ross, 'Household Accounts of Elizabeth Berkeley, Countess of Warwick, 1420–21', *Transactions of the Bristol and Gloucestershire Archaeological Society* LXX, 1951, pp. 86–90, 96.
9 Davies, 'The Medieval Warhorse', p. 5.
10 Edward Bateson, *History of Northumberland* I (Newcastle, 1893). p. 37; H. Ellis (ed.), 'A memoir of the state of Norham Castle in the time of Henry VIII', *Archaeologia* XVII, 1814, p. 203.
11 H. M. Colvin (ed.), *The History of the King's Works* (London, 1963–82), II, p. 919.
12 Davies, 'The Medieval Warhorse', pp. 7–11.
13 M. W. Thompson, *Pickering Castle* (London, 1958), pp. 10, 21–22.
14 Herbert J. Hewitt, *The Horse in Medieval England* (London, 1983), p. 27.
15 Thomas Wright (ed.), 'Letters relating to the suppression of the monasteries', *Camden Society* XXVI, 1843, p. 158.
16 A. B. Whittingham, 'Bury St Edmunds Abbey', *Architectural Journal* CVIII, 1951, p. 177.
17 John H. Harvey (ed.), *William Worcestre: Itineraries* (Oxford, 1969), pp. 218–19, 221; R. B. Pugh (ed.), *Victoria County History of England*: *Warwickshire* VIII (London, 1981), pp. 457– 58; Mark Girouard, 'The Smythson Collection of the Royal Institute of British Architects', *Architectural History* 5, 1962, 1/21. I am grateful to Dr John Goodall for drawing my attention to these stables. *The Victoria County History of England* argues convincingly that the Beauchamp stables survived until the Civil War.
18 W. H. St John Hope, *Windsor Castle* I (London, 1913), pp. 39, 41, 111; Francis Grose, *Antiquities of England and Wales* (London, 1776), unpaginated; Whittingham, 'Bury St Edmunds Abbey', pp. 169–86; William Brown, 'Description of the Buildings of Twelve Yorkshire Priories at the Reformation', *Yorkshire Archaeological Journal* IX, 1886, pp. 197–215, 321–33.
19 Anthony Emery, *Dartington Hall* (Oxford, 1970), p. 155; Peter Drewett, 'The Archiepiscopal Palace at Croydon', *Archaeological Journal* CXXVIII, 1971, pp. 162–65; John Willis Clark, *The Architectural History of the University of Cambridge and the Colleges of Cambridge and Eton* (Cambridge, 1886), p. 356.

20 J. T. Fowler (ed.), 'Cistercian Statutes', *Yorkshire Archaeological Journal* IX, 1886, p. 341; *Victoria County History of England: North Riding* II (London, 1923), p. 25.

21 A. J. Taylor, *Minster Lovell Hall* (London, 1958), p. 18.

22 J. T. Fowler (ed.), *Rites of Durham* (Durham, 1903), p. 91.

23 Joan Greatrex, 'St Swithun's Priory in the later Middle Ages', in John Crook (ed.), *Winchester Cathedral: Nine Hundred Years 1093–1993* (Winchester, 1993), fig. 11.1. I am grateful to Dr John Crook for sharing his knowledge of this building and for his cautious endorsement of its probable identification as a medieval stable. The identification of substantial excavated remains at Reading Abbey as stables is similarly reliant on circumstantial evidence of later documented usage (John Hawkes, 'Excavations on the site of the Reading Abbey Stables, 1983', *Berkshire Archaeological Journal* LXXIII, 1986–90, pp. 67–87).

24 Essex Record Office, D/DmhPI; Anthony Emery, *The Greater Houses of England and Wales II: East Anglia, Central England and Wales* (Cambridge, 2000), p. 445; N. W. Alcock *et al.*, 'Maxstoke Castle, Warwickshire', *Archaeological Journal* CXXXV, 1978, p. 228.

25 Faulkener, 'Castle Planning in the Fourteenth Century', pp. 225–30, fig. 8. Faulkener's assumption that the stables must be within the body of the castle is not questioned by Anthony Emery. See Anthony Emery, *Greater Medieval Houses of England and Wales, 1300–1500* I (Cambridge, 1996), fig. 74.

26 F. A. Faulkener, 'Some Medieval Archiepiscopal Palaces', *Archaeological Journal* CXXVII, 1970, p. 132; W. Hutchinson, 'The Ancient Manor House of the Bishop of Durham at Howden, Yorkshire', *Yorkshire Archaeological Journal* IX, 1886, p. 389.

27 Colvin, *The History of the King's Works* I, p. 551; H. M. Colvin, 'Four Medieval Building Contracts', *Architectural History* II, 1959, p. 25; Colvin, *The History of the King's Works* II, p. 997; F. C. E. Erwood, 'The Praemonstratensian Abbey of Langley, Co. Norfolk', *British Archaeological Association Journal* n.s. XXVIII, 1922, pp. 101-3; M. W. Thompson, *Kenilworth Castle* (London, 1977), pp. 23–24; W. D. Simpson (ed.), 'The Building Accounts of Tattershall Castle, 1434–72', *Lincoln Record Society* LV, 1960, p. 65.

28 Thomas Arnold (ed.), 'Memorials of St Edmund's Abbey', I, *Rolls Society* XCVI, 1890, p. 297; Hope, *Windsor Castle*, I, p. 93; Colvin, 'Four Medieval Building Contracts', p. 25; Graham J. Dawson, 'The Black Prince's Palace at Kennington, Surrey', *British Archaeological Reports* XXVI, 1976, p. 56; Brown, 'Twelve Yorkshire Priories at the Reformation', IX, 1886, pp. 197–215, 321–33; P. A. Rahtz, 'Excavations at King John's Hunting Lodge, Writtle, Essex, 1955–57', *Society for Medieval Archaeology Monograph Series 3*, 1969, p. 8; J. C. Hodgson, *The History of Northumberland* V (Newcastle, 1899), p. 51.

29 R. Wilcox, 'Excavations at Farleigh Hungerford Castle, Somerset, 1973–76', *Somerset Archaeological and Natural History Society* CXXIV, 1979/80, p. 98–99, fig. 5; Susan Young *et al.*, 'Medieval Britain and Ireland in 1982', *Medieval Archaeology* 27, 1983, p. 211; Norman Drinkwater, 'The Old Deanery, Salisbury', *Antiquaries Journal* XLIV, 1964, p. 58; J. H. Harvey, 'Great Milton, Oxfordshire, and Thorncroft, Surrey: The Building Accounts of Two Manor Houses of the Late Fifteenth Century', *British Archaeological Association Journal* 3rd ser. XVIII, 1955, p. 50; Henry Littlehales, 'The Medieval Records of a London City Church', *Early English Text Society* CXXV, 1904, p. 29; David Dymond, 'A 15th-Century Building Contract from Suffolk', *Vernacular Architecture* 9, 1978, pp. 10–11.

30 W. H. St John Hope and J. Bilson, 'Architectural Description of Kirkstall Abbey', *Thoresby Society* XVI, 1907, p. 62; Wilcox, 'Excavations at Farleigh Hungerford Castle, pp. 98–99.

31 D. J. Cathcart King and J. Clifford Perks, 'Carew Castle, Pembrokeshire', *Archaeological Journal* CXIX, 1962, p. 274; John Leland, *The Itinerary*, ed. L. T. Smith I, i (London, 1964), p. 79; A. E. Preston, 'The Demolition of Reading Abbey', *Berkshire Archaeological Journal* XXXIX, 1935, p. 137.

32 Sir Thomas Phillipps (ed.), 'Survey of the Manor and Forest of Clarendon, Wiltshire, 1272', *Archaeologica* XXV, 1834, p. 155.

33 Clement Gardet (ed.), *Le Livre d'Heures du Duc Louis de Savoie* (Annecy, 1959) pl. VI; M. Meiss, *French Painting in the Time of Jean de Berry: The Boucicaut Master* (London, 1968), figs 90, 339.

34 Hope, *Windsor Castle*, pp. 93, 102; Dymond, 'A 15th-Century Building Contract from Suffolk', pp. 10–11.

35 John Plummer, *The Last Flowering* (New York and London, 1982), pl. 73; Meiss, *French Painting in the Time of Jean de Berry*, figs 339.

36 Gaston Phoebus, *The Hunting Book*, ed. Gabriel Bise (London, n.d.), p. 58; Bibliotheque National, MS Francais 66.

37 Emery, *Dartington Hall*, p. 145; H. Gordon Slade, 'Broughton Castle, Oxfordshire', *Archaeological Journal* CXXXV, 1978, p. 152; T. H. Turner, *Some Account of Domestic Architecture from the Conquest to the 13th Century* (Oxford, 1851), pp. 202, 231.

CHAPTER 2

1 W. B. Rye, *England as Seen by Foreigners in the Days of Elizabeth and James the First* (London, 1865), pp. 45, 258; Anne Clifford, *The Diary of Lady Anne Clifford*, ed. Vita Sackville-West ed. (London, 1923), p. 78.

2 Thomas Blundevil, *The Fower Chiefyst Offices Belonging to Horsemanshippe* (London, 1565), p. 10.

3 H. M. Colvin (ed.), *The History of the King's Works* (London, 1963–82), III, pp. 80, 80n; IV, p. 221, fig. 11 (London, 1975 and 1982). Loseley Hall, Surrey, was built in the 1560s for £1,600 19s 7d, see Malcolm Airs, *The Tudor and Jacobean House: A Building History* (Stroud, 1995), p. 99.

4 Simon Thurley, *The Royal Palaces of Tudor England* (New Haven and London, 1993), pp. 70–72, figs 93–94.

5 Public Record Office, MR 1/329 (1). The survey of 1694 is reproduced in John Bold, *Greenwich* (New Haven and London, 2000), fig. 130. The roofline of the stables can be seen in Anthonis van Wyngaerde's drawing of Greenwich of *c.*1558 reproduced in Thurley, *Royal Palaces of Tudor England*, fig. 67.

6 I am grateful to Dr Simon Thurley for drawing my attention to the significance of the Greenwich stables and for providing me with transcripts of accounts and copies of plans.

7 I am grateful to Dr Simon Thurley for drawing these stables to my attention and for passing on details about the relevant building accounts.

8 Jean-Marie Pérouse de Montclos, *Fontainbleau* (London, 1998), pp. 165–86, 228–31.

9 Peter Ellis (ed.), 'The Elizabethan Gardens

and Leicester's Stables at Kenilworth Castle', *Birmingham and Warwickshire Archaeological Society Transactions*, 1995, p. 99.

10 William Dugdale, *The Antiquities of Warwickshire* (London, 1730), pp. 243, 249.

11 Ellis, 'The Stables at Kenilworth Castle', p. 115.

12 John Harris, *The Artist and the Country House* (London, 1979) pl. 155.

13 Gervase Markham, *Cavelarice or The English Horseman* (London, 1607) V, pp. 3–4; Thomas de Grey, *The Compleat Horseman and Expert Farrier* (London, 1639), pp. 1–2.

14 de Grey, *The Compleat Horseman*, p. 10.

15 Harris, *The Artist and the Country House*, pl. 83; Giles Worsley, 'The Osterley Stables', in Tim Knox and Anthea Palmer (eds), *Aspects of Osterley* (High Wycombe, 2000), pp. 27–34.

16 John Dent, *The Quest for Nonsuch* (London, 1962), pp. 291–92; Jennifer C. Ward, *Old Thorndon Hall* (Chelmsford, 1972), pp. 6–7; John Humphreys (ed.), 'The Elizabethan Estate Book of Grafton Manor near Bromsgrove', *Transactions of the Birmingham Archaeological Society* XLIV, 1918, p. 39; Basil Stallybrass, 'Bess of Hardwick's Buildings and Building Accounts', *Archaeologia* LXIV, 1913, p. 375.

17 Blundeville, *Horsemanshippe*, p. 10; de Grey, *The Compleat Horseman*, p. 11; Survey of Mr Henry Smythe's lands at Cossam 1602, Longleat Archives no. 6711, quoted in F. J. Ladd, *Architects at Corsham Court* (Bradford-on-Avon, 1978), p. 8; Caroline Stanley-Millson and John Newman, 'Blickling Hall: The Building of a Jacobean Mansion', *Architectural History* XXIX, 1986, pp. 33–35.

18 John Gage, *The History and Antiquities of Hengrave Hall in Suffolk* (London, 1822), pp. 17, 21–22, pl. 1.

19 John Heward and Robert Taylor, *The Country Houses of Northamptonshire*, (Swindon, 1996), fig. 321.

20 Edward Roberts and Pat Grover, 'Elizabethan Riding Stables at Chawton and their Context', *Proceedings of the Hampshire Field Club Archaeological Society* LII, 1997, pp. 151–64.

21 Hampshire Record Office, 5M 53/1562–66; Grover, 'Elizabethan Riding Stables at Chawton', pp. 158–61, figs 5–7.

22 I am grateful to Dr Mark Girouard for alerting me to these stables and providing me with photographs.

23 Christopher Hussey, 'Brocklesby Park, Lincolnshire', *Country Life*, 24 February 1934.

24 Newman, 'Blickling Hall', p. 34.

25 Gordon Nares, 'Whitmore Hall, Staffordshire', *Country Life*, 6 June 1957.

26 Blundeville, *Horsemanshippe*, p. 12.

27 Carlo Pedretti, *Leonardo Architect* (London, 1986), p. 260; Blundeville, *Horsemanshippe*, p. 12.

28 Victoria and Albert Museum Research Department archives, Ingatestone Hall inventory of 1600; Newman, 'Blickling Hall', pp. 33–35.

29 R. T. Gunther, *The Architecture of Sir Roger Pratt* (Oxford, 1928), pp. 152–53; *The Wren Society* V (London, 1928), pl. XXXI.

30 It has been suggested that the stables may date from repairs carried out after damage caused by a siege during the Civil War, but similarities in the detail of the woodwork to earlier examples, particularly to the church of St Lawrence, Folke, Dorset, which was rebuilt in 1628, suggest that they probably date from Arnold's work. I am grateful to Dr Mark Girouard for his comments on this.

31 Roger North, *Of Building*, ed. Howard Colvin and John Newman (Oxford, 1981), pp. 95–96.

32 Sidney Herbert, *The Pembroke Diaries 1750–94* (London, 1954), p. 104.

33 Public Record Office, WORK, 6/63.

34 British Library, King's Maps 66.24.0-s, J. Mariette, *Plan General des grandes escuries que le Roy faict bastir à Versailles du dessein de Manssart*.

35 François de Garsault, *Le Nouveau Parfait Marechal* (Paris, 1741), p. 92; Charles Etienne Briseux, *L'Art de bâtir des maisons de campagne* I (Paris, 1761) pp. 8–10.

36 Salomon Kleiner, *Suite des residences memorables d'Eugène François, Duc de Savoye et Piemont* (Augsburg, 1738), IX/4; Georgina Masson, *Italian Villas and Palaces* (London, 1959), p. 83, pl. 70.

37 Christopher Clifford, *The Schoole of Horsemanship* (London, 1585), p. 53; Gunther, *Sir Roger Pratt*, pp. 99, 152–53; North, *Of Building*, pp. 95–96.

38 Clifford, *The Schoole of Horsemanship*, pp. 58–61.

39 Robert Plot, *The Natural History of Oxfordshire* (Oxford, 1677), p. 263.

40 William Horman, *Vulgaria* (London, 1519), pp. 170–71.

41 Captain Hammond, 'Relation of a short survey of the Western Counties 1635', ed. L. G. Wickham Legg, *Camden Society* 3rd ser. LII, 1936, p. 36.

42 Thomas Tusser, *Five Hundreth Points of Good Husbandry*, ed. William Mavor (London, 1812), pp. 5, 10.

43 Gage, *Hengrave Hall*, p. 8; M. E. Blackman (ed.), 'Ashley House Building Accounts 1602–7', *Surrey Record Series* XXIX, 1977, p. 37; Historic Manuscripts Commission Report on the Manuscripts of Lord de L'Isle and Dudley, H. M. C. 77 IV (London, 1942), p. 302; Stanley-Millson and Newman, 'Blickling Hall', pp. 33–35.

44 M. A. Havinden (ed.), 'Household and Family Inventories in Oxford, 1550–90', *Oxford Record Society* XLIV, 1965, p. 144; F. G. Emmison, 'Jacobean Household Inventories', *Bedfordshire Historical Record Society* XX, 1938, p. 119.

45 Blundeville, *Horsemanshippe*, p. 10; *Dictionary of the National Biography*, Supplement, I, 1901, p. 421.

46 Clifford, *The Schoole of Horsemanship*, pp. 58–61.

47 Gervase Markham, *Cavalerice or The English Horseman* V, pp. 3–4; de Grey, *The Compleat Horseman*, pp. 3–4; Nicholas Morgan, *The Perfection of Horsemanship* (London, 1609), pp. 150–51; de Grey, *The Compleat Horseman*, pp. 12–13.

48 Society of Antiquaries, Proclamations, 8, 226, *A Proclamation for the Preservation of Grounds for the Making of Salt–Peeter*, 1624.

49 John Nichols, *The History and Antiquities of the County of Leicestershire* I (London, 1795), p. 290.

50 Sir William Brereton, 'Travels in Holland, the United Provinces, England, Scotland and Ireland, 1634–5', *Chetham Society* I, 1844, p. 141.

51 Balthazar Gerbier, *A Brief Discourse Concerning the Three Chief Principles of Magnificent Building* (London, 1662), pp. 31–35; Gunther, *Sir Roger Pratt*, p. 224.

52 Blundeville, *Horsemanshippe*, p. 12; Nicholas Morgan, *The Perfection of Horsemanship* (London, 1609), p. 152; de Grey, *The Compleat Horseman*, pp. 6–11; Markham, *Cavelarice or The English Horseman* V, pp. 3–4.

53 Hampshire Record Office, 5M53/1558–64.

54 de Grey, *The Compleat Horseman*, pp. 10–11.

55 Thomas Pedcock (ed.), 'Inventories Made for Sir William and Sir Thomas Fairfax', *Archaeologia* XLVIII, 1884, p. 150.

56 Clifford, *The Schoole of Horsemanship*, p. 61.

57 Davies, 'The Medieval Warhorse', pp. 11–12; Anthony Dent, *Horses in Shakespeare's England* (London, 1987), p. 11.

58 John Stow, *A Survay of London* (London, 1598), p. 65; Fynes Moryson, *An Itinerary* III, (London, 1617), p. 62.

59 Keith Chivers, *The Shire Horse* (London, 1976), p. 12; John Stow, quoted by Henry Humpherus, *History of the Origin and Progress of the Company of Watermen and Lightermen* I, (London, 1887), pp. 141–42.

60 Gage, *Hengrave Hall*, p. 198; William Sandys, 'Copy of the Inventory of Archbishop Parker's Goods at the Time of his Death', *Archaeologia* XXX, 1844, pp. 24–25; Catherine Parsons, 'Horseheath and its Owners', *Cambridge Antiquarian Society* XLI, 1943–47, p. 8.

61 Transcript of the inventory of Chenies Manor, Buckinghamshire, 1585, in Victoria and Albert Museum; Ward, 1972, p. 6; William Greenwell, 'Durham Wills and Inventories II', *Surtees Society* XXXXVIII, 1860, p. 228; Blackman, 'Ashley House, Building Accounts', p. 37; G. R. Batho, 'Household Papers of Henry Percy, 9th Earl of Northumberland (1564–1632)', *Camden Society* 3rd series, XCIII, 1962, pp. 103, 124.

62 Julian Munby, *Medieval Carriages and the Origins of the Coach: The Archaeology of the European Transport Revolution* (forthcoming); Colvin, *History of the King's Works* III, pp. 79–80.

63 Johannes Kip and Leonard Knyff, *Britannia Illustrata* (London, 1707), p. 59.

64 Colvin, *History of the King's Works* IV, p. 141; Kip and Knyff, *Britannia Illustrata*, p. 80; Harris, *The Artist and the Country House*, fig. 67.

65 C. L. Kingsford, 'Essex House, Formerly Leicester House and Exeter Inn', *Archaeologia* LXXIII, 1922–23, pp. 46–47; William Ridgeway, *The Origin and Influence of the Thoroughbred Horse* (Cambridge, 1905), p. 379.

66 Clifford, *The Diary of Lady Anne Clifford*, p. 20; Vita Sackville-West, *Knole and the Sackvilles* (London, 1922), p. 72; F. W. Steer, 'Inventory of Anne, Countess of Dorchester', *Notes and Queries* CXCVIII, 1953, p. 471.

67 J. O. Halliwell, *Ancient Inventories of Furniture, Pictures, Tapestry, Plate etc* (London, 1854), p. 112; printed in Guidebook to Stapleford Park, Leicestershire, n.d. (copy in the National Monuments Record); Ham Inventory, *Furniture History* XVI, 1980, microfiche.

68 Harris, *The Artist and the Country House*, p. 54.

69 Edward Martin, Timothy Easton and Ian McKechnie, 'Conspicuous display: the extraordinary garden and buildings of a minor gentry family in mid-Suffolk', *Proceedings of the Suffolk Institute of Archaeology and History* XXXVIII, part 1, 1993, pp. 56–74.

CHAPTER 3

1 The terminology associated with the riding house has led to confusion. The earliest drawing for such a building, that of 1609 for the riding house at St James's Palace, is inscribed by Robert Smythson 'The Princes Ridinge Place'. However, his son John Smythson inscribes his 1622 drawing of the riding house at Welbeck Abbey as 'The Ridinge House' and 'riding house' is the common terminology throughout the seventeenth and eighteenth centuries, and much of the nineteenth century. In the nineteenth century, the term 'riding' school was also used, as it is today. This has led to confusion with commercial riding schools (institutions rather than buildings), as, for instance, in the Department of Culture, Media and Sport's List of Buildings of Special Architectural and Historic Interest. It has also led to confusion between the riding house (a covered, barn-like structure), the exercising ring (a circular or similar shaped covered riding ring usually with open sides) and the covered ride (a long, straight enclosed space usually with turning areas at the end), all of which are sometimes described as riding schools but should be seen as separate building types. For that reason I have used the term riding house consistently. The French term manege was also used, particularly in the eighteenth century. This can refer to an open or covered space.

2 F. R. de la Gueriniere, *Ecole de Cavalerie* (Paris, 1733), p. 62.

3 Baldassare Castiglione, *The Book of the Courtier*, trans. Thomas Hoby (London, 1900), p. 54.

4 Gervase Markham, *Countrey Contentments in Two Bookes: The First Containing the Whole Art of Riding Great Horses* (London, 1615), p. 35.

5 Castiglione, *The Book of the Courtier*, p. 54.

6 William Shakespeare, *The Merchant of Venice*, I, II, 36.

7 André Montheilet, 'A History of Academic Equitation', in J. and L. P. Froissard (eds), *The Horseman's International Book of Reference* (London, 1980), p. 99.

8 William Shakespeare, *Venus and Adonis*, l. 260.

9 Mathilde Windisch-Graetz, *The Spanish Riding School* (London, 1956), p. 5.

10 Calendar of State Papers, Venice, II (London, 1867), p. 162.

11 Sebastian Giustinian, *Four Years at the Court of Henry VIII*, I, trans. Rawdon Brown (London, 1854), p. 90.

12 Giustinian, *The Court of Henry VIII*, p. 102.

13 Calendar of State Papers, Venice, II (London, 1867), p. 175.

14 Calendar of State Papers, Foreign and Domestic, Henry VIII III, I (London, 1867), p. 64.

15 William Cavendish, Duke of Newcastle, *A New Method and Extraordinary Invention to Dress Horses* (London, 1667), p. 2.

16 Morgan, *The Perfection of Horsemanship* (London, 1609), preface.

17 Thomas Elyot, *The Boke Named the Governour* (London, 1531) p. 60.

18 Richard Berenger, *The History and Art of Horsemanship* II (London, 1771), pp. 164–65. It has been suggested that the 'Hospice' at Ansty Manor, Wiltshire, was designed as a riding house either in the 1570s or 1590s. This seems unlikely; see Giles Worsley, 'A History and Catalogue of the British Riding House', *Transactions of the Ancient Monuments Society* 47, 2002, p. 72.

19 Claudio Corte, *The Art of Riding*, trans. Thomas Bedingfield (London, 1584), preface.

20 Thomas Blundevil, *The Fower Chiefyst Offices Belonging to Horsemanshippe* (London, 1565), preface.

21 C. M. Prior, *The Royal Studs of the Sixteenth and Seventeenth Centuries* (London, 1935), pp. 13–14, 19.

22 Colvin, *History of the King's Works* V, p. 163.

23 Philip Sidney, *The Defence of Poesie*, ed. A. Feuillerat (Cambridge, 1923), p. 3.

24 Philip Sidney, *Astrophel and Stella*, XLIX.

25 Philip Sidney, *The Correspondence of Sir Philip*

Sidney and Hubert Languet, ed. W. A. Bradley (Boston, 1912), p. 224.

26 Cavendish, *A New Method*, p. 2.

27 Markham, *Cavelarice, or the English Horseman*, p. 25.

28 Elyot, *A Boke Named the Governour*, p. 60.

29 Humphry Gilbert, 'Copy of the plan proposed to Queen Elizabeth by Sir Humphry Gilbert for instituting a London Academy', *Archaeologia*, 1827, pp. 508–20.

30 William Higford, *Institutions or Advice to his Grandson* (London, 1658), pp. 69–71.

31 William Shakespeare, *Henry VIII*, v, iii, 21–24.

32 Blundeville, *Horsemanshippe*, p.18v.

33 Cavendish, *A New Method*, p. 3.

34 H. de Terrebasse, *Antoine de Pluvinel* (Lyon, 1911), p. 13.

35 Walter Liedtke, *The Royal Horse and Rider* (New York, 1989), p. 23.

36 J. W. Stoye, *English Travellers Abroad* (London, 1952), pp. 49–51, but see Calendar of State Papers, France, 6 April 1611; André Joubert, 'Les Gentilhommes étrangers . . . à l'Académie d'équitation d'Angers au XVII siècle', *Revue d'Anjou* 26, 1893, pp. 9–22.

37 British Library, Harl. MS 1760 f. 24.

38 Maximilian de Bethune, *Memoires of Maximilian de Bethune, Duc de Sully* II (London, 1745), p. 261.

39 Prior, *The Royal Studs*, pp. 73, 75.

40 Society of Antiquaries, *A Collection of Ordinances and Regulations for the Government of the Royal Household* (London, 1790), p. 334.

41 Cavendish, *A New Method*, p. 5.

42 Stoye, *English Travellers Abroad*, pp. 49–51, 59; Hatfield Papers, SFP I, 70, 123.

43 Calendar of State Papers, Domestic, 1623–35 (London, 1859), p. 295.

44 John Evelyn, *The Diary of John Evelyn*, ed. J. S. de Beer I, (London, 1955), p. 54.

45 Stoye, *English Travellers Abroad*, p. 119.

46 Cavendish, *A New Method*, p. 4; Stoye, *English Travellers Abroad*, p. 124.

47 W. H., *The True Picture and Relation of Prince Henry* (Leyden, 1636), pp. 26–27.

48 Charles Cornwallis, *A Discourse of the Most Illustrious Prince, Henry, Late Prince of Wales* (London, 1641), p. 16.

49 John Cloake, *Palaces and Parks of Richmond and Kew* I: *The Palaces of Shene and Richmond* (Chichester, 1995), p. 181.

50 Calendar of State Papers, Venice, XII, London, 1905, p. 142.

51 Calendar of State Papers, Venice, XII, London, 1905, p. 194.

52 Stoye, *English Travellers Abroad*, p. 49; Hatfield Papers, SFP I, 69.

53 Roy Strong, *Henry Prince of Wales* (London, 1986), p. 42.

54 Margaret Cavendish, *The Life of . . . William Cavendish, Duke, Marquess and Earl of Newcastle* (London, 1667), p. 20.

55 Edward Hyde, *History of the Rebellion*, ed. W. Dunn Macray III (Oxford, 1888), p. 381.

56 Cavendish, *A New Method*, p. 41.

57 Ben Jonson, *The Under-Wood*, LIII.

58 Historical Manuscripts Commission 6, London, 1877, p. 230.

59 Robert Burton, *Anatomy of Melancholy* II (London, 1800), pp. 406–7.

60 Henry Slingsby, *Diary*, ed. D. Parsons (London, 1836), pp. 265, 312.

61 C. R. L. Fletcher (ed.), 'Dr Wallis's Letter against Mr Maidwell, 1700', *Oxford Historical Society* V, 1885, p. 318.

62 Historical Manuscripts Commission 29, II, (London, 1893), p. 125.

63 Berenger, *Art of Horsemanship*, II, pp. 164–65.

64 Landesbibliothek Kassel, MS Hass. 68, f. 79v.; information from Professor K. J. Holtgen.

65 Public Records Office, E317 Surrey 46.

66 Cloake, *Palaces of Shene and Richmond*, p. 181.

67 Girouard, 'The Smythson Collection of the Royal Institute of British Architects', p. 76.

68 Lieutenant Hammond, 'Relation of a short survey of the Western Counties 1635', ed. L. G. Wickham Legg, *Camden Society* 3rd ser. LII, 1936, p. 37.

69 Lucy Worsley, 'Riding Houses and Horses: William Cavendish's Architecture for the Art of Horsemanship', *Architectural History* 45, 2002, pp. 194–229.

70 Giles Worsley, 'Inigo Jones and the Hatfield Riding House', *Architectural History* 45, 2002, pp. 88–95.

71 Anon., *The Workes of King Charles the Martyr* I (London, 1662), p. 3.

72 Anon., *King Charles the Martyr* I, p. 115.

73 Colvin, *History of the King's Works* 1982, IV, pp. 162–64, 244–45.

74 W. D. Compton, *History of the Comptons* (London, 1930), p. 54.

75 L. Mostyn and T. A. Glenn, *History of the Family of Mostyn* (London, 1925), p. 116.

76 Sir Sidney Lee (ed.), *Dictionary of the National Biography* LI (London, 1897), pp. 154–55.

77 Hilary Ballon, *Louis le Vau: Mazarin's College, Colbert's Revenge*, (Princeton, 1999), p. 14.

78 Catherine Skeel, 'Ludlow Castle in 1631', *Shropshire Archaeological and Natural History Transactions* 4th ser., V, 1915, p. 123.

79 François Boudon and Jean Blécon, *Le Château de Fontainbleau de François Ier to Henri IV* (Paris, 1998), fig. 19.

80 Pierre Dan, *Le Trésor des merveilles de la maison royale de Fontainbleau*, (n.p., 1642), p. 29.

81 Jacques Hillairet, *Le Palais de Tuilleries* (Paris, 1965), p. 20, fig. 16; Henri Sauval, *Histoires et recherches des antiquités de la ville de Paris* II (Paris, 1724), p. 59.

82 Pascal Lievaux, 'Le Manege et les écuries du roi aux Tuileries', in Patrice Franchet d'Espery (ed.), *François Robichon de la Gueriniere* (Paris, 2000), pp. 31–34, 43.

83 Sauval, *Histoire et recherches des antiquités de la ville de Paris* II (Paris, 1724), p. 59.

84 Lievaux also suggests that a plan from the reign of Henri IV for linking the Louvre and the Tuileries – dated by Hilary Ballon to 1594–95 (Hilary Ballon, *The Paris of Henri IV* (New York, 1991), fig. 10) – includes a proposed riding house. However, the dotted lines inside this building are a convention used in delineating tennis courts, as can be seen in Androuet du Cerceau's plans of Coucy, Amboise and Anet (David Thomson (ed.), *Les Plus Excellents Bastiments de France par J.-A. du Cerceau* (Paris, 1988) pp. 74, 178, 182, 264, 268).

85 Sauval, *Histoires et recherches des antiquités de la ville de Paris* pp. 41, 498; Hillairet, *Le Palais de Tuilleries*, p. 24.

86 Moriz Dreger, *Baugeschichte der k.k. Hofburg in Wien* (Wien, 1914) 123, figs. 114–16, 153–54; Mathilde Windisch-Graetz, *The Spanish Riding School* (London, 1956), pp. 16, 29.

87 De Grey, *The Compleat Horseman*, preface.

88 Huntingdon Library, San Marino, California, Bridgwater MSS 8117.

89 Balthazar de Monconys, *Journal des Voyages* II (Lyon, 1666), p. 35.

90 Rolf Loeber, *Biographical Dictionary of Architects in Ireland* (London, 1981), pp. 34–35.

91 Worsley, 'Catalogue of the British Riding House', pp. 55, 76n. 5.

92 Evelyn, *Diary* IV, pp. 257–58.

93 H. Ross Williamson, *Four Stuart Portraits* (London, 1949), pp. 26–40, 143–50. A plan associated with Newcastle's house in

Clerkenwell shows a college-like building with a series of independent suites of rooms around a courtyard, a stable for twenty-four horses, unusually large for London, and an open manege with post and rails. If this was a scheme by Newcastle to promote his love of *haute école* it was never carried through. Balliol College Archives, MS B.21.24; H. Ross Wilkinson, *Four Stuart Portraits* (London, 1949), pp. 26–40, 143–50; Worsley, 'Riding Houses and Horses', pp. 194–229.

94 W. H. Manchée, 'The Fouberts and their Royal Academy', *Huguenot Society Proceedings* XVI, 1941, pp. 79–82; Evelyn, *Diary*, I, pp. 87–88, IV, pp. 257–58, 290.

95 Evelyn, *Diary* IV, pp. 257–8, 400.

96 Manchée, 'The Fouberts and their Royal Academy', p. 82.

97 Lambeth Palace Library, Fifty New Churches Papers, MS 2750.45.

98 Lewis Maidwell, *A Scheme for a Public Academy* (London, 1700); *The Survey of London* XXXI (London, 1963), pp. 177–80.

99 Maidwell, *A Scheme for a Public Academy*; F. H. W. Sheppard (ed.), *Survey of London* XXXI: *The Parish of St James Westminster* (London, 1963), pp. 177–80.

100 *Twenty Five Actions of the Manage House, Engrav'd by Josephus Sympson From the Original Drawings by Mr John Vanderbank* (London, 1729), preface.

101 Maidwell, *A Scheme for a Public Academy*; Sheppard, *The Parish of St James Westminster*, pp. 177–80.

CHAPTER 4

1 Sir Balthazar Gerbier, *A Brief Discourse Concerning the Three Chief Principles of Magnificent Building* (London, 1662), pp. 31–32; I am grateful to Dr Robert Oresko for advice on the identity of Prince Francesco Thomaso.

2 Gerbier, *The Three Chief Principles of Magnificent Building*, pp. 31–32.

3 Giles Worsley, 'Courtly Stables and Their Implications for Seventeenth-Century English Architecture', *Georgian Group Journal* XIII, 2003, pp. 114–40.

4 G. R. Batho (ed.), 'The Percies at Petworth', *Sussex Archaeological Collections* XCV, 1957, pp. 18–19; Thomas Fuller, *The History of the Worthies of England* (London, 1662), p. 100; Daniel Defoe, *A Tour tho' the Whole Island of Great Britain* I (London, 1724), Letter II, p. 67; John Harris, *The Artist and the Country House* (London, 1979), p. 84, fig. 85.

5 H. M. Colvin (ed.), *The History of the King's Works* (London, 1963–82), IV, pp. 274–75.

6 Mark Girouard (ed.), 'The Smythson Collection of Architectural Drawings', *Architectural History* V, 1962, pp. 30, 67.

7 Public Record Office, E317 Herts., no. 24 f. 3.

8 Fuller, *The History of the Worthies of England*, p. 346.

9 Howard Colvin, *Biographical Dictionary of British Architects* (New Haven and London, 1995), pp. 397, 559, 561.

10 Pearl Finch, *History of Burley-on-the-Hill* (London, 1901), pp. 101–2; Anne Blandamer, 'The Duke of Buckingham's House at Burley on the Hill', *Rutland Record*, XVIII, 1998, pp. 349–60.

11 Finch, *History of Burley-on-the-Hill*, I, pp. 6–7; Blandamer, 'The Duke of Buckingham's House', p. 355, fig. 2.

12 Jones also used similar gables on his design of around 1633–34 for the west front of St Paul's Cathedral and on his design for the elevation of a seven-window house with an iron pergola, see John Harris and Gordon Higgott, *Inigo Jones: Complete Architectural Drawings* (London, 1989), pp. 88–89, 241–43, 260–1.

13 Colvin, *The History of the King's Works* IV, p. 277.

14 Thomas Wentworth, *Letters and Despatches of Thomas Wentworth, Earl of Strafford*, ed. W. Knowler, II (London, 1739), p. 118.

15 Harris, *The Artist and the Country House*, fig. 278.

16 Giles Worsley, *Classical Architecture in Britain: The Heroic Age* (New Haven and London, 1995), pp. 8–15.

17 Dianne Duggan, 'The Fourth Side of the Covent Garden Piazza: New Light on the History and Significance of Bedford House', *British Art Journal* III.3, Autumn 2002, pp. 53–65, figs 1, 9.

18 Adjacent stables for Cecil House, though clearly impressive for their day, were about 70 ft long and lacked the generous courtyard in front of the Bedford House stables, see Jill Husselby and Paula Henderson, 'Location, Location, Location! Cecil House in the Strand', *Architectural History* XLV, 2002, pp. 161–93, figs 3–5.

19 John Aubrey, *A Natural History of Wiltshire*, ed. John Britton (London, 1847), p. 87.

20 John Bold, *Wilton House and English Palladianism* (London, 1988), p. 34. Bold does not address the specific date of the stables.

21 Evelyn, *Diaries* III, p. 114.

22 Colvin, *Biographical Dictionary of British Architects*, pp. 298–99.

23 Harris, *The Artist and the Country House*, fig. 89; William Page (ed.), *The Victoria County History of England: Buckinghamshire* II (London, 1908), p. 283; Nikolaus Pevsner and Elizabeth Williamson, *The Buildings of England: Buckinghamshire* (London, 1994), p. 422.

24 Public Record Office, E351/3251.

25 Isaac Ware, *Designs of Inigo Jones and Others* (London, 1735), pls 52–53. One copy, owned by John Carr, is in Sir John Soane's Museum. The other, owned by Thomas Worsley, surveyor-general of the Office of Works from 1761 to 1778, is at Hovingham Hall, Yorkshire.

26 West Yorkshire Archives Service, Leeds, TN/EA/13/74 ff. 34–35; Malcolm Airs, 'Inigo Jones, Isaac de Caus and the Stables at Holland House', *Georgian Group Journal* XIII, 2003, pp. 141–60.

27 J. T. Smith, *Sixty-Two Additional Plates to Smith's Antiquities of Westminster* (London, 1809), p. 5.

28 David Howarth, *Lord Arundel and his Circle* (New Haven and London, 1985), pp. 185, 224n. 52.

29 Harris, *Inigo Jones*, pp. 84, 126–28, 308–9, figs 41, 107.

30 Harris, *Inigo Jones*, pp. 101–2, fig. 34.

31 Mark Girouard, 'The Smythson Collection of the Royal Institute of British Architects', *Architectural History* V, 1962, III/15 (6). Vertue recorded an inscription of 1625 on the stables. (Katherine A Esdaile, the Earl of Ilchester ad Sir Henry M. Hake; Vertue's notebooks II, *Walpole Society* XX, 1932, p. 32.)

32 Worsley, 'Riding Houses and Horses', pp. 208, 210.

33 Girouard, 'The Smythson Collection, III/2.

34 Worsley, 'Riding Houses and Horses', pp. 199, 208, fig. 17.

35 Gerbier, *The Three Chief Principles of Magnificent Building*, pp. 31–35.

36 David R. Coffin, *The Villas in the Life of Renaissance Rome* (Princeton, 1979), pl. 60; Carlo Pedretti, *Leonardo Architect* (London, 1986), pp. 258–63, fig. 385.

37 North, *Of Building*, pp. 95–96.
38 Clifford, *The Schoole of Horsemanship*, p. 53.
39 Markham, *Cavalerice or The English Horseman*, v, pp. 3–4; de Grey, *The Compleat Horseman*, p. 13.
40 North, *Of Building*, pp. 72, 135.
41 Celia Fiennes, *Journeys*, ed. Christopher Morris, (London, 1947), p. 153.
42 John Newman, 'Hugh May, Clarendon and Cornbury', in John Bold and Edward Chaney (ed.), *English Architecture Public and Private* (London, 1993), pp. 80–87, fig. 34.
43 Loeber, *Architects in Ireland*, pp. 60–61.
44 R. T. Gunther, *The Architecture of Sir Roger Pratt* (Oxford, 1928), pp. 152–54, 161–65.
45 Johannes Kip and Leonard Knyff, *Britannia Illustrata* (London, 1707), p. 74; Harris, *The Artist and the Country House*, figs 115, 128.
46 Joseph Taylor, *A Journey to Edenborough* (Edinburgh, 1903), p. 159; Nora Hardwick, *A Diary of the Journey Through the North of England Made by William and John Blathwayt of Dyrham Park in 1703* (Dursley, 1977), p. 28; Kip and Knyff, *Britannia Illustrata*, p. 41; H. M. Colvin *et al.*, 'Architectural Drawings from Lowther Castle', *Architectural History Monograph* no. 2, 1980, no. 12, fig. 7b.
47 *The Wren Society* v, London, 1928, pl. XXXII.
48 *The Wren Society* v, London, 1928, pl. XXXI.
49 Colvin, *The History of the King's Works* v, p. 65; *Wren Society* XII, pl. XXVI.
50 Colvin, *The History of the King's Works* IV, IV, pp. 274–75, v, p. 205, fig. 11.
51 Kip and Knyff, *Britannia Illustrata*, pp. 3, 36, 41, 62, 71.
52 Harris, *The Artist and the Country House*, figs 71–72.
53 I am grateful to Mr John Harris for the dating of these stables.
54 *Victoria County History of England: Warwickshire* VIII (London, 1969), pp. 460–62.
55 Giles Worsley, 'Inept or Innovative', *Country Life*, 4 February 1988.
56 Colvin, *Biographical Dictionary of British Architects*, p. 950.
57 John Newman, *The Buildings of Wales: Gwent/Monmouthshire* (London, 2000), pp. 291–92.
58 George Lipscomb, *The History and Antiquities of the County of Buckinghamshire* I (London, 1831) pp. 556–57.
59 North, *Of Building*, p. 95.

CHAPTER 5

1 Christopher Simon Sykes, *Private Palaces* (London, 1985), p. 100.
2 G. S. Thomson, *Life in a Noble Household, 1641–1700* (London, 1984), p. 163–64.
3 John Schofield (ed.), 'The London Surveys of Ralph Treswell', *London Topographical Record* 135, 1987, p. 26–27.
4 H. M. Colvin (ed.), *The History of the King's Works* (London, 1963–82), v, fig. 22; C. L. Kingsford, 'Essex House, Formerly Leicester House and Exeter Inn', *Archaeologia* LXXIII (1922–23), pp. 1–54, pl. 1, fig. 3; C. L. Kingsford, 'Bath Inn or Arundel House', *Archaeologia* LXXII (1921–22), pp. 243–77, pls XLVI–XLVII.
5 Jill Husselby and Paula Henderson, 'Location, Location, Location! Cecil House in the Strand', *Architectural History* XLV, 2002, figs 3–4; *The Survey of London* XXXVI, 1970, p. 36, fig. 45, pl. 1; Dianne Duggan, 'The fourth side of the Covent Garden Piazza: New light on the history and significance of Bedford House', *British Art Journal* III.3, Autumn 2002, figs 1, 9.
6 *The Survey of London* XVIII, 1937, pp. 13–14; *The Survey of London* XX, 1940, p. 89.
7 Mews can be found in other large Georgian developments such as Edinburgh New Town and the principal colonial towns of America, but these all derive from the London model. Nowhere else is there such a large concentration of mews.
8 Perhaps surprisingly, Sir John Summerson's *Georgian London* (London, 2003), still the most comprehensive account of the development of London in this period, does not address the issue of the mews. Nor does the most recent relevant study of London town planning, Elizabeth McKellar's *The Birth of Modern London: The Development and Design of the City 1660–1720* (Manchester and New York, 1999). Such books as have been published on London's mews, particularly Barbara Rosen and Wolfgang Zuckermann, *The Mews of London* (London, 1982) and Sebastian Deckker, *Mews Style* (London, 1998) do not examine its origins.
9 Fynes Moryson, *An Itinerary* (London, 1617), III, p. 62.
10 There is no detailed study that examines the arrival and impact of the coach in Britain, though see Lawrence Stone, 'The Residential Development of the West End of London in the 17th Century', in Barbara C. Malament (ed.), *After the Reformation* (Manchester, 1980), pp. 177–80.
11 Sykes, *Private Palaces*, p. 45.
12 Summerson, *Georgian London*, pp. 10, 15–21.
13 A close study of Ogilby and Morgan's City of London map of 1676, Ralph Hyde (ed.), *The A to Z of Restoration London*, London Topographical Society CXLV, 1992, proves the novelty of this building type. In particular, Moorfields, which is cited by Jules Lubbock as the model for the London square, was not laid out with mews, see Jules Lubbock, *The Tyranny of Taste* (New Haven and London, 1995), pp. 30–33.
14 John Harris, Stephen Orgel and Roy Strong, *The King's Arcadia: Inigo Jones and the Stuart Court* (London, 1973), pp. 185.
15 Dianne Duggan, 'London the Ring, Covent Garden the Jewell of that Ring: New Light on Covent Garden', *Architectural History* XLIII, pp. 1–22.
16 *The Survey of London* XXXVI, 1970, pp. 169, 180.
17 John Strype, *A Survey of the Cities of London and Westminster* (London, 1720), II, bk. VI, p. 93.
18 *The Survey of London* XXXVI, 1970, p. 240.
19 Duggan, 'London the Ring', figs 5 and 6.
20 *The Survey of London* XXXVI, 1970, pp. 72–73, pl. 1; John Bruce, 'Observations on a lease of Two Houses in the Piazza, Covent Garden, granted to Sir Edmund Verney, AD 1634', *Archaeologia* XXXV, 1835, pp. 194–201.
21 *The Survey of London* XXXVI, 1970, pp. 72–73, pl. 4.
22 Sir Laurence Gomme and Philip Norman (eds), *Survey of London* v, 1914, p. 38–39. Hollar's engraving, in the British Library, is reproduced in Felix Barker and Peter Jackson, *London: 2000 Years of a City and its People* (London, 1974), pp. 136–37.
23 R. Latham and W. Matthew, *The Diary of Samuel Pepys* (London, 1971), IV, pp. 17–18.
24 Sir Laurence Gomme and Philip Norman (eds), *The Survey of London* III (London, 1912), p. 26.
25 Penelope Hunting, 'The Survey of Hatton Garden in 1694 by Abraham Arlidge', *London Topographical Record* XXV, 1985, pp. 87–105; Hyde, *The A to Z of Restoration London*.
26 Bedford Archives, Middlesex Estates Survey 1795.

27 *The Survey of London* XXIX, 1960, pp. 76, 295; *The Survey of London* XXX, 1960, pp. 5, 6.

28 Patricia Croot, 'Before and after Drury House: the development of a Suburban House 1250–1800', *London Topographical Society* XXVIII, 2001, pp. 36–39.

29 I am grateful to Mr Frank Kelsall for drawing my attention to this.

30 *The Survey of London* XXXIX, 1977, pp. 110–11, fig. 4.

31 *The Survey of London* XXXIX, 1977, pp. 12–17, pls 2–3, figs 4, 9.

32 Philips Vingboons, *Afbeelsels der voornaemste gebouwen* II (Amsterdam, 1674), pp. 39, 44, 51–52.

33 Sarah Churchill, *Letters of a Grandmother*, ed. G. S. Thomson (London, 1943), p. 122; 'The "Best Turned" House of the Duke of Bedford', *Georgian Group Journal* VI, 1996, pp. 63–73.

34 John Rocque, *A Plan of the Cities of London and Westminster and Borough of Southwark*, 1747; John Woolfe and James Gandon, *Vitruvius Britannicus* IV (London, 1767), pp. 19, 67; James Paine, *Elevations and Sections of Noblemen and Gentlemen's Houses* I (London, 1767), pp. 26–32; *The Survey of London* XXXII, 1963, p. 369, pls 111–14

35 John Kip and Leonard Knyff, *Britannia Illustrata* (London, 1707), p. 29.

36 Colen Campbell, *Vitruvius Britannicus* (London, 1715–22), I, pp. 43, 39.

37 *The Survey of London* XX, 1940, pl. 55; *The Survey of London* XL, London, 1977, fig. 55; Royal Institute of British Architects, *Catalogue of the Drawings Collection*: B (1972), Barry, Sir C., Stafford House, 7–8; *Building News*, 13 May 1881; *The Builder*, 17 February, 1883; Royal Institute of British Architects, *Catalogue of the Drawings Collection*: B (1972), Barry, Sir C., Bridgwater House, 92–94.

38 Isaac Ware, *A Complete Body of Architecture* (London, 1756), pp. 422, 426, pls 59–60.

39 *The Survey of London* XIV, 1931, p. 62; *The Survey of London* XXXII, 1963, pp. 407–12, pl. 50.

40 Nottingham University Library, Portland Papers, PwK 5165–8; 1st ser. 25 in Ordnance Survey map, 1870.

41 Terry Friedman, *James Gibbs* (New Haven and London, 1984), pp. 202–4, pls 226–29.

42 *The Survey of London* XXXIX, 1977, pp. 12–17, pl. 2–3, figs 4, 9; *The Survey of London* XXXIX, 1977, p. 197; Ware, *A Complete Body of Architecture*, pp. 422, 426.

43 British Library, Add. MSS 22267 f. 69.

44 John Sheffield, Duke of Buckingham, *The Works* II (London, 1729), p. 261.

45 British Library, Add. MS 5238 f. 4.

46 *The Survey of London* XXXIX, 1977, p. 197.

47 Anne Riches, 'London Mews Arches', *Architectural Review* 151, April 1972, pp. 233–36; Dekker, *Mews Style*, pp. 58–79.

48 A. J. Younger, *The Making of Classical Edinburgh* (Edinburgh, 1966), figs 15, 28.

49 Younger, *Classical Edinburgh*, pp. 209, 216–19, fig. 40.

50 Daniel Defoe, *A Tour thro' the Whole Island of Great Britain* II (London, 1724) Letter III, 53; Edward Clarke, *A Tour Through the South of England, Wales and Part of Ireland* (London, 1793), p. 148.

51 J. Pigott Smith, *Map of the Town of Brighton* (Brighton, 1824 and 1830).

52 William Gordon, *The Horse World of London* (London, 1983), p. 101.

53 Byng Giraud, *Stable Building and Stable Fitting* (London, 1891), p. 7.

54 Daniel J. Olsen, *Town Planning in London* (London, 1964), pp. 129–30.

55 Rosen and Zuckermann, *The Mews of London*, p. 13.

56 Gordon Toplis, 'The History of Tyburnia', *Country Life*, 29 November 1973.

57 F. M. L. Thompson, *Hampstead, Building a Borough, 1650–1964* (London, 1974), pp. 278–81.

58 *The Survey of London* XXXVIII, 1975, pp. 68, 324.

59 F. M. L. Thompson, 'Nineteenth-century Horse Sense', *Economic History Review* n.s. 29, 1976, p. 77.

60 S. Sidney, *The Book of the Horse* (London, 1873), pp. 120–21.

61 Thompson, *Hampstead*, pp. 278–81.

62 *The Survey of London* XXXVIII, 1975, p. 324.

CHAPTER 6

1 Andor Gomme, *Smith of Warwick* (Stamford, 2000), pp. 101–8, fig. 40.

2 John Hervey, *Lord Hervey and his Friends*, ed. Earl of Ilchester (London, 1950), p. 71.

3 Historic Manuscripts Commission XLII, (London, 1897), p. 85.

4 Colen Campbell, *Vitruvius Britannicus* (London, 1715–22), III, pp. 27–28; Isaac Ware, *The Plans, Elevations and Sections . . . of Houghton in Norfolk* (London, 1735), pp. 29–32; Historic Manuscripts Commission XLII pp. 85–86; Cambridge University Library, CH (H) Sir Robert Walpole Correspondence 899, 925, 957; CH (H) Account Books 33, 35, 47.

5 Isaac Ware, *A Complete Body of Architecture* (London, 1756), p. 406.

6 Giles Worsley, 'Rokeby Park, Yorkshire', *Country Life*, 19–26 March 1987; Historic Manuscripts Commission 42, pp. 85–86.

7 Lawrence Stone and Jeanne C. Fawtier Stone, *An Open Elite?: England 1540–1880* (Oxford, 1984), p. 311.

8 Raymond Carr, *English Fox Hunting: A History* (London, 1976), p. 25; Historic Manuscripts Commission 42, p. 85.

9 *The Spectator*, 13 July 1711.

10 Carr, *English Fox Hunting*, pp. 29–30.

11 George Burrows, *Gentleman Charles: A History of Foxhunting* (London, 1951), p. 29.

12 Robert Atkyns, *The Ancient and Present State of Glostershire* (London, 1712), p. 618–19; Samuel Rudder, *A New History of Gloucestershire* (Cirencester, 1779), p. 621.

13 Eric Gee, *Bishopthorpe Palace: An Architectural History* (York, 1983), pp. 39–40.

14 John Harris, *The Artist and the Country House* (London, 1979), fig. 83.

15 Johannes Kip and Leonard Knyff, *Britannia Illustrata* (London, 1707), p. 71.

16 Humphry Repton, *Fragments on the Theory and Practice of Landscape Gardening* (London, 1816), pp. 21–22.

17 John Summerson, *Architecture in Britain: 1530–1830* (Harmondsworth, 1979), p. 333.

18 Robert Morris, *An Essay in Defence of Ancient Architecture* (London, 1728), p. 85.

19 Vitruvius, *The Ten Books of Architecture*, trans. Morris Hicky Morgan (New York, 1960), p. 13.

20 Summerson, *Architecture in Britain: 1530–1830* (Harmondsworth, 1979), p. 346.

21 Robert Morris, *Lectures on Architecture* (London, 1734), p. 161.

22 Sarah Churchill, *Letters of a Grandmother*, ed. G. S. Thomson (London, 1943), p. 53.

23 Churchill, *Letters of a Grandmother*, p. 53.

24 Campbell, *Vitruvius Britannicus*, II, pl. 66; Giles Worsley, 'Ecclesiastical Stable', *Country Life*, 1 October 1987; G. L. M. Goodfellow, 'Colen Campbell's Last Years', *Burlington Magazine* CXI, April 1969, pp. 145–46.

25 Howard Colvin, *Biographical Dictionary of*

British Architects (New Haven and London, 1995), p. 669.

26 Duke of Bedford papers, Letters of Sarah, Duchess of Marlborough, II, 57.

27 There is also a much smaller version with a bulging Tuscan portico in timber at Alderston House, East Lothian. This is believed to date from about 1750–60, though architecturally it could be earlier. I am grateful to Mr Bob Heath for drawing it to my attention.

28 James Paine, *Elevations and Sections of Noblemen and Gentlemen's Houses* I (London, 1767), pls I–VII.

29 Giles Worsley, 'The Osterley Stables', in Tim Knox and Anthea Palmer (eds), *Aspects of Osterley* (High Wycombe, 2000), pp. 32–33.

30 Churchill, *Letters of a Grandmother*, pp. 77–78.

31 David R. Coffin, *The Villa in the Life of Renaissance Rome* (Princeton, 1979), pl. 60; Carlo Pedretti, *Leonardo Architect* (London, 1986), pp. 258–63, fig. 385.

32 Francesco di Giorgio Martini, *Trattato di architettura civile e militare*, ed. Carlo Promis, II (Torino, 1841), pp. 50–52; Domenico Rossi, *Architettura civile* III (Roma, 1702), p. 73.

33 John Cornforth, 'Vaux le Vicomte, Ile-de-France', *Country Life*, 9 January 1986, fig. 8; R. Blomfield, *History of French Architecture 1661–1774* II (London, 1921), pp. 73–76.

34 Pearl Finch, *History of Burley-on-the-Hill* (London, 1901), pp. 101–2

35 British Library, King's Maps 66.24.0-s, J. Mariette, *Plan General des Grandes Escuries que le Roy a Faict Bastir à Versailles du Dessein de Monsieur Manssart*; Germain Boffrand, *Livre d'architecture* (Paris, 1745), pl. XXIV.

36 Thomas Blundeville, *The Fower Chiefyst Offices Belonging to Horsemanshippe* (London, 1565), p. 12v.

37 Curzon MSS, Samuel Wyatt to Lord Scarsdale, 22 October 1765, cited by John Martin Robinson, *Georgian Model Farms* (Oxford, 1983), p. 11.

38 Girouard, 'The Smythson Collection', III/15 (6).

39 Finch, *History of Burley-on-the-Hill*, pp. 101–2.

40 Giles Worsley, 'Hovingham Hall, Yorkshire', *Country Life*, 15 and 22 September 1994.

41 Frances Askham, *The Gay Delavals* (London, 1955), p. 126. I am grateful to Lord Hastings for this reference.

42 John Wood, *A Description of Bath* II (London, 1765), pp. 427–29.

43 *Dictionary of the Welsh National Biography* (London, 1959), p. 335.

44 Anon., 'The Building of Winslow Hall', *Records of Buckinghamshire* XI/7, 1926, p. 420.

45 William Gibson, *The True Method of Dieting Horses* (London, 1721), p. 91.

46 John Lawrence, *A Philosophical and Practical Treatise on Horses* II (London, 1796), pp. 41, 49.

47 Salomon Kleiner, *Suite des residences memorables d'Eugene Francois, Duc de Savoye et Piemont* (Augsburg, 1738), IX/4; Wolfgang Gotz, *Marstalle des Barock* (1964), pls 19, 23.

48 Francois de Garsault, *Le Nouveau Parfait Marechal* (Paris, 1741), pp. 94–95; Colvin, *History of the Kings Works* V, p. 102.

49 J. Britton, *The Beauties of England and Wales* XVII (London, 1812), p. 451.

50 I am grateful to Dr Eileen Harris for her suggestion that these stables are by Thomas Wright. Though no documentary proof has been found, Wright was closely connected to the fourth Earl of Berkeley (who succeeded in 1736 and died in 1755) through Norborne Berkeley, Lord Botetort and the Duke of Beaufort. A design survives by him for a garden building for Lord Berkeley at Whitley Park, Ham, Gloucestershire. Alterations to Berkeley Castle at this date may also be by Wright.

51 Charles McKean and David Walker, *Dundee* (Edinburgh, 1984), p. 14.

52 Penicuik papers, James Adam sketchbook 1/6. Photograph in the National Monuments Record for Scotland.

53 Gervase Jackson-Stops, 'When Adam Delved', *Country Life*, 5 March 1987.

54 Giles Worsley, 'Stables with Steeples', *Country Life*, 10 September 1987.

55 Alistair Rowan, 'Penicuik House, Mid Lothian', *Country Life*, 15 August 1968.

56 Sir John Soane's Museum, vols XXXV, 43–47; XXXI, 48–51; XXXVI, 71–72.

57 Unless the curious amalgam of styles at Rosebery and Farm in Midlothian could be described as Gothic.

CHAPTER 7

1 Richard Berenger, *The History and Art of Horsemanship* (London, 1771), pp. 212–14.

2 Percy Archives, Alnwick Castle Archives, X.II.I Box 5e. The riding house is not mentioned in the relevant volume of *The Survey of London* and it is not certain when it was demolished.

3 F. H. W. Sheppard (ed.), *The Survey of London* XL: *The Grosvenor Estate in Mayfair*, 1980, p. 185. The Horse Guards occupied these stables until about 1784 when they were taken over by a coachmaker.

4 Sheppard, *The Survey of London* XL: *The Grosvenor Estate in Mayfair*, 1980, p. 330. The stables ceased to be used as a barracks later in the eighteenth century but the riding house survived until the formation of Balfour Mews in 1898.

5 Westminster City Archives, St George's, Hanover Square ratebooks C241–43, C296–99; Marylebone ratebooks 1753. Both stables can be seen in John Roque's map of London (Ralph Hyde, *The A to Z of Georgian London* (Lympne Castle, 1981), pp. 5, 19), which claims to have been published in 1746, though they were not built for five and seven years respectively. This suggests the copy of Roque's map published by Hyde had been updated since its original publication.

6 Edward Topham, *The Life of the Late John Elwes Esq* (London, 1790), p. 29. The change in size can be seen by comparing Rocque's map of 1746 with Horwood's of 1792.

7 West Yorkshire Archives Service, Leeds, NHC 12/2/1.

8 Edward McParland, *Public Architecture in Ireland 1680–1760* (New Haven and London, 2001), p. 125.

9 Berenger, *Art of Horsemanship*, preface.

10 Westminster City Archives, St George's Hanover Square ratebooks, C129; Marylebone ratebooks, 1740.

11 Giles Worsley, 'A History and Catalogue of the British Riding House', *Transactions of the Ancient Monuments Society* 47, 2003, pp. 33–78.

12 Topham, *John Elwes*, p. 2.

13 Mountstuart, Bute Archives; information from Mr Francis Russell.

14 S. C. Herbert (ed.), *Henry, Elizabeth and George (1734–80), Letters and Diaries of Henry Tenth Earl of Pembroke and his Circle* (London, 1939), *passim*.

15 André Montheilhet, 'A History of Academic Equitation', in J. and L. P. Froissard (eds), *The Horseman's International Book of Reference* (London, 1980), p. 106.

16 Inscribed on an equestrian portait of Blomberg at Hovingham Hall, Yorkshire.

17 Henry Angelo, *Reminiscences of Henry Angelo* I (London, 1828), p. 35; Charlotte Papandiek, *Court and Private Life in the Time of Queen Charlotte* I (London, 1887), p. 49.

18 *Dictionary of the National Biography*, IV (London, 1885), pp. 326–27.

19 Wiltshire Record Office, Pembroke Papers, H1/4, H3/14b, H3/15, H3/16, H3/17; Christopher Hussey, 'The Gardens of Wilton House, Wiltshire', *Country Life*, 1 August 1963; *The Survey of London* XIII, London, 1930, pp. 168–70, pl. 68.

20 Hester Lynch Thrale, *Thraliana*, ed. K. C. Balderston (Oxford, 1942), p. 737.

21 Giles Worsley, 'Hovingham Hall, Yorkshire'.

22 Howard Colvin (ed.), *The History of the King's Works* (London 1973 and 1976), V, p. 137, VI, pp. 303–7; John Harris, *Sir William Chambers* (London, 1970), p. 219, pl. 120.

23 Hovingham Hall, Worsley MSS, ZON 13/3/114; *The Survey of London* XXXIX, 1977, pl. 13a; *The Survey of London* XI, 1980, p. 240, pl. 67a; West Yorkshire Archives Service, Leeds, NHC 12/2/3.

24 Derbyshire Record Office, Harpur Crewe Papers, D 2375 M 171/56 (information from Sir Howard Colvin); Howard Colvin, *Calke Abbey* (London, 1985), pp. 45, 106.

25 S. Shaw, *A Tour to the West of England in 1788* (London, 1789), p. 601; Stratfield Saye, MS Recollections by Mr Brooke (information from Mrs Joan Wilson).

26 Daniel Lysons, *The Environs of London* II (London, 1790), p. 197; George Colman, *Random Records* I (London, 1830), p. 280; Thomas Faulkener, *The History and Antiquities of Brentford, Ealing and Chiswick* (London, 1845), pp. 484–85; Warwick Draper, *Chiswick* (London, 1923), p. 145.

27 BL Add. MSS 32, 736 f. 571.

28 Horace Walpole, *Memoirs of the Reign of George III*, I (London, 1894), p. 29.

29 BL, Add. MSS 41197 f. 101.

30 Francis Askham, *The Gay Delavals* (London, 1955), p. 139.

31 Hovingham Hall, Worsley MSS, ZON 13/3/192.

32 Royal Institute of British Architects, *Catalogue of the Drawings Collection*: L–N (London, 1973), W. Newton, [82].

33 Strickland Freeman, *The Art of Horsemanship, Altered and Abbreviated According to the Principles of the Late Sir Sidney Medows* (London, 1806), p. 2.

34 Hovingham Hall, Worsley MSS, ZON 17/2/22.

35 Hovingham Hall, Worsley MSS, ZON 17.2/122–25, 142–43, 167–70.

36 Hovingham Hall, Worsley MSS, ZON 13/3/4–5; Royal Institute of British Architects, *Drawings Catalogue*, G–K (Lon-don, 1973), [4], sketch book II, f. 53v.

37 West Yorkshire Archives Service, Leeds, NHC 12/2/1.

38 Francis Place, *The British Carpenter* (London, 1735) (information from Sir Howard Colvin).

39 Mark Girouard, 'The Smythson Collectionof the Royal Institute of British Architects', *Architectural History* 5, 1962, p. 76.

40 Harris, *Inigo Jones*, p. 180, fig. 47; Sir John Soane's Museum, VI/3/1&6; West Yorkshire Archives Service, Leeds, NHC 12/2/3.

41 Westminster Public Library, Gardner Collection 39/30; Sir John Soane's Museum, LXII/3/6.

42 Royal Institute of British Architects, *Catalogue of the Drawings Collection*: G–K (London, 1973), [4], sketch book II, f.53v.

43 Gloucester County Record Office, D1245 FF38/A/4; Eileen Harris, 'New Light on Wyatt at Fawley', *Architectural History* 27, 1984, pp. 265–67, fig. 2b.

44 H. D. Roberts, *A History of the Royal Pavilion* (London, 1939), pp. 57–65; J. B. Nichols, *Illustrations of Her Majesty's Palace at Brighton* (London, 1838), pp. 15–17, pls II, XXIX, XXXI, XXXII.

45 Freeman, *The Art of Horsemanship*, p. 8.

46 Topham, *John Elwes*, p. 29; West Yorkshire Archives Service, Leeds, NHC 12/2/1.

47 Royal Institute of British Architects, *Catalogue of the Drawings Collection*: L–N (London, 1973), W. Newton, [82].

48 Hovingham Hall, Worsley MSS, ZON 13/3/5.

49 Royal Institute of British Architects, *Catalogue of the Drawings Collection*: C–F (London, 1972), T. Cundy, p. 63.

50 Freeman, *The Art of Horsemanship*, p. 4.

51 Christopher Hussey, 'The Gardens of Wilton House, Wiltshire', figs 6 and 7.

52 Hugo Arnot, *The History of Edinburgh* (Edinburgh, 1788), p. 424.

53 Freeman, *The Art of Horsemanship*, p. 245.

54 S. C. Herbert (ed.), *Pembroke Papers 1780–94: Letters and Diaries of Henry Tenth Earl of Pembroke and his Circle* (London, 1950), pp. 113–14.

55 Antoine de Pluvinel, *Maneige royale* (Paris, 1623).

56 William Cavendish, *Methode et invention nouvelle de dresser les chevaux* (Antwerp, 1658).

57 Erika Neubauer, *Wiener Barockgarten* (Dortmund, 1980), pp. 164–67.

58 Sir John Soane's Museum, VIII/2/2–3.

59 Angelo, *Reminscences* I, p. 438.

60 *The Survey of London* XXXI, 1963, pp. 177–80.

61 Angelo, *Reminiscences* I, p. 143.

62 Angelo, *Reminiscences*, *passim*; *The Survey of London* XXXXIIII, 1966, pp. 145–46.

63 Sir John Soane's Museum 21/116, 38/50–54; William Edgar, *The Plan of the City and Castle of Edinburgh* (1765); British Library, King's Maps XLIX/59; Arnot, *The History of Edinburgh*, pp. 423–25; James Grant, *Cassell's Old and New Edinburgh* II (London, 1882), pp. 334–35; A. T. Bolton, *The Architecture of James and Robert Adam* I (London, 1922), pp. 60–64.

64 Angelo, *Reminiscences* I, p. 438.

65 James Boswell, *Life of Johnson* II (Oxford, 1934), pp. 424, 527–28; Thrale, *Thraliana* I, pp. 116–19; Samuel Johnson, *Letters*, ed. R. W. Chapman, I (Oxford, 1952), pp. 10, 14–18, 108–9, II, 111, III, 263.

66 British Library King's Maps XXX.11.1; *The Survey of London* XXXVII, London, 1973, p. 276.

67 James Boswell, *Boswell in Extremis: 1776–78*, ed. C. McC. Weiss (London, 1921), pp. 305–6; *The Survey of London* XLI, London, 1986, 180–83; East Riding of Yorkshire Record Office, DDGR 42/27/37.

68 Lawrence, *Treatise on Horses* I (London, 1796), p. 333; John Adams, *The Analysis of Horsemanship* I (London, 1805), p. xvi.

69 John Lawrence, *The History and Delineation of the Horse in all its Varieties* (London, 1809), pp. 203–4.

70 Berenger, *Art of Horsemanship*, pp. 212–14.

71 Lawrence, *Treatise on Horses* I, p. 333.

72 West Yorkshire Archives Service, Leeds, NHC 12/2/1; Topham *John Elwes*, p. 29.

73 West Yorkshire Archives Service, Leeds, NHC 12/2/2; *The Survey of London* XLI, London, 1986, pp. 121–23.

74 *A New and Correct Plan of the City of Bath* (1772); *The Original Bath Guide* (nd).

75 J. Pigott Smith, *Map of the Town of Brighton* (1824–25).

76 This riding house was subsequently converted into a warehouse, part of it being a bacon smokery. In the late 1970s,

attempts were made to convert it into an arts centre.

77 *The Original Bath Guide* (n.d.).

78 *Survey of London* XXIII, London, 1951, pp. 70–72; Giles Worsley, 'Before the Big Top', *Country Life*, 3 December 1987.

79 John Blackner, *The History of Notingham* (London, 1815), p. 51. The Nottingham riding house was demolished in 1926.

80 For a fuller account of the design of permanent circus buildings and hippodromes see Worsley, 'Before the Big Top'.

81 M. A. Richardson, *Local Historians Table Talk* II (Newcastle, 1842), p. 237; John Latimer, *The Annals of Bristol in the Eighteenth Century* (Frome, 1893), p. 492; C. J. Broadbent, *Annals of the Liverpool Stage* (Liverpool, 1908), p. 179; J. C. Dibdin, *The Annals of the Edinburgh Stage* (Edinburgh, 1888), pp. 332–33; J. L. Hodgkinson and R. Pogson, *The Early Manchester Theatre* (London, 1969), p. 139; British Library, King's Maps XLIX/64.1, John Ainslie, *Plan of the Old and New Town of Edinburgh and Leith* (1804); J. A. Erredge, *History of Brightelmstone* (Brighton, 1862), p. 215; David Laurie, *A Proposal for Erecting Public Markets in the Gorbals* (1810), II, map, q; Colvin, *Biographical Dictionary of British Architects*, pp. 465, 998.

82 Hovingham Hall, Worsley MSS, uncatalogued, Elizabeth Chivers, *Journal of a Tour Made in 1805*.

83 Clifford Musgrave, *Life in Brighton* (London, 1970), p. 309.

84 Public Record Office, WO3/4; PRO, MPF 317. According to Dr P. B. Boyden of the National Army Museum, there were two (Lieutenant) Colonel Blaguieres in the army in 1773. James Blaguiere was with the thirteenth Dragoons at Gort, Loughrea and Portumna. John Blaguiere was with the seventeenth (Light) Dragoons at Kilkenny, Ros and Carrick.

85 Herbert, *Pembroke Papers 1780–94*, pp. 38, 45, 59, 125–26; Christopher Gilbert, 'Ingram's Palace, York', *Leeds Art Calendar* 71, 1972, pp. 26–31.

86 T. H. McK. Clough, 'The Riding School of the Rutland Fencible Cavalry', *Rutland Record* 15, 1995, pp. 213–27. Richard Horwood's map of London of 1792–99.

87 James Douet, *British Barracks 1600–1914* (London, 1998), pp. 77–78. Douet's account lists a number of military riding houses, although it does not give a comprehensive account of the subject.

88 John Greenacombe, *Survey of London* XLV: *Knightsbridge*, London, 2000, pp. 64–65, pl. 4a, 7. This riding house was replaced by a considerably larger riding house in 1878, which was itself demolished and replaced by an even larger riding house in 1967–70.

89 Douet, *British Barracks*, pp. 77–78.

90 Colvin, *Biographical Dictionary of British Architects*, p. 111; Douet, *British Barracks*, p. 5.

91 In the series of barrack plans for the South-Western, Aldershot, London, South-Eastern and Eastern Districts for 1859–67 held in the British Library Map Room, the following barracks had riding houses: Dorchester, Aldershot West, East and South Barracks, Croydon (infantry barracks); Hounslow; Hyde Park; Regent's Park; St John's Wood; Windsor; Brighton Pavilion; Colchester; Ipswich; Norwich; Shorncliffe. Of cavalry barracks only Trowbridge Barracks, which housed only 87 horses, and those at Hampton Court and Kensington Palace did not have riding houses.

92 C. B. Ewart, 'Military Riding Schools', *Professional Papers of the Royal Engineers*, 1862, paper 11, p. 181; Douet, *British Barracks*, p. 136.

93 Margaret Richardson (ed.), *Catalogue of the Drawings Collection of the Royal Institute of British Architects: C–F* (Farnborough, 1972), p. 56, T. Cundy, record book p. 59; M. H. Hayes, *Stable Management and Exercise* (London, 1900), p. 250.

94 Douet, *British Barracks*, pp. 107, 133. There are also surviving riding houses, now converted to other uses, at the former Hillsborough Barracks in Sheffield, built in 1848–53 and in Northumberland Road, Newcastle, where John Dobson is believed to have been responsible for the Northumberland Yeomanry's riding house of 1849, see H. Pease, *History of the Northumberland Yeomanry (Hussars)* (London, 1924), p. 12.

CHAPTER 8

1 Raymond Carr, *English Fox Hunting: A History* (London, 1976), p. 40; George Burrows, *Gentleman Charles: A History of Foxhunting* (London, 1951), p. 29.

2 Peter Willett, *An Introduction to the Thoroughbred* (London, 1875), pp. 28, 31.

3 Burrows, *A History of Foxhunting*, p. 67.

4 Burrows, *A History of Foxhunting*, p. 67.

5 Carr, *English Fox Hunting*, pp. 72, 77.

6 G. E. Collins, *History of the Brocklesby Hounds* (London, 1902), pp. 246–47.

7 T. G. T. Paget, *History of the Althorp and Pytchley Hunt* (London, 1937), p. 70.

8 Carr, *English Fox Hunting*, pp. 72, 77.

9 William Taplin, *The Gentleman's Stable Directory* (London, 1791; 13th ed., 1796); William Augustus Osbaldiston, *The British Sportsman or Nobleman, Gentleman and Farmer's Dictionary* (London, 1792); J. Hunter, *A Complete Dictionary of Farriery and Horsemanship* (Birmingham, 1796); John Lawrence, *A Philosophical and Practical Treatise on Horses* (London, 1798; 3rd ed., 1810); James White, *A Compendium of the Veterinary Art with Observations on . . . the Construction of Stables* (London, 1802; 19th ed., 1861); R. W. Dickson, *Practical Architecture* (London, 1805); Delabre Pritchett Blaine, *A Domestic Treatise on the Diseases of Horses and Dogs . . . Including the Natural Management as Stabling . . . etc* (London, 1803; 4th ed., 1810); Robert Lugar, *The Country Gentleman's Architect . . . with Plans for Stables* (London, 1807); John Lawrence, *The History and Delineation of the Horse in all its Varieties* (London, 1809); Thomas Peall, *Observations Chiefly Practical on Some of the More Common Diseases of the Horse* (London, 1814); William Flint, *A Treatise on the Breeding, Training and Management of Horses* (Hull, 1815); Richard Lawrence, *The Complete Farrier* (London, 1816); William Henry Scott, *British Field Sports* (London, 1818).

10 Taplin, *The Gentleman's Stable Directory*, p. 187.

11 Lawrence, *The History and Delineation of the Horse*, p. 153.

12 Giles Worsley, 'The Osterley Stables', in Tim Knox and Anthea Palmer (eds), *Aspects of Osterley* (High Wycombe, 2000), p. 32.

13 Lawrence, *The History and Delineation of the Horse*, p. 154.

14 John Lawrence, *The Horse in all its Varieties and Uses* (London, 1829), pp. 94–95.

15 Lawrence, *The History and Delineation of the Horse*, p. 153.

16 Sir John Soane Museum, 64.3.86, 64.3.98, 64.4.61, Dance Cabinet 2.9.9r.

17 Lincolnshire County Record Office,

Ancaster MS, 3 Anc 5/104/12; Royal Institute of British Architects, *Catalogue of the Drawings Collection*: C–F, Cundy, T. Senr, pp. 55–61; Derek Linstrum (ed.), *Catalogue of the Drawings Collection of the Royal Institute of British Architects*: *The Wyatt Family* (Farnborough, 1974), Sir Jeffry Wyatville, [11]; 2, [1], 399.

18 Lugar, *The Country Gentleman's Architect*, pl. 15.

19 Scott, *British Field Sports*, pp. 335–37.

20 Royal Institute of British Architects, *Catalogue of the Drawings Collection*: C–F, Donthorn, *passim*.

21 J. C. Loudon, *An Encylopaedia of Cottage, Farm and Villa Architecture* (London, 1836), p. 383.

22 Nicholas Morgan, *The Perfection of Horsemanship* (London, 1609), p. 151; Markham, *Cavelarice*, pp. 6–7.

23 Physiologus (pseud.), Thomas Tryon, *The Countryman's Companion* (London, [?1684]) pp. 12–13.

24 William Gibson, *The True Method of Dieting Horses* (London, 1721), p. 89; John Reeves, *The Art of Farriery* (London, 1758), p. 135; John Mills, *A Treatise on Cattle* (London, 1776), p. 113.

25 Gibson, *The True Method of Dieting Horses*, p. 89.

26 William Ellis, *The Modern Husbandman* III (London, 1750), p. 171; William Ellis, *Ellis's Husbandry Abridged and Methodized* II (London, 1772), pp. 297–98.

27 Taplin, *The Gentleman's Stable Directory* II, p. 195.

28 Lawrence, *Treatise on Horses*, II, p. 46.

29 John Stewart, *Stable Economy* (Edinburgh, 1838), p. 35.

30 Lawrence, *Treatise on Horses*, II, 191.

31 William Roper, *On the Nature and Management of the Horse in Health and Disease* (London, 1844), pp. 6–7.

32 Mills, *A Treatise on Cattle*, p. 113.

33 Taplin, *The Gentleman's Stable Directory*, p. 201.

34 White, *A Compendium of the Veterinary Art*, p. 222.

35 Lugar, *The Country Gentleman's Architect*, p. 18; Lawrence, *The Complete Farrier*, p. 54; A. Lawson, *The Modern Farrier* (Newcastle, 1825), pp. 142–43.

36 Sir John Soane's Museum, 64.235, 237, 239; 67.16.

37 A. E. Richardson et al., *Southill: A Regency House* (London, 1951), fig. 4; Derek Linstrum (ed.), *Catalogue of the Drawings Collection of the Royal Institute of British Architects*: *The Wyatt Family* (Farnborough, 1974), Sir Jeffry Wyatville, [11], 2.

38 Tattersall, *Sporting Architecture*, p. 33.

39 Sir John Soane's Museum, 64.235, 237, 239; 67.16.

40 Lawrence, *Treatise on Horses*, p. 45; William Ward, *A New Treatise on the Method of Breeding, Breaking and Training Horses* (Edinburgh, 1771), p. 52; Sir John Soane's Museum, Dance Cabinet, Dr.2.9.9R.

41 Mark Girouard (ed.), 'The Smythson Collection of the Royal Institute of British Architects', *Architectural History* 5, 1962, pp. 67, 134; R. T. Gunther, *The Architecture of Sir Roger Pratt* (Oxford, 1928), pp. 99, 152–53.

42 Giles Worsley, 'Thomas Worsley 1710–1778', B. A. thesis, University of Oxford 1982, p. 36.

43 No trace survives, for instance, to suggest that the Darley Arabian, bred by Richard Darley (d. 1706), another Yorkshire landowner at Aldby Park, one of the three key Arabian stallions from which the English thoroughbred is descended, was handsomely housed.

44 Paget Toynbee (ed.), 'Horace Walpole's Journal of Visits to Country Seats', *Walpole Society* XVI, 1928, pp. 35, 267.

45 Ward, *A New Treatise on the Method of Breeding, Breaking and Training Horses*, pp. 5–6.

46 Tattersall, *Sporting Architecture*, p. 14.

47 William Cavendish, Marquis of Newcastle, *Methode et Invention Nouvelle de dresser les Chevaux* (Antwerp, 1658), pl. 11.

48 H. M. Colvin (ed.), *The History of the King's Works* (London, 1963–82), VI, p. 337; Public Record Office, WORK 19/4/6–8, 19/68.

49 Tattersall, *Sporting Architecture*, pp. 6–25.

50 Charles Knight, *London* VI (London, 1844), p. 363.

51 Tattersall, *Sporting Architecture*, p. 10.

52 Anon, 'Sandringham, the Royal Home of Sport', reprinted from *Land and Sport Illustrated* (London, 1906), pp. 28–32; M. and H. D. Dossenbach and Kohler, *Great Stud-Farms of the World* (London, 1978), pp. 120–21.

53 J. F. Blakeborough, *Sykes of Sledmere* (London, 1929), p. xiii.

54 Tattersall, *Sporting Architecture*, pp. 18–19.

55 Colvin, *Biographical Dictionary of British Architects* (New Haven and London, 1995) p. 558.

56 C. P. Lewis, 'John Chapman's Maps of Newmarket', *Proceedings of the Cambridge Antiquarian Society* LXXX, 1991, pp. 68–78, pls 1, 3.

57 Sheffield Archives, WWM.R.200b.

58 Nancy Briggs, *John Johnson* (Chelmsford, 1991), p. 132.

59 Laura Thompson, *Newmarket from James I to the Present Day* (London, 2000), p. 99.

60 R. Darvill, *A Treatise on the Care, Treatment and Training of the English Race Horse* (London, 1828), p. 2.

61 Tattersall, *Sporting Architecture*, pp. 42–43.

62 Lawrence, *Treatise on Horses* II, p. 41; Darvill, *Training of the English Race Horse* I, pp. 69–70, pl. 1.

63 Darvill, *Training of the English Race Horse*, I, pp. 2, 6.

64 Frank Siltzer, *Newmarket* (London, 1923), p. 86.

65 Sir John Soane's Museum, 41.71.3; 59.153; 64.3.86; 64.4.61, 64. The rides at Tyringham and Mulgrave were not built.

66 Colvin, *Biographical Dictionary of British Architects*, p. 445.

67 Irish Architectural Archive, 32/40 R18.

68 Lawrence, *Treatise on Horses* II, p. 47.

69 John Loudon, *An Encylopaedia of Cottage, Farm and Villa Architecture* (London 1833), pp. 965–66; Thomas Lloyd, *The Lost Houses of Wales* (London, 1986), p. 39.

70 Historic Landscape Management and Anne Riches, *Brocklesby Park Historic Landscape Survey* (n.d.), pp. 25, 43.

71 Shakespeare Memorial Trust, Stratford, Leigh Papers, DR 18/17/17/18.

72 Peter Meadows, *Joseph Bonomi, Architect, 1739–1808* (London, 1988), pp. 27–28.

73 Tattersall, *Sporting Architecture*, p. 43.

74 Worsley, 'The British Riding House', pp. 33–78.

75 Worsley, 'The British Riding House', p. 49.

76 Worsley, 'The British Riding House', pp. 57, 66.

77 Anon., *An Historical and Topographical Account of the Town of Woburn* (Woburn, 1818), pp. 60–81.

78 Worsley, 'The British Riding House', pp. 54, 56, 57–58, 63–64.

79 Worsley, 'The British Riding House', pp. 64, 70.

80 William Kent, *The Designs of Inigo Jones* II (London, 1727), pls 8–9.

81 Richard Hewlings, 'Wakefield Lodge and

other houses of the second Duke of Grafton', *Georgian Group Journal* III, 1993, pp. 43–61.

82 T. J. Bennett, 'Charlton and the Charlton Hunt', *Sussex Archaeological Collections* XV, 1863, pp. 76–78.

83 Lionel Edwards, *Famous Foxhunters* (London, 1932), p. 10; Nimrod (*pseud.*), C. J. Apperley, *Nimrod's Sporting Tours* (London, 1835), p. 22.

84 Apperley, *Nimrod's Sporting Tours*, pp. 372–73; J. H. Crosby, *Ignatius Bonomi of Durham, Architect* (Durham, 1987), p. 79.

85 Jack Brownlow, *Melton Mowbray: Queen of the Shires* (Wymondham, 1980), p. 96.

86 Brownlow, *Melton Mowbray*, pp. 155–56.

87 Lugar, *The Country Gentleman's Architect*, pl. 15.

88 J. E. Eardley-Wilmot, *Reminiscences of the Late Thomas Assheton Smith* (London, 1860), p. 71; Tattersall, *Sporting Architecture*, p. 38.

89 Tattersall, *Sporting Architecture*, p. 58; *Victoria County History of England*: *Leicestershire* III (London, 1955), p. 275.

90 Tattersall, *Sporting Architecture*, p. 31; *Victoria County History of England*: *Leicestershire*, III, p. 271.

91 Nimrod (*pseud.*), C. J. Apperley, *Hunting Reminiscences* (London, 1835), pp. 10, 232.

92 Castor (*pseud.*), Lord North, *A Century of Foxhunting with the Warwickshire Hounds* (Banbury and London, 1891) pp. 104–5; Tattersall, *Sporting Architecture*, p. 58; Eardley-Wilmot, *Reminiscences of the late Thomas Assheton Smith*, p. 71.

CHAPTER 9

1 F. M. L. Thompson, 'Nineteenth-century Horse Sense', *Economic History Review* n.s. 29, 1976, p. 14.

2 F. M. L. Thompson, 'Victorian England: The Horse-Drawn Society', inaugural lecture at Bedford College, and 'Nineteenth-Century Horse Sense', pp. 60–81.

3 Susanna Wade Martins, *Historic Farm Buildings* (London, 1991), 172–73.

4 Arthur Young, *A General View of the Agriculture of the County of Norfolk* (London, 1804), pp. 479–80, cited in Martins, *Historic Farm Buildings*, p. 172.

5 E. J. T. Collins, 'The Farm Horse Economy of England and Wales in the early tractor age 1900–40', in F. M. L. Thompson (ed.), *Horses in European Economic History* (Reading, 1983).

6 John Hudson, 'On the construction of farm buildings', *Journal of the Royal Agricultural Society of England* II, 1850, pp. 280–82, cited in Martins, *Historic Farm Buildings*, p. 172.

7 Matthew Brettingham, *The Plans and Elevations of the Late Earl of Leicester's House at Holkham* (London, 1773), pl. 65.

8 John Martin Robinson, *Georgian Model Farms* (Oxford, 1983), p. 132.

9 Burton Constable archives, design by Lancelot Brown for Burton Constable Park, *circa* 1772, formerly on display at Burton Constable.

10 Robinson, *Georgian Model Farms*, pp. 164, 149, 162.

11 Robinson, *Georgian Model Farms*, p. 162; Edward MacParland, *James Gandon* (London, 1985) p. 131, figs 138, 140–1.

12 Designs at Rednock, copies in the National Monuments Record for Scotland.

13 Susanna Wade Martins, *The English Model Farm: Building the Agricultural Ideal, 1700–1914* (Bollington, 2002), p. 179, figs 100, 103.

14 Martins, *The English Model Farm*, pp. 56, 58, 86, 130, 132.

15 Martins, *Historic Farm Buildings*, pp. 177–79.

16 Martins, *The English Model Farm*, p. 132, fig. 65.

17 William Flint, *A Treatise on the Breeding, Training and Management of Horses* (Hull, 1815), pp. 46–47.

18 Eurwyn William, *The Historical Farm Buildings of Wales* (Edinburgh, 1986), p. 135.

19 R. W. Dickson, *Practical Agriculture* (London, 1805), p. 50.

20 Chatsworth MSS 561.0, James Paine to the Duke of Devonshire, 3 May 1758.

21 Nigel Crowe, *Canals* (London, 1994), p. 96.

22 *Victoria Country History*: *City of York* (London, 1964), p. 477.

23 Anthony Bird, *Roads and Vehicles* (London, 1973), p. 126.

24 William Bayzand, 'Coaching in and out of Oxford from 1820 to 1840', *Oxford Historical Society* (1905) Collectanea IV, p. 270.

25 *Victoria County History of England*: *Oxford* IV (London, 1964), p. 438.

26 E. M. Jope and W. A. Pantin, 'The Clarendon Hotel, Oxford', *Oxoniensia* XXIII, 1958, p. 128.

27 Jope and Pantin, 'The Clarendon Hotel, Oxford', pp. 88–91, 128, pls XVIB, XVII; H. E. Salter, 'A Cartulary of the Hospital of St John the Baptist', *Oxford Historical Society* LXVI, 1914, pl. VI.

28 Wiltshire Record Office, Wilton Archives, F 5/2, Sir Matthew Decker, 'An Account of a Journey Done into Hartford, Cambridgeshire, Suffolk, Norfolk and Essex' (transcript courtesy of Mr John Cornforth).

29 R. S. Surtees, *Hawbuck Grange* (London, 1847), pp. 94, 125.

30 John Graves, *The History and Antiquities of Cleveland* (Carlisle, 1808), p. 126.

31 Anon., 'A Tour Through Derbyshire', *Ackermann's Repository of the Arts* X, 1813, p. 322.

32 Ivan Hall, *Georgian Buxton* (Derby, 1984), pp. 25–29, 33.

33 William Macritchie, *Diary of a Tour through Great Britain in 1795* (London, 1897), p. 54.

34 Hall, *Georgian Buxton*, pp. 25–29, 33.

35 Anon., 'Ingleby Arncliffe', *Yorkshire Archaeological Journal* XVI, 1900–2, p. 123.

36 Thompson, 'Nineteenth-century Horse Sense', p. 13; Bryan Holden, *The Long Haul: The Life and Times of the Railway Horse* (London, 1985), p. 18.

37 Douet, *British Barracks*, pp. 14–42.

38 *Survey of London* XVI, London, 1935, pp. 8–9.

39 H. M. Colvin (ed.), *The History of the King's Works* (London, 1963–82), V, p. 186.

40 Sir John Soane's Museum, 111.40.

41 *Survey of London* XVI, London, 1935, pp. 10–13, pl. 22.

42 John Rocque, *A Plan of the Cities of London and Westminster and Borough of Southwark*, 1747; Giles Worsley, 'The British Riding House', p. 38.

43 Douet, *British Barracks*, pp. 60–62.

44 Douet, *British Barracks*, pp. 61–71.

45 George Paget, Marquess of Anglesey, *History of the British Cavalry* I (London, 1973), pp. 128–29.

46 James Douet, *British Barracks 1600–1914* (London, 1988), pp. 58–59, 86–88.

47 *Victoria County History of England*: *City of York* (London, 1964), pp. 541–42; Royal Commission on Historic Monuments, *York* IV (London, 1975), p. 46.

48 Douet, *British Barracks*, p. 88.

49 Douet, *British Barracks*, pp. 104, 106–11.

50 Douet, *British Barracks*, pp. 130–33, 136–37.

51 Stewart, *Stable Economy*, p. 36.

52 General Report of the Commission appointed for Improving the Sanitary Condition of Barracks and Hospitals (London, 1861), p. 19.

53 General Report (London, 1861), p. 162.

54 General Report (London, 1861), pp. 81, 252.
55 Douet, *British Barracks*, pp. 139–41, 145.
56 John Bright, *Pit Ponies* (London, 1986), p. 19.
57 Jennifer Tann, 'Horse Power 1780–1880', in Thompson, *Horses in European Economic History*, pp. 21–26.
58 A. J. Parkinson, *Report on the Dowlais Stables Prepared by the National Monuments Record (Wales)* 1985; J. A. Owen, *History of the Dowlais Ironworks* (Dowlais, 1977), pp. 24, 35.
59 T. E. Coleman, *Stable Sanitation and Construction* (London, 1897), pp. 3–4.
60 William Gordon, *The Horse World of London* (London, 1893), pp. 24, 26, 32, 67.
61 *Civil Engineer and Architect's Journal*, October 1837.
62 Alfred Barnard, *The Noted Breweries of Great Britain and Ireland* (London, 1889), pp. 264–65, 290–91, 319–20, 363–64.
63 H. C. Moore, *Omnibuses and Cabs, their Origins and History* (London, 1902), p. 12.
64 Thompson, 'Nineteenth-century Horse Sense', p. 14.
65 Gordon, *The Horse World of London*, p. 10.
66 Gordon, *The Horse World of London*, p. 10.
67 Byng Giraud, *Stable Building and Stable Fittings* (London, 1891) pls 32–54.
68 Roger Whitehouse, *A London Album: Early Photographs Recording the History of the City and its People from 1840 to 1915* (London, 1980), fig. 227.
69 Charles Dudley, 'Equine London', in G. R. Sims (ed.), *Living London* II (London, 1901–3), pp. 160–61.
70 Gordon, *The Horse World of London*, pp. 53, 63, 80, 119, 136.
71 Giraud, *Stable Building*, p. 19; *The Builder*, 15 April 1876; *The Builder*, 19 April 1884.
72 Gordon, *The Horse World of London*, p. 19.
73 *The Builder*, 10 July 1880.
74 Thompson, 'Nineteenth-century Horse Sense', p. 61.
75 Anne Grimshaw, *The Horse: A Bibliography of British Books 1851–1976* (London, 1982), p. 92.
76 Thompson, 'Nineteenth-century Horse Sense', p. 14.
77 Gordon, *The Horse World of London*, p. 46.
78 *The Builder*, 15 April 1876.
79 *The Builder*, 19 April 1884.
80 Angus Wilson, *London's Industrial Heritage* (Newton Abbot, 1967), p. 90.
81 Thompson, 'Nineteenth-century Horse Sense', p. 13.
82 Holden, *The Long Haul*, p. 20.
83 Gordon, *The Horse World of London*, pp. 52, 60.
84 Gordon, *The Horse World of London*, p. 63.
85 Bright, *Pit Ponies*, pp. 52–58; Holden, *The Long Haul*, p. 46.
86 Gordon, *The Horse World of London*, pp. 117, 136–37.
87 Horace Smith, *A Horseman Through Six Reigns* (London, 1955), pp. 29–30.

CHAPTER 10

1 A bibliography of stable theorists can be found in Giles Worsley, 'The Design and Development of the Stable and Riding House in Great Britain from the 13th Century to 1914', Ph. D. thesis, Courtauld Institute of Art, London University, 1989.
2 Edward Mayhew, *The Illustrated Horse Management Containing Descriptive Remarks Upon . . . Stables* (London, 1864), p. 326.
3 Thomas Knightley, *Stable Architecture* (London, 1862), p. iii.
4 George Tattersall, *Stable Architecture* (London, 1841), p. 33; William Haycock, *The Gentleman's Stable Manual* (London, 1859), pp. 19–20.
5 R. S. Surtees, *Mr Sponge's Sporting Tour* (London, 1853), p. 30.
6 S. Sidney, *The Book of the Horse* (London, 1873), pp. 498–503. Lord Stamford's very large stables at Bradgate near Groby in Leicestershire survive, though derelict.
7 William Miles, *General Remarks on Stables and Examples of Stable Fittings* (London, 1860), pp. 8–9; Edward Richerson, *The General Management of the Hunter; Explaining How the Stable Should be Constructed* (London, 1904), p. 9.
8 John Birch, *Examples of Stables, Hunting Boxes, Kennels, Racing Establishments etc* (Edinburgh and London, 1892), pp. 22–23.
9 Byng Giraud, *Stable Building and Stable Fitting* (London, 1891), p. 38; Richerson, *The General Management of the Hunter*, p. 10.
10 F. W. G. Fitzwygram, *Horses and Stables* (London, 1886), p. 15.
11 Miles, *General Remarks on Stables and Examples of Stable Fittings*, p. 28; Knightley, *Stable Architecture*, p. 4; T. E. Coleman, *Stable Sanitation and Construction* (London, 1897), p. 93.
12 Coleman, *Stable Sanitation and Construction* , p. 158.
13 Tattersall, *Sporting Architecture*, p. 36.
14 Edward Mayhew, *The Illustrated Horse Management*, p. 337.
15 Miles, *General Remarks on Stables*, p. 39; Fitzwygram, *Horses and Stables*, p. 28.
16 Fitzwygram, *Horses and Stables*, p. 28; J. Wortley Axe, *The Horse, Horsemen and Stable Management* (London, 1907), p. 324.
17 *The Builder*, 13 February 1875. The stables were rebuilt to form galleries for the Wallace Collection.
18 Miles, *General Remarks on Stables*, p. 2.
19 Coleman, *Stable Sanitation and Construction*, p. 146.
20 *The Builder*, 12 December 1863.
21 Mark Girouard, *The Victorian Country House* (New Haven and London, 1979), p. 38n.
22 J. H. Walsh and J. I. Lupton, *The Horse in the Stable and Field* (London, 1861), p. 190.
23 Royal Institute of British Architects, unpublished catalogue of Devey drawings.
24 Fitzwygram, *Horses and Stables*, p. 29.
25 *The Builder*, January 16 1864.
26 Tattersall, *Sporting Architecture*, pp. 32–33.
27 Robert Kerr, *The Gentleman's House; or How to Plan English Residences from the Parsonage to the Palace* (London, 1864), p. 296.
28 H. M. Colvin (ed.), *The History of the King's Works* (London, 1963–82), IV, p. 162; John Cole, *A Full and More Particular Account of the Late Fire, with Several Losses at Newmarket* (London, 1683).
29 R. S. Surtees, *Hawbuck Grange* (London, 1847), pp. 90, 125–26.
30 J. C. Loudon, *An Encyclopaedia of Cottage, Farm and Villa Architecture* (London, 1836), p. 367; Surtees, *Mr Sponge's Sporting Tour*, p. 30.
31 Miles, *General Remarks on Stables*, p. 56, pl. 18.
32 George Armagate, *The Horseowner and Stableman's Companion* (London, 1869), p. 21.
33 Birch, *Examples of Stables*, pp. 17–19; Coleman, *Stable Sanitation and Construction*, p. 89.
34 E. L. Tarbuck, *The Builder's Practical Director* (Leipzig, Dresden and London, *c.*1856), 197; Miles, *General Remarks on Stables*, pp. 43–44; Knightley, *Stable Architecture*, p. 8.
35 Knightley, *Stable Architecture*, p. 6.
36 Hermann Muthesius, *Das Englische Haus* (Berlin, 1904/05; London, 1979), p. 104.
37 Anthony Trollope, *The Small House at Allington* (Oxford, 1980), p. 195.
38 Jack Brownlow, *Melton Mowbray: Queen of the Shires* (Wymondham, 1980), p. 148.

39 Christopher Hussey, 'Earnshill, Somerset II', *Country Life*, 20 October 1960.

40 Christopher Hussey, 'Brocklesby Park, Lincolnshire', *Country Life*, 24 Februrary 1934.

41 Brownlow, *Melton Mowbray*, p. 148.

42 Lupton, *The Horse in the Stable and the Field*, pp. 266–69; *Dublin Builder* I, January 1859, pp. 7, 60, September 1859, p. 112.

43 Knightley, *Stable Architecture*, p. 24; John Gamgee, *Plain Rules for the Stable* (London, 1866), pp. 25–28.

44 William Cavendish-Bentick, sixth Duke of Portland, *Memoirs of Racing and Hunting* (London, 1935), pp. 86–87.

45 Matthew Hayes, *Stable Management and Exercise* (London, 1900), pp. 249–50.

46 Other examples include Welbeck Abbey, Nottinghamshire, where the stables of the fifth Duke of Portland, who was considered one of the best judges of horses in the country, held 94 horses at his death in 1879 and Crabbet Park where Wilfred Scawen Blunt established a famous stud of Arabian horses in the 1870s. Giles Worsley, 'A History and Catalogue of the British Riding House', *Transactions of the Ancient Monuments Society*, 2003, pp. 49, 50, 57, 66.

47 Examples include Stapleford Park, Leicestershire, Hornby Castle, Yorkshire, Firle Place, Sussex, and Crofton Hall, Cumbria. Worsley, 'A History and Catalogue of the British Riding House', pp. 50, 54, 56, 57–58, 60, 63–64.

48 Other examples include Syon House, Middlesex (circa 1819); Hamilton Palace, Lanarkshire (1837); Windsor Castle, Berkshire (1839); Grimston Park, Yorkshire (1840); Trentham Park, Staffordshire (1840), Crom Castle, County Fermanagh (1860); Blenheim Palace, Oxfordshire (*c.*1850–60).

49 Worsley, 'A History and Catalogue of the British Riding House', pp. 47, 50–51, 53, 45–55, 59–60, 62, 64, 65, 66–67.

50 Worsley, 'A History and Catalogue of the British Riding House', pp. 52.

51 Worsley, 'A History and Catalogue of the British Riding House', pp. 51, 52, 53, 65, 66.

52 *The Builder*, 3 January 1903.

53 Humphry Repton, *An Enquiry into the Changes of Taste in Landscape Gardening* (London, 1806), p. 31.

54 Humphry Repton, *Fragments on the Theory and Practice of Landscape Gardening* (London, 1816), p. 38.

55 Mark Girouard, *The Victorian Country House* (New Haven and London, 1979), pp. 194–204, pl. XIV, figs 184–89.

56 James Lees-Milne, *Mottisfont Abbey* (London, 1978), p. 18.

57 Thompson, 'Nineteenth-century Horse Sense', p. 75.

58 Tattersall, *Sporting Architecture*, p. 58; *Victoria County History of England: Leicestershire* III (London, 1955), p. 275.

59 Lionel Edwards, *Famous Foxhunters* (London, 1932), p. 86.

60 *Victoria County History of England: Rutland* I (London, 1908), p. 304.

61 T. F. Dale, *The Eighth Duke of Badminton and the Badminton Hunt* (London, 1901), pp. 115–16.

62 C. F. Somerset, Duke of Beaufort, and Mowbray Morris, *Hunting* (London, 1885), pp. 92, 95.

63 Edwards, *Famous Foxhunters*, pp. 62–64; C. F. Somerset, Duke of Beaufort, and Mowbray Morris, *Hunting* (London, 1885), pp. 94–96.

64 J. L. Randall, *A History of the Meynell Hounds and Country* (London, 1901), pp. 115–16.

65 Knightley, *Stable Architecture*, p. 13; Birch, *Examples of Stables*, pp. 59–60, pls 27–28.

66 Brownlow, *Melton Mowbray*, p. 186.

67 Brownlow, *Melton Mowbray*, p. 96; Birch, *Examples of Stables*, pp. 59–60, pls 27–28.

68 Birch, *Examples of Stables*, pp. 48–49, pl. 20.

69 Brownlow, *Melton Mowbray*, p. 209.

70 Giles Worsley, 'Stapleford Park, Leicestershire,' *Country Life*, 23 June 1988.

71 Knightley, *Stable Architecture*, pls 9–12.

72 H. C. Howard, Earl of Suffolk, and W. G. Craven, *Racing and Steeple Chasing* (London, 1886), pp. 108–11.

73 Howard and Craven, *Racing and Steeple Chasing*, p. 109.

74 Byng Giraud, *Stable Building and Stable Fitting* (London, 1891), p. 2.

75 J. Wortley Axe, *The Horse, Its Treatment in Health and Disease* III (London, 1907), p. 319. The situation was not so different in the 1980s at the Jockey Club in Hong Kong, where the stables were air-conditioned, but not the jockeys' quarters.

76 Hermann Muthesius, *Das Englische Haus* (Berlin, 1904/05; London, 1979), p. 102.

77 *The Builder*, 31 December 1910.

78 Clive Aslet, 'Manderston, Berwickshire – III', *Country Life*, 1 March 1979.

Adams, John, *The Analysis of Horsemanship* (London, 1805)
Aemilianus, Palladius Junius Moderatus, *Of Husbandrie* (Early English Text Society, London, 1873–79)
Alberti, Leon Battista, *Ten Books on Architecture*, trans. Giacomo Leoni (London, 1726)
Anon., *The Gentleman's New Jockey* (London, 1687; further edns 1691, 1721)
Anon., *The Sportsman's Dictionary* (London, 1778; fifth edn 1807)
Apperley, C. J. [Nimrod, pseud.], *Nimrod's Sporting Tours* (London, 1835)
Armatage, George, *The Horseowner and Stableman's Companion* (London, 1869)
A. S., *The Gentleman's Complete Jockey* (London, 1782)
Astley, John, *The Art of Riding* (London, 1584)
Axe, J. Wortley, *The Horse, its Treatment in Health and Disease* (London, 1907)
Barnard, Alfred, *The Noted Breweries of Great Britain and Ireland* (London, 1889)
Berenger, Richard (ed.), *The History and Art of Horsemanship* (London, 1771)

BIBLIOGRAPHY

Birch, John, *Country Architecture* (London, 1874)
Birch, John, *The Architecture of Stables and Country Mansions* (London, 1883)
Birch, John, *Examples of Stables, Hunting Boxes, Kennels, Racing Establishments etc* (Edinburgh and London, 1892)
Bird, Anthony, *Roads and Vehicles* (London, 1973)
Blaine, Delabre Pritchett, *A Domestic Treatise on the Diseases of Horses and Dogs . . . including likewise the natural management of stabling . . .* (London, 1803; fourth edn 1810)
Blondel, J. F., *De la distribution des maisons de plaisance* (Paris, 1737)
Blondel, J. F., *Cours d'architecture* (Paris, 1773)
Blundevil, Thomas, *A New Booke Containing the Arte of Ryding, and Breaking Grete Horses* (London, 1560)
Blundevil, Thomas, *The Fower Chiefyst Offices Belonging to Horsemanshippe* (London, 1565; further edns 1570, 1580, 1597, 1609)
Boorde, Andrew, *Hereafter Foloweth a Compendyous Regyment or a Dyetary of Helth* (London, 1542)
Bosvile, Godfrey, *Horses, Horsemen and Stable Management* (London, 1908)
Bourgelat, C., *A New System of Horsemanship*, trans. Richard Berenger (London, 1754)
Bright, John, *Pit Ponies* (London, 1986)
Briseux, Charles Etienne, *Architecture moderne* (Paris, 1728)
Briseux, Charles Etienne, *L'Art de batir des maisons de campagne* (Paris, 1761)
Burrows, George, *Gentleman Charles: A History of Foxhunting* (London, 1951)
Campbell, Colen, *Vitruvius Britannicus* (London, 1715–25)
Caracciolo, Pasquale, *La Gloria del cavallo* (Venice, 1567)
Carr, Raymond, *English Fox Hunting: A History* (London, 1976)
Cato, Marcus Porcius, *On Agriculture* (London, 1934)
Cavendish, William, *Methode et invention nouvelle de dresser les chevaux* (Antwerp, 1658)
Cavendish, William, Duke of Newcastle, *A New Method and Extraordinary Invention to Dress Horses* (London, 1667)
Cavendish, William, Duke of Newcastle, and Gaspard de Saunier, *A General System of Horsemanship* (London, 1743)
Chivers, Keith, *The Shire Horse* (London, 1976)
Clark, James, *A Treatise on the Prevention of Diseases Incidental to Horses* (Edinburgh, 1781, second edn 1790)

Clifford, Christopher, *The Schoole of Horsemanship* (London, 1585)
Coleman, T. E., *Stable Sanitation and Construction* (London, 1897)
Columella, Lucius Junius Moderatus, *Of Husbandry* (London, 1745)
Colvin, H. M. (edn), *The History of the King's Works* (London, 1963–82)
Colvin, Howard, *Biographical Dictionary of British Architects* (New Haven and London, 1995)
Conrads, Norbert, *Ritterakademien der fruhen Neuzeit: Bildung als Standesprivileg im 16 und 17 Jahrhundert* (Gottingen, 1982)
Corte, Claudio, *The Art of Riding*, trans. Thomas Bedingfield (London, 1584)
Crowe, Nigel, *Canals* (London, 1994)
Dale, T. F., *The Stable Handbook* (London and New York, 1907)
Davies, R. W., 'Training Grounds of the Roman Cavalry', *Archaeological Journal* cxxv, 1968, pp. 73–100
d'Aviler, A. C., *Explication des termes d'architecture* (Paris, 1691)
d'Eisenberg, R., *Description du manege moderne* (Paris, 1727)
de Garsault, François, *Le Nouveau Parfait Marechal* (Paris, 1741?; sixteenth edn 1805)
de Grey, Thomas, *The Compleat Horseman and Expert Ferrier* (London, 1639; fifth edn. 1684)
Dekker, Sebastian, *Mews Style* (London, 1998)
de la Broue, Salomon, *Le Cavalerice francois* (Paris, 1602)
de la Gueriniere, F. R., *Ecole de cavalerie* (Paris, 1733)
Dent, Anthony, *Horses in Shakespeare's England* (London, 1987)
de Pluvinel, Antoine, *Maneige royale* (Paris, 1623)
de Solleysel, Jacques, *The Compleat Horseman*, trans. William Hope (London, 1696; fourth edn 1729)
Dickson, R. W., *Practical Architecture* (London, 1805)
Douet, James, *British Barracks 1600–1914* (London, 1998)
Dudley, Charles, 'Equine London', in G. R. Sims (ed.), *Living London* (London, 1901–3)
Edwards, Peter, *The Horse Trade of Tudor and Stuart England* (Cambrige, 1988)
Emery, Anthony, *Greater Medieval Houses of England and Wales, 1300–1500* (Cambridge, 1996–2000)
E. R., *The Experienced Farrier* (London, 1678; fourth edn 1720)
Fairfax, T., *The Compleat Sportsman or Country Gentleman's Recreation* (London, 1758; further edns 1762, 1764, 1765, 1795)
Fillis, James, *Breaking and Riding*, trans. M. H. Hayes (London, 1902)
Fitzwygram, Frederick, *Lectures on Horses and Stables* (London, 1862)
Fitzwygram, Frederick, *Horses and Stables* (London, 1869; fifth edn 1901)
Flint, William, *A Treatise on the Breeding, Training and Management of Horses* (Hull, 1815)
Freeman, Strickland, *The Art of Horsemanship, Altered and Abbreviated According to the Principles of the Late Sir Sidney Medows* (London, 1806)
Gambrill, Richard V. N., and James C. Mackenzie, *Sporting Stables and Kennels* (London, 1935)
Gamgee, John, *Plain Rules for the Stable* (London, 1866)
Garland, J. A. [Jorrocks, pseud.], *The Private Stable: Its Establishment, Management and Appointments* (Boston, 1899)
Gerbier, Balthazar, *A Brief Discourse Concerning the Three Chief Principles of Magnificent Building* (London, 1662)
Gerbier, Balthazar, *Counsel and Advise to all Builders* (London, 1663)
Gibson, William, *The True Method of Dieting Horses* (London, 1721; third edn 1731)
Giraud, Byng, *Stable Building and Stable Fitting* (London, 1891)
Girouard, Mark, *The Victorian Country House* (New Haven and London, 1979)
Girouard, Mark, *Robert Smythson and the Elizabethan Country House* (New Haven and London, 1983)
G. M., *The Gentleman's Accomplished Jockey with Compleat Horseman and Approved Farrier* (London, 1722)
Goodall, Daphne, *A History of Horse Breeding* (London, 1977)

Gordon, William, *The Horse World of London* (London, 1893)
Grimshaw, Anne, *The Horse: A Bibliography of British Books 1851–1976* (London, 1982)
Grisone, Federico, *Gli Ordini di Cavalcare* (Naples, 1550)
Guillet, George, *The Gentleman's Dictionary* (London, 1705)
Gunther, R. T. (ed.), *The Architecture of Sir Roger Pratt* (London, 1928)
Handler, Hans, and Erich Lessing, *The Spanish Riding School in Vienna* (London, 1972)
Harris, John, *The Artist and the Country House* (London, 1979)
Haycock, William, *The Gentleman's Stable Manual* (London, 1859)
Hayes, Matthew, *The Points of the Horse* (London, 1893)
Hayes, Matthew, *Among Men and Horses* (London, 1894)
Hayes, Matthew, *Stable Management and Exercise* (London, 1900; sixth edn 1968)
Herbert, Henry, Earl of Pembroke, *A Method of Breaking Horses and Teaching Soldiers How to Ride Designed for Use of the Army* (London, 1761)
Hewitt, Herbert J., *The Horse in Medieval England* (London, 1983)
Holden, Bryan, *The Long Haul: The Life and Times of the Railway Horse* (London, 1985)
Hope, William, *A Supplement of Horsemanship to the first part of the Parfait Mareschal* (Edinburgh, 1696)
Horlock, K. V. [Scrutator, pseud.], *Practical Lessons on Hunting* (London, 1865)
Horman, William, *Vulgaria* (London, 1519)
Hunt, Vere D., *The Horse and his Master: With Hints on Breeding, Breaking, Stable Management* (London, 1859)
Hunter, J., *A Complete Dictionary of Farriery and Horsemanship* (Birmingham, 1796)
Hyland, Ann, *The Warhorse 1250–1600* (Stroud, 1998)
Hyland, Ann, *The Horse in the Middle Ages* (Stroud, 1999)
Kerr, Robert, *The Gentleman's House; or How to Plan English Residences from the Parsonage to the Palace* (London, 1864; third edn 1871)
Knightley, Thomas, *Stable Architecture* (London, 1862)
Knyff, Leonard, *Nouveau Theatre de la Grande Bretagne* (London, 1708)
Lambton, Lucinda, *Beastly Buildings* (London, 1985)
Lawrence, John, *A Philosophical and Practical Treatise on Horses* (London, 1798; third edn 1810)
Lawrence, John, *The History and Delineation of the Horse* (London, 1809)
Lawrence, John, *The Horse in All its Varieties and Uses* (London, 1829)
Lawrence, Richard, *The Complete Farrier* (London, 1816)
Lawson, A., *The Modern Farrier* (Newcastle, 1820; seventh edn 1825)
Liedtke, Walter, *Royal Horse and Rider* (New York, 1989)
Loehhnyss, G. E., *Della Cavalleria* (?1699)
Loudon, John Claudius, *An Encyclopedia of Cottage, Farm and Villa Architecture* (London, 1833)
Loudon, John Claudius, *An Encyclopedia of Agriculture* (London, 1844)
Lugar, Robert, *The Country Gentleman's Architect . . . with Plans for Stables* (London, 1807)
Mackay-Smith, Alexander, 'Riding in the Renaissance and in the Baroque Period', in Walter Liedtke, *Royal Horse and Rider* (New York, 1989), pp. 86–91
Markham, Gervase, *A Discourse of Horsemanshippe* (London, 1593)
Markham, Gervase, *Cavelarice or the English Horseman: Book Five, or that Part of the Arte which Containeth the Office of the Keeper Groome of the Stable* (London, 1607; another edition, 1617)
Markham, Gervase, *Countrey Contentments in Two Bookes: The First Containing the Whole Art of Riding Great Horses* (London, 1615)
Martins, Susanna Wade, *Historic Farm Buildings* (London, 1991)
Martins, Susanna Wade, *The English Model Farm: Building the Agricultural Ideal, 1700–1914* (Bollington, 2002)
Mayhew, Edward, *The Illustrated Horse Management Containing Descriptive Remarks Upon . . . Stables* (London, 1864; second edn 1876)

Miles, William, *General Remarks on Stables and Examples of Stable Fittings* (London, 1860; tenth edn 1888)
Mills, John, *A Treatise on Cattle, Shewing the Most Approved Methods of Breeding, Rearing and Fitting for use . . . Horses etc* (London, 1776)
Montheilet, André, 'A History of Academic Equitation', in J. and L. P. Froissard (eds), *The Horseman's International Book of Reference* (London, 1980), pp. 97–116
Moore, H. C., *Omnibuses and Cabs, their Origins and History* (London, 1902)
Morgan, Nicholas, *The Perfection of Horsemanship* (London, 1609)
Morris, Robert, *An Essay in Defence of Ancient Architecture* (London, 1728)
Morris, Robert, *Lectures on Architecture* (London, 1734)
Muthesius, Hermann, *Das Englische Haus* (Berlin, 1904/5; London, 1979)
North, Roger, *Of Building*, ed. H. M. Colvin and John Newman (Oxford, 1981)
Olsen, Donald J., *Town Planning in London: The Eighteenth and Nineteenth Centuries* (New Haven and London, 1982)
Osbaldiston, William Augustus, *The British Sportsman or Nobleman, Gentleman and Farmer's Dictionary* (London, 1792)
Paine, James, *Plans, Elevations and Sections of Noblemen and Gentlemen's Houses* (London, 1767)
Peall, Thomas, *Observations Chiefly Practical on Some of the More Common Diseases of the Horse* (London, 1814)
Physiologus (Thomas Tryon, pseud.), *The Countryman's Companion* (London, ?1684)
Pollio, Marcus Vitruvius, *Ten Books on Architecture*, trans. M. H. Morgan (New York, 1960)
Powell, Christopher, *Stables and Stable Blocks* (Princes Risborough, 1991)
Poynter, F. N. L., 'A Bibliography of Gervase Markham', *Oxford Bibliographic Society* n.s. IX, 1962
Prior, C. M., *The Royal Studs of the Sixteenth and Seventeenth Centuries* (London, 1935)
Renatus, Vegetius, *Of the Distempers of Horses* (London, 1748)
Repton, Humphry, *An Enquiry into the Changes of Taste in Landscape Gardening* (London, 1806)
Repton, Humphry, *Fragments on the Theory and Practice of Landscape Gardening* (London, 1816)
Richardson, Charles James, *Picturesque Designs for Mansions, Villas and Lodges* (London, 1870)
Richerson, Edward Sherlock, *The General Management of the Hunter: Explaining How the Stable Should be Constructed* (London, 1904)
Riches, Anne, 'London Mews Arches', *Architectural Review* 151, April 1972, pp. 233–36
Ridgeway, William, *The Origin and Influence of the Thoroughbred Horse* (Cambridge, 1905)
Robinson, John Martin, *Georgian Model Farms* (Oxford, 1983)
Roche, Daniel (ed.), *Les Écuries royales du XVI au XVIII siècle* (Paris, 1998)
Roche, Daniel, and Daniel Reytier (eds), *Voitures, chevaux et attelages du XVI au XIX siècle* (Paris, 2000)
Roper, William, *On the Nature and Management of the Horse in Health and Disease . . . with Instructions for Stabling* (London, 1844)
Rosen, Barbara, and Wolfgang Zuckermann, *The Mews of London* (London, 1982)
Royal Institute of British Architects, *Drawings Catalogue* (Farnborough, 1969–89)
Sadler, Julius Trousdale, Jnr and Jacquelin D. J. Sadler, *American Stables: An Architectural Tour* (New York, 1981)
Savot, Louis, *L'Architecture françoise* (Paris, 1624; second edn 1685)
Scamozzi, Vincenzo, *L'Idea della architettura universale* (Venice, 1615)
Scott, William Henry, *British Field Sports* (London, 1818)
Shaw, Vera Kemball, *The Encyclopedia of the Stable* (London, 1909)
Sidney, S., *The Book of the Horse* (London, 1873)
Smith, Peter C., *The Design and Construction of Stables* (London, 1967)
Somerset, C. F., Duke of Beaufort, and Mowbray Morris, *Hunting* (London, 1885)
Stewart, John, *Stable Economy* (Edinburgh, 1838; sixth edn 1854)
Sturgis, Russell, *A Dictionary of Architecture and Building* (London, 1902)

Taplin, William, *The Gentleman's Stable Directory* (London, 1791; thirteenth edn 1796)

Tarbuck, E. L., *The Builder's Practical Director* (Leipzig, Dresden and London, 1857)

Tattersall, George, *Sporting Architecture* (London, 1841)

Thompson, F. M. L., 'Nineteenth-Century Horse Sense', *Economic History Review* n.s. 29, 1976, pp. 60–81

Thompson, F. M. L. (ed.), *Horses in European Economic History* (Reading, 1983)

Thurley, Simon, *The Royal Palaces of Tudor England: Architecture and Court Life, 1460–1547* (New Haven and London, 1993)

Tusser, Thomas, *Five Hundreth Points of Good Husbandry* (London, 1573; ed. William Mavor, London, 1812)

Vanderbank, John, and Gaspard de Saunier, *Twenty Five Actions of the Manege Horse* (London, 1729)

Waistell, Charles, *Designs for Agricultural Buildings* (London, 1827)

Wallis, Thomas, *The Farrier's and Horseman's Complete Dictionary* (London, 1759; third edn 1775)

Walsh, J. H., and J. I. Lupton, *The Horse in the Stable and the Field* (London, 1861)

Ward, William, *A New Treatise on the Method of Breeding, Breaking and Training Horses* (Edinburgh, 1776)

Ware, Isaac, *A Complete Body of Architecture* (London, 1756)

Warth, Keith, *Stables and other Equestrian Buildings: A Guide to Design and Construction* (London, 1997)

White, James, *A Compendium of the Veterinary Art with Observations on the Construction of Stables* (London, 1802; nineteenth edn 1861)

Willett, Peter, *An Introduction to the Thoroughbred* (London, 1875)

William, Eurwyn, *The Historical Farm Buildings of Wales* (Edinburgh, 1986)

Williams, Dorian (ed.) *The Wilton House 'Riding School'* (London, 1978)

Wilson, Angus, *London's Industrial Heritage* (Newton Abbot, 1967)

Worsley, Giles, 'The Design and Development of the Stable and Riding House in Great Britain from the Thirteenth Century to 1914', Ph. D. thesis, Courtauld Institute of Art, London University, 1989.

Worsley, Giles, *Classical Architecture in Britain: The Heroic Age* (New Haven and London, 1995)

Worsley, Giles, 'A Courtly Art: The History of "Haute Ecole" in England', *The Court Historian* 6.1, May 2001, pp. 29–47

Worsley, Giles, 'A History and Catalogue of the British Riding House', *Transactions of the Ancient Monuments Society*, 2003, pp. 33–78

Worsley, Giles, 'Courtly Stables and their Implications for Seventeenth-Century English Architecture', *Georgian Group Journal* XIII, 2003, pp. 114–40

Worsley, Lucy, 'Riding Houses and Horses: William Cavendish's Architecture for the Art of Horsemanship', *Architectural History* XLV, 2002, pp. 194–229

Xenophon, *The Art of Horsemanship*, in Richard Berenger, *The History and Art of Horsemanship* I (London, 1771), pp. 219–70

Young, Alan, *Tudor and Jacobean Tournaments* (London, 1987)

INDEX